Business Intelligence

Business Intelligence: Data Mining and Optimization for Decision Making

Carlo Vercellis

Politecnico di Milano, Italy.

A John Wiley and Sons, Ltd., Publication

This edition first published 2009
© 2009 John Wiley & Sons Ltd

Registered office
John Wiley & Sons Ltd, The Atrium, Southern Gate, Chichester, West Sussex, PO19 8SQ, United Kingdom

2 2010

For details of our global editorial offices, for customer services and for information about how to apply for permission to reuse the copyright material in this book please see our website at www.wiley.com.

Library of Congress Cataloging-in-Publication Data

Vercellis, Carlo.
   Business intelligence : data mining and optimization for decision making / Carlo Vercellis.
      p. cm.
   Includes bibliographical references and index.
   ISBN 978-0-470-51138-1 (cloth) – ISBN 978-0-470-51139-8 (pbk. : alk. paper)
1. Decision making–Mathematical models. 2. Business intelligence. 3. Data mining. I. Title.
   HD30.23.V476 2009
   658.4′038–dc22

                            2008043814

A catalogue record for this book is available from the British Library.

ISBN: 978-0-470-51138-1 (Hbk)
ISBN: 978-0-470-51139-8 (Pbk)

Typeset in 10.5/13pt Times by Laserwords Private Limited, Chennai, India

Contents

Preface

Since the 1990s, the socio-economic context within which economic activities are carried out has generally been referred to as the *information and knowledge society*. The profound changes that have occurred in methods of production and in economic relations have led to a growth in the importance of the exchange of intangible goods, consisting for the most part of transfers of information. The acceleration in the pace of current transformation processes is due to two factors. The first is *globalization*, understood as the ever-increasing interdependence between the economies of the various countries, which has led to the growth of a single *global economy* characterized by a high level of integration. The second is the new *information technologies*, marked by the massive spread of the Internet and of wireless devices, which have enabled high-speed transfers of large amounts of data and the widespread use of sophisticated means of communication.

In this rapidly evolving scenario, the wealth of development opportunities is unprecedented. The easy access to information and knowledge offers several advantages to various actors in the socio-economic environment: *individuals*, who can obtain news more rapidly, access services more easily and carry out on-line commercial and banking transactions; *enterprises*, which can develop innovative products and services that can better meet the needs of the users, achieving competitive advantages from a more effective use of the knowledge gained; and, finally, the *public administration*, which can improve the services provided to citizens through the use of e-government applications, such as on-line payments of tax contributions, and e-health tools, by taking into account each patient's medical history, thus improving the quality of healthcare services.

In this framework of radical transformation, methods of governance within complex organizations also reflect the changes occurring in the socio-economic environment, and appear increasingly more influenced by the immediate access to information for the development of effective action plans. The term *complex organizations* will be used throughout the book to collectively refer to a diversified set of entities operating in the socio-economic context, including enterprises, government agencies, banking and financial institutions, and non-profit organizations.

The adoption of low-cost massive data storage technologies and the wide availability of Internet connections have made available large amounts of data that have been collected and accumulated by the various organizations over the years. The enterprises that are capable of transforming data into *information* and *knowledge* can use them to make quicker and more effective decisions and thus to achieve a competitive advantage. By the same token, on the public administration side, the analysis of the available information enables the development of better and innovative services for citizens. These are ambitious objectives that technology, however sophisticated, cannot perform on its own, without the support of competent minds and advanced analysis methodologies.

Is it possible to extract, from the huge amounts of data available, knowledge which can then be used by *decision makers* to aid and improve the governance of the enterprises and the public administration?

Business intelligence may be defined as a set of mathematical models and analysis methodologies that systematically exploit the available data to retrieve information and knowledge useful in supporting complex decision-making processes.

Despite the somewhat restrictive meaning of the term *business*, which seems to confine the subject within the boundaries of enterprises, business intelligence systems are aimed at companies as well as other types of complex organizations, as mentioned above.

Business intelligence methodologies are interdisciplinary and broad, spanning several domains of application. Indeed, they are concerned with the representation and organization of the decision-making process, and thus with the field of decision theory; with collecting and storing the data intended to facilitate the decision-making process, and thus with data warehousing technologies; with mathematical models for optimization and data mining, and thus with operations research and statistics; finally, with several application domains, such as marketing, logistics, accounting and control, finance, services and the public administration.

We can say that business intelligence systems tend to promote a scientific and rational approach to managing enterprises and complex organizations. Even the use of an electronic spreadsheet for assessing the effects induced on the budget by fluctuations in the discount rate, despite its simplicity, requires on the part of decision makers a mental representation of the financial flows.

A business intelligence environment offers decision makers information and knowledge derived from data processing, through the application of mathematical models and algorithms. In some instances, these may merely consist of the calculation of totals and percentages, while more fully developed analyses make use of advanced models for optimization, inductive learning and prediction.

In general, a model represents a selective abstraction of a real system, designed to analyze and understand from an abstract point of view the operating behavior of the real system. The model includes only the elements of the system deemed relevant for the purpose of the investigation carried out. It is worth quoting the words of Einstein on the subject of model development: 'Everything should be made as simple as possible, but not simpler.'

Classical scientific disciplines, such as physics, have always made use of mathematical models for the abstract representation of real systems, while other disciplines, such as operations research, have dealt with the application of scientific methods and mathematical models to the study of artificial systems, such as enterprises and complex organizations.

'The great book of nature', as Galileo wrote, 'may only be read by those who know the language in which it was written. And this language is mathematics.' Can we apply also to the analysis of artificial systems this profound insight from one of the men who opened up the way to modern science?

We believe so. Nowadays, the mere intuitive abilities of decision makers managing enterprises or the public administration are outdone by the complexity of governance of current organizations. As an example, consider the design of a marketing campaign in dynamic and unpredictable markets, where however a wealth of information is available on the buying behavior of the consumers. Today, it is inconceivable to leave aside the application of advanced inferential learning models for selecting the recipients of the campaign, in order to optimize the allocation of resources and the redemption of the marketing action.

The interpretation of the term *business intelligence* that we have illustrated and that we intend to develop in this book is much broader and deeper compared to the narrow meaning publicized over the last few years by many software vendors and information technology magazines. According to this latter vision, business intelligence methodologies are reduced to electronic tools for querying, visualization and reporting, mainly for accounting and control purposes. Of course, no one can deny that rapid access to information is an invaluable tool for decision makers. However, these tools are oriented toward business intelligence analyses of a *passive* nature, where the decision maker has already formulated in her mind some criteria for data extraction. If we wish business intelligence methodologies to be able to express their huge strategic potential, we should turn to *active* forms of support for decision making, based on the systematic adoption of mathematical models able to transform data not only into *information* but also into *knowledge*, and then knowledge into actual competitive advantage. The distinction between passive and active forms of analysis will be further investigated in Chapter 1.

One might object that only simple tools based on immediate and intuitive concepts have the ability to prove useful in practice. In reply to this objection, we cannot do better than quote Vladimir Vapnik, who more than anyone has contributed to the development of inductive learning models: 'Nothing is more practical than a good theory.'

Throughout this book we have tried to make frequent reference to problems and examples drawn from real applications in order to help readers understand the topics discussed, while ensuring an adequate level of methodological rigor in the description of mathematical models.

Part I describes the basic components that make up a business intelligence environment, discussing the structure of the decision-making process and reviewing the underlying information infrastructures. In particular, Chapter 1 outlines a general framework for business intelligence, highlighting the connections with other disciplines. Chapter 2 describes the structure of the decision-making process and introduces the concept of a decision support system, illustrating the main advantages it involves, the critical success factors and some implementation issues. Chapter 3 presents data warehouses and data marts, first analyzing the reasons that led to their introduction, and then describing on-line analytical processing analyses based on multidimensional cubes.

Part II is more methodological in character, and offers a comprehensive overview of mathematical models for pattern recognition and data mining. Chapter 4 describes the main characteristics of mathematical models used for business intelligence analyses, offering a brief taxonomy of the major classes of models. Chapter 5 introduces data mining, discussing the phases of a data mining process and their objectives. Chapter 6 describes the activities of data preparation for business intelligence and data mining; these include data validation, anomaly detection, data transformation and reduction. Chapter 7 provides a detailed discussion of exploratory data analysis, performed by graphical methods and summary statistics, in order to understand the characteristics of the attributes in a dataset and to determine the intensity of the relationships among them. Chapter 8 describes simple and multiple regression models, discussing the main diagnostics for assessing their significance and accuracy. Chapter 9 illustrates the models for time series analysis, examining decomposition methods, exponential smoothing and autoregressive models. Chapter 10 is entirely devoted to classification models, which play a prominent role in pattern recognition and learning theory. After a description of the evaluation criteria, the main classification methods are illustrated; these include classification trees, Bayesian methods, neural networks, logistic regression and support vector machines. Chapter 11 describes association rules and the Apriori algorithm. Chapter 12 presents the best-known clustering models: partition methods, such as K-means and K-medoids, and hierarchical methods, both agglomerative and divisive.

Part III illustrates the applications of data mining to relational marketing (Chapter 13), models for salesforce planning (Chapter 13), models for supply chain optimization (Chapter 14) and analytical methods for performance assessment (Chapter 15).

Appendix A provides information and links to software tools used to carry out the data mining and business intelligence analyses described in the book. Preference has been given to *open source* software, since in this way readers can freely download it from the Internet to practice on the examples given. By the same token, the datasets used to exemplify the different topics are also mostly taken from repositories in the public domain. Appendix B includes a short description of the datasets used in the various chapters and the links to sites that contain these as well as other datasets useful for experimenting with and comparing the analysis methodologies.

Bibliographical notes at the end of each chapter, highly selective as they are, highlight other texts that we found useful and relevant, as well as research contributions of acknowledged historical value.

This book is aimed at three main groups of readers. The first are students studying toward a master's degree in economics, business management or other scientific disciplines, and attending a university course on business intelligence methodologies, decision support systems and mathematical models for decision making. The second are students on doctoral programs in disciplines of an economic and management nature. Finally, the book may also prove useful to professionals wishing to update their knowledge and make use of a methodological and practical reference textbook. Readers belonging to this last group may be interested in an overview of the opportunities offered by business intelligence systems, or in specific methodological and applied subjects dealt with in the book, such as data mining techniques applied to relational marketing, salesforce planning models, supply chain optimization models and analytical methods for performance evaluation.

At Politecnico di Milano, the author leads the research group *MOLD – Mathematical modeling, optimization, learning from data*, which conducts methodological research activities on models for inductive learning, prediction, classification, optimization, systems biology and social network analysis, as well as applied projects on business intelligence, relational marketing and logistics. The research group's website, *www.mold.polimi.it*, includes information, news, in-depth studies, useful links and updates.

A book free of misprints is a rare occurrence, especially in the first edition, despite the efforts made to avoid them. Therefore, a dedicated area for errata and corrigenda has been created at *www.mold.polimi.it*, and readers are welcome to contribute to it by sending a note on any typos that they might find in the text to the author at *carlo.vercellis@polimi.it*.

I wish to express special thanks to Carlotta Orsenigo, who helped write Chapter 10 on classification models and discussed with me the content and the organization of the remaining chapters in the book. Her help in filling gaps, clarifying concepts, and making suggestions for improvement to the text and figures was invaluable.

To write this book, I have drawn on my experience as a teacher of graduate and postgraduate courses. I would therefore like to thank here all the many students who through their questions and curiosity have urged me to seek more convincing and incisive arguments.

Many examples and references to real problems originate from applied projects that I have carried out with enterprises and agencies of the public administration. I am indebted to many professionals for some of the concepts that I have included in the book: they are too numerous to name but will certainly recognize themselves in some statements, and to all of them I extend a heartfelt thank-you.

All typos and inaccuracies in this book are entirely my own responsibility.

Part I

Components of the decision-making process

1

Business intelligence

The advent of low-cost data storage technologies and the wide availability of Internet connections have made it easier for individuals and organizations to access large amounts of data. Such data are often heterogeneous in origin, content and representation, as they include commercial, financial and adminis- trative transactions, web navigation paths, emails, texts and hypertexts, and the results of clinical tests, to name just a few examples. Their accessibility opens up promising scenarios and opportunities, and raises an enticing question: is it possible to convert such data into information and knowledge that can then be used by decision makers to aid and improve the governance of enterprises and of public administration?

Business intelligence may be defined as a set of mathematical models and analysis methodologies that exploit the available data to generate information and knowledge useful for complex decision-making processes. This opening chapter will describe in general terms the problems entailed in business intelli- gence, highlighting the interconnections with other disciplines and identifying the primary components typical of a business intelligence environment.

1.1 Effective and timely decisions

In complex organizations, public or private, decisions are made on a continual basis. Such decisions may be more or less critical, have long- or short-term effects and involve people and roles at various hierarchical levels. The ability of these *knowledge workers* to make decisions, both as individuals and as a community, is one of the primary factors that influence the performance and competitive strength of a given organization.

Business Intelligence: Data Mining and Optimization for Decision Making   C. Vercellis
© 2009 John Wiley & Sons, Ltd

Most knowledge workers reach their decisions primarily using easy and intuitive methodologies, which take into account specific elements such as experience, knowledge of the application domain and the available information. This approach leads to a stagnant decision-making style which is inappropriate for the unstable conditions determined by frequent and rapid changes in the economic environment. Indeed, decision-making processes within today's organizations are often too complex and dynamic to be effectively dealt with through an intuitive approach, and require instead a more rigorous attitude based on analytical methodologies and mathematical models. The importance and strategic value of analytics in determining competitive advantage for enterprises has been recently pointed out by several authors, as described in the references at the end of this chapter. Examples 1.1 and 1.2 illustrate two highly complex decision-making processes in rapidly changing conditions.

Example 1.1 – Retention in the mobile phone industry. The marketing manager of a mobile phone company realizes that a large number of customers are discontinuing their service, leaving her company in favor of some competing provider. As can be imagined, low customer loyalty, also known as customer *attrition* or *churn*, is a critical factor for many companies operating in service industries. Suppose that the marketing manager can rely on a budget adequate to pursue a customer retention campaign aimed at 2000 individuals out of a total customer base of 2 million people. Hence, the question naturally arises of how she should go about choosing those customers to be contacted so as to optimize the effectiveness of the campaign. In other words, how can the probability that each single customer will discontinue the service be estimated so as to target the best group of customers and thus reduce churning and maximize customer retention? By knowing these probabilities, the target group can be chosen as the 2000 people having the highest churn likelihood among the customers of high business value. Without the support of advanced mathematical models and data mining techniques, described in Chapter 5, it would be arduous to derive a reliable estimate of the churn probability and to determine the best recipients of a specific marketing campaign.

Example 1.2 – Logistics planning. The logistics manager of a manufacturing company wishes to develop a medium-term logistic-production plan. This is a decision-making process of high complexity which includes, among other choices, the allocation of the demand originating from different market areas to the production sites, the procurement of raw materials and purchased parts from suppliers, the production planning of the plants and the distribution of end products to market areas. In a typical manufacturing company this could well entail tens of facilities, hundreds of suppliers, and thousands of finished goods and components, over a time span of one year divided into weeks. The magnitude and complexity of the problem suggest that advanced optimization models are required to devise the best logistic plan. As we will see in Chapter 14, optimization models allow highly complex and large-scale problems to be tackled successfully within a business intelligence framework.

The main purpose of business intelligence systems is to provide knowledge workers with tools and methodologies that allow them to make *effective* and *timely* decisions.

Effective decisions. The application of rigorous analytical methods allows decision makers to rely on information and knowledge which are more dependable. As a result, they are able to make better decisions and devise action plans that allow their objectives to be reached in a more effective way. Indeed, turning to formal analytical methods forces decision makers to explicitly describe both the criteria for evaluating alternative choices and the mechanisms regulating the problem under investigation. Furthermore, the ensuing in-depth examination and thought lead to a deeper awareness and comprehension of the underlying logic of the decision-making process.

Timely decisions. Enterprises operate in economic environments characterized by growing levels of competition and high dynamism. As a consequence, the ability to rapidly react to the actions of competitors and to new market conditions is a critical factor in the success or even the survival of a company.

Figure 1.1 illustrates the major benefits that a given organization may draw from the adoption of a business intelligence system. When facing problems such as those described in Examples 1.1 and 1.2 above, decision makers ask themselves a series of questions and develop the corresponding analysis. Hence, they examine and compare several options, selecting among them the best decision, given the conditions at hand.

If decision makers can rely on a business intelligence system facilitating their activity, we can expect that the overall quality of the decision-making process will be greatly improved. With the help of mathematical models and algorithms, it is actually possible to analyze a larger number of alternative

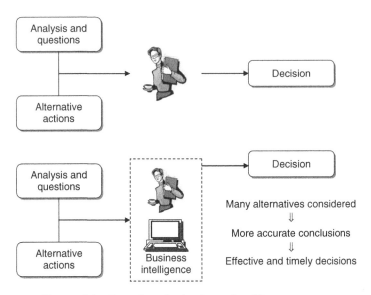

Figure 1.1   Benefits of a business intelligence system

actions, achieve more accurate conclusions and reach effective and timely decisions. We may therefore conclude that the major advantage deriving from the adoption of a business intelligence system is found in the increased *effectiveness* of the decision-making process.

1.2   Data, information and knowledge

As observed above, a vast amount of data has been accumulated within the information systems of public and private organizations. These data originate partly from internal transactions of an administrative, logistical and commercial nature and partly from external sources. However, even if they have been gathered and stored in a systematic and structured way, these data cannot be used directly for decision-making purposes. They need to be processed by means of appropriate extraction tools and analytical methods capable of transforming them into information and knowledge that can be subsequently used by decision makers.

The difference between *data*, *information* and *knowledge* can be better understood through the following remarks.

Data. Generally, data represent a structured codification of single primary entities, as well as of transactions involving two or more primary entities. For example, for a retailer data refer to primary entities such as customers, points of sale and items, while sales receipts represent the commercial transactions.

Information. Information is the outcome of extraction and processing activities carried out on data, and it appears meaningful for those who receive it in a specific domain. For example, to the sales manager of a retail company, the proportion of sales receipts in the amount of over €100 per week, or the number of customers holding a loyalty card who have reduced by more than 50% the monthly amount spent in the last three months, represent meaningful pieces of information that can be extracted from raw stored data.

Knowledge. Information is transformed into knowledge when it is used to make decisions and develop the corresponding actions. Therefore, we can think of knowledge as consisting of information put to work into a specific domain, enhanced by the experience and competence of decision makers in tackling and solving complex problems. For a retail company, a sales analysis may detect that a group of customers, living in an area where a competitor has recently opened a new point of sale, have reduced their usual amount of business. The knowledge extracted in this way will eventually lead to actions aimed at solving the problem detected, for example by introducing a new free home delivery service for the customers residing in that specific area. We wish to point out that knowledge can be extracted from data both in a *passive* way, through the analysis criteria suggested by the decision makers, or through the *active* application of mathematical models, in the form of inductive learning or optimization, as described in the following chapters.

Several public and private enterprises and organizations have developed in recent years formal and systematic mechanisms to gather, store and share their wealth of knowledge, which is now perceived as an invaluable intangible asset. The activity of providing support to knowledge workers through the integration of decision-making processes and enabling information technologies is usually referred to as *knowledge management*.

It is apparent that business intelligence and knowledge management share some degree of similarity in their objectives. The main purpose of both disciplines is to develop environments that can support knowledge workers in decision-making processes and complex problem-solving activities. To draw a boundary between the two approaches, we may observe that knowledge management methodologies primarily focus on the treatment of information that is usually unstructured, at times implicit, contained mostly in documents, conversations and past experience. Conversely, business intelligence systems are based on structured information, most often of a quantitative nature and usually organized in a database. However, this distinction is a somewhat fuzzy one: for example, the ability to analyze emails and web pages through text mining methods progressively induces business intelligence systems to deal with unstructured information.

1.3   The role of mathematical models

A business intelligence system provides decision makers with information and knowledge extracted from data, through the application of mathematical models and algorithms. In some instances, this activity may reduce to calculations of totals and percentages, graphically represented by simple histograms, whereas more elaborate analyses require the development of advanced optimization and learning models.

In general terms, the adoption of a business intelligence system tends to promote a scientific and rational approach to the management of enterprises and complex organizations. Even the use of a spreadsheet to estimate the effects on the budget of fluctuations in interest rates, despite its simplicity, forces decision makers to generate a mental representation of the financial flows process.

Classical scientific disciplines, such as physics, have always resorted to mathematical models for the abstract representation of real systems. Other disciplines, such as operations research, have instead exploited the application of scientific methods and mathematical models to the study of artificial systems, for example public and private organizations. Part II of this book will describe the main mathematical models used in business intelligence architectures and decision support systems, as well as the corresponding solution methods, while Part III will illustrate several related applications.

The rational approach typical of a business intelligence analysis can be summarized schematically in the following main characteristics.

- First, the objectives of the analysis are identified and the performance indicators that will be used to evaluate alternative options are defined.

- Mathematical models are then developed by exploiting the relationships among system control variables, parameters and evaluation metrics.

- Finally, *what-if* analyses are carried out to evaluate the effects on the performance determined by variations in the control variables and changes in the parameters.

Although their primary objective is to enhance the effectiveness of the decision-making process, the adoption of mathematical models also affords other advantages, which can be appreciated particularly in the long term. First, the development of an abstract model forces decision makers to focus on the main features of the analyzed domain, thus inducing a deeper understanding of the phenomenon under investigation. Furthermore, the knowledge about the domain acquired when building a mathematical model can be more easily transferred in the long run to other individuals within the same organization, thus allowing a sharper preservation of knowledge in comparison to empirical decision-making processes. Finally, a mathematical model developed for a specific decision-making task is so general and flexible that in most cases it can be applied to other ensuing situations to solve problems of similar type.

1.4   Business intelligence architectures

The architecture of a business intelligence system, depicted in Figure 1.2, includes three major components.

Data sources. In a first stage, it is necessary to gather and integrate the data stored in the various primary and secondary sources, which are heterogeneous in origin and type. The sources consist for the most part of data belonging to operational systems, but may also include unstructured documents, such as emails and data received from external providers. Generally speaking, a major effort is required to unify and integrate the different data sources, as shown in Chapter 3.

Data warehouses and data marts. Using extraction and transformation tools known as *extract, transform, load* (ETL), the data originating from the different sources are stored in databases intended to support business intelligence analyses. These databases are usually referred to as *data warehouses* and *data marts*, and they will be the subject of Chapter 3.

Business intelligence methodologies. Data are finally extracted and used to feed mathematical models and analysis methodologies intended to support decision makers. In a business intelligence system, several decision support applications may be implemented, most of which will be described in the following chapters:

- multidimensional cube analysis;
- exploratory data analysis;

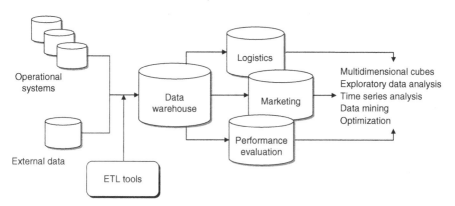

Figure 1.2   A typical business intelligence architecture

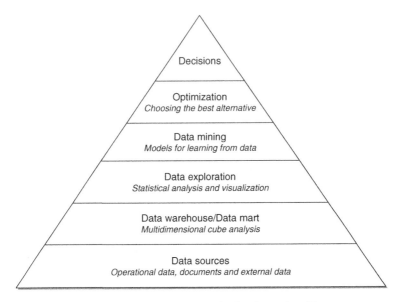

Figure 1.3   The main components of a business intelligence system

- time series analysis;
- inductive learning models for data mining;
- optimization models.

The pyramid in Figure 1.3 shows the building blocks of a business intelligence system. So far, we have seen the components of the first two levels when discussing Figure 1.2. We now turn to the description of the upper tiers.

Data exploration. At the third level of the pyramid we find the tools for performing a *passive* business intelligence analysis, which consist of query and reporting systems, as well as statistical methods. These are referred to as passive methodologies because decision makers are requested to generate prior hypotheses or define data extraction criteria, and then use the analysis tools to find answers and confirm their original insight. For instance, consider the sales manager of a company who notices that revenues in a given geographic area have dropped for a specific group of customers. Hence, she might want to bear out her hypothesis by using extraction and visualization tools, and then apply a statistical test to verify that her conclusions are adequately supported by data. Statistical techniques for exploratory data analysis will be described in Chapters 6 and 7.

Data mining. The fourth level includes *active* business intelligence methodologies, whose purpose is the extraction of information and knowledge from data.

These include mathematical models for pattern recognition, machine learning and data mining techniques, which will be dealt with in Part II of this book. Unlike the tools described at the previous level of the pyramid, the models of an active kind do not require decision makers to formulate any prior hypothesis to be later verified. Their purpose is instead to expand the decision makers' knowledge.

Optimization. By moving up one level in the pyramid we find optimization models that allow us to determine the best solution out of a set of alternative actions, which is usually fairly extensive and sometimes even infinite. Example 1.2 shows a typical field of application of optimization models. Other optimization models applied in marketing and logistics will be described in Chapters 13 and 14.

Decisions. Finally, the top of the pyramid corresponds to the choice and the actual adoption of a specific decision, and in some way represents the natural conclusion of the decision-making process. Even when business intelligence methodologies are available and successfully adopted, the choice of a decision pertains to the decision makers, who may also take advantage of informal and unstructured information available to adapt and modify the recommendations and the conclusions achieved through the use of mathematical models.

As we progress from the bottom to the top of the pyramid, business intelligence systems offer increasingly more advanced support tools of an active type. Even roles and competencies change. At the bottom, the required competencies are provided for the most part by the information systems specialists within the organization, usually referred to as *database administrators*. Analysts and experts in mathematical and statistical models are responsible for the intermediate phases. Finally, the activities of decision makers responsible for the application domain appear dominant at the top.

As described above, business intelligence systems address the needs of different types of complex organizations, including agencies of public administration and associations. However, if we restrict our attention to enterprises, business intelligence methodologies can be found mainly within three departments of a company, as depicted in Figure 1.4: marketing and sales; logistics and production; accounting and control. The applications of business intelligence described in Part III of this volume will be precisely devoted to these topics.

1.4.1 Cycle of a business intelligence analysis

Each business intelligence analysis follows its own path according to the application domain, the personal attitude of the decision makers and the available analytical methodologies. However, it is possible to identify an ideal cyclical

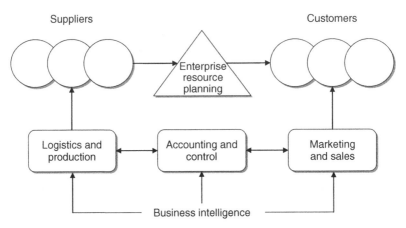

Figure 1.4   Departments of an enterprise concerned with business intelligence systems

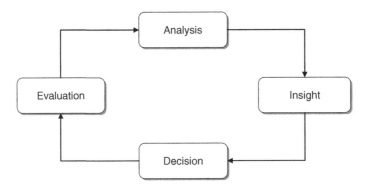

Figure 1.5   Cycle of a business intelligence analysis

path characterizing the evolution of a typical business intelligence analysis, as shown in Figure 1.5, even though differences still exist based upon the peculiarity of each specific context.

Analysis. During the analysis phase, it is necessary to recognize and accurately spell out the problem at hand. Decision makers must then create a mental representation of the phenomenon being analyzed, by identifying the critical factors that are perceived as the most relevant. The availability of business intelligence methodologies may help already in this stage, by permitting decision makers to rapidly develop various paths of investigation. For instance, the exploration of data cubes in a multidimensional analysis, according to different logical views as described in Chapter 3, allows decision makers to modify their hypotheses flexibly and rapidly, until they reach an interpretation scheme that they deem satisfactory. Thus, the first phase in the business intelligence cycle leads decision makers to ask several questions and to obtain quick responses in an interactive way.

Insight. The second phase allows decision makers to better and more deeply understand the problem at hand, often at a causal level. For instance, if the analysis carried out in the first phase shows that a large number of customers are discontinuing an insurance policy upon yearly expiration, in the second phase it will be necessary to identify the profile and characteristics shared by such customers. The information obtained through the analysis phase is then transformed into knowledge during the insight phase. On the one hand, the extraction of knowledge may occur due to the intuition of the decision makers and therefore be based on their experience and possibly on unstructured information available to them. On the other hand, inductive learning models may also prove very useful during this stage of analysis, particularly when applied to structured data.

Decision. During the third phase, knowledge obtained as a result of the insight phase is converted into decisions and subsequently into actions. The availability of business intelligence methodologies allows the analysis and insight phases to be executed more rapidly so that more effective and timely decisions can be made that better suit the strategic priorities of a given organization. This leads to an overall reduction in the execution time of the *analysis–decision–action–revision* cycle, and thus to a decision-making process of better quality.

Evaluation. Finally, the fourth phase of the business intelligence cycle involves performance measurement and evaluation. Extensive metrics should then be devised that are not exclusively limited to the financial aspects but also take into account the major performance indicators defined for the different company departments. Chapter 15 will describe powerful analytical methodologies for performance evaluation.

1.4.2   Enabling factors in business intelligence projects

Some factors are more critical than others to the success of a business intelligence project: *technologies*, *analytics* and *human resources*.

Technologies. Hardware and software technologies are significant enabling factors that have facilitated the development of business intelligence systems within enterprises and complex organizations. On the one hand, the computing capabilities of microprocessors have increased on average by 100% every 18 months during the last two decades, and prices have fallen. This trend has enabled the use of advanced algorithms which are required to employ inductive learning methods and optimization models, keeping the processing times within a reasonable range. Moreover, it permits the adoption of state-of-the-art graphical visualization techniques, featuring real-time animations. A further relevant enabling factor derives from the exponential increase in the capacity of mass storage devices, again at decreasing costs, enabling any organization to store terabytes of data for business intelligence systems. And network connectivity, in the form of *Extranets* or *Intranets*, has played a primary role in the diffusion within organizations of information and knowledge extracted from business intelligence systems. Finally, the easy integration of hardware and software purchased by different suppliers, or developed internally by an organization, is a further relevant factor affecting the diffusion of data analysis tools.

Analytics. As stated above, mathematical models and analytical methodologies play a key role in information enhancement and knowledge extraction from the data available inside most organizations. The mere visualization of the data according to timely and flexible logical views, as described in Chapter 3, plays a relevant role in facilitating the decision-making process, but still represents a passive form of support. Therefore, it is necessary to apply more advanced models of inductive learning and optimization in order to achieve active forms of support for the decision-making process.

Human resources. The human assets of an organization are built up by the competencies of those who operate within its boundaries, whether as individuals or collectively. The overall knowledge possessed and shared by these individuals constitutes the *organizational culture*. The ability of knowledge workers to acquire information and then translate it into practical actions is one of the major assets of any organization, and has a major impact on the quality of the decision-making process. If a given enterprise has implemented an advanced business intelligence system, there still remains much scope to emphasize the personal skills of its knowledge workers, who are required to perform the analyses and to interpret the results, to work out creative solutions and to devise effective action plans. All the available analytical tools being equal, a company employing human resources endowed with a greater mental agility and willing to accept changes in the decision-making style will be at an advantage over its competitors.

1.4.3   Development of a business intelligence system

The development of a business intelligence system can be assimilated to a project, with a specific final objective, expected development times and costs, and the usage and coordination of the resources needed to perform planned

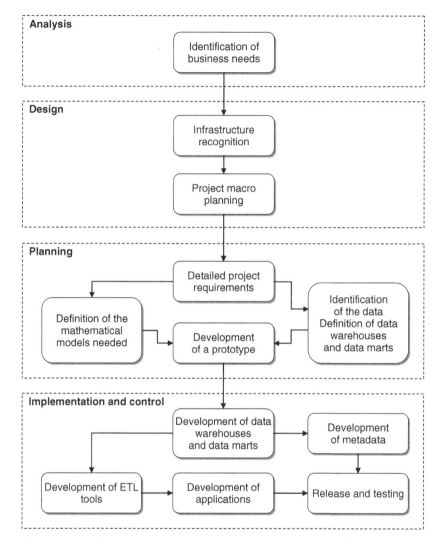

Figure 1.6   Phases in the development of a business intelligence system

activities. Figure 1.6 shows the typical development cycle of a business intelligence architecture. Obviously, the specific path followed by each organization
might differ from that outlined in the figure. For instance, if the basic information structures, including the data warehouse and the data marts, are already in
place, the corresponding phases indicated in Figure 1.6 will not be required.

Analysis. During the first phase, the needs of the organization relative to the
development of a business intelligence system should be carefully identified.
This preliminary phase is generally conducted through a series of interviews of

knowledge workers performing different roles and activities within the organization. It is necessary to clearly describe the general objectives and priorities of the project, as well as to set out the costs and benefits deriving from the development of the business intelligence system.

Design. The second phase includes two sub-phases and is aimed at deriving a provisional plan of the overall architecture, taking into account any development in the near future and the evolution of the system in the mid term. First, it is necessary to make an assessment of the existing information infrastructures. Moreover, the main decision-making processes that are to be supported by the business intelligence system should be examined, in order to adequately determine the information requirements. Later on, using classical project management methodologies, the project plan will be laid down, identifying development phases, priorities, expected execution times and costs, together with the required roles and resources.

Planning. The planning stage includes a sub-phase where the functions of the business intelligence system are defined and described in greater detail. Subsequently, existing data as well as other data that might be retrieved externally are assessed. This allows the information structures of the business intelligence architecture, which consist of a central data warehouse and possibly some satellite data marts, to be designed. Simultaneously with the recognition of the available data, the mathematical models to be adopted should be defined, ensuring the availability of the data required to feed each model and verifying that the efficiency of the algorithms to be utilized will be adequate for the magnitude of the resulting problems. Finally, it is appropriate to create a system prototype, at low cost and with limited capabilities, in order to uncover beforehand any discrepancy between actual needs and project specifications.

Implementation and control. The last phase consists of five main sub-phases. First, the data warehouse and each specific data mart are developed. These represent the information infrastructures that will feed the business intelligence system. In order to explain the meaning of the data contained in the data warehouse and the transformations applied in advance to the primary data, a *metadata* archive should be created, as described in Chapter 3. Moreover, ETL procedures are set out to extract and transform the data existing in the primary sources, loading them into the data warehouse and the data marts. The next step is aimed at developing the core business intelligence applications that allow the planned analyses to be carried out. Finally, the system is released for test and usage.

Figure 1.7 provides an overview of the main methodologies that may be included in a business intelligence system, most of which will be described

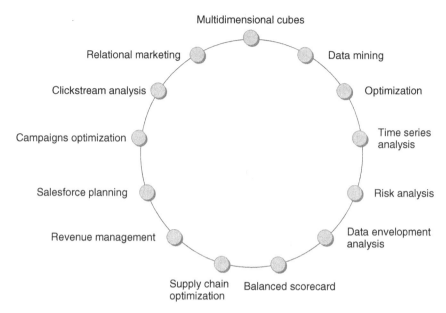

Figure 1.7   Portfolio of available methodologies in a business intelligence system

in the following chapters. Some of them have a methodological nature and can be used across different application domains, while others can only be applied to specific tasks.

1.5   Ethics and business intelligence

The adoption of business intelligence methodologies, data mining methods and decision support systems raises some ethical problems that should not be overlooked. Indeed, the progress toward the information and knowledge society opens up countless opportunities, but may also generate distortions and risks which should be prevented and avoided by using adequate control rules and mechanisms. Usage of data by public and private organizations that is improper and does not respect the individuals' right to privacy should not be tolerated. More generally, we must guard against the excessive growth of the political and economic power of enterprises allowing the transformation processes outlined above to exclusively and unilaterally benefit such enterprises themselves, at the expense of consumers, workers and inhabitants of the Earth ecosystem.

However, even failing specific regulations that would prevent the abuse of data gathering and invasive investigations, it is essential that business intelligence analysts and decision makers abide by the ethical principle of respect for the personal rights of the individuals. The risk of overstepping the boundary between correct and intrusive use of information is particularly high within the relational marketing and web mining fields, described in Chapter 13. For example, even if disguised under apparently inoffensive names such as 'data enrichment', private information on individuals and households does circulate, but that does not mean that it is ethical for decision makers and enterprises to use it.

Respect for the right to privacy is not the only ethical issue concerning the use of business intelligence systems. There has been much discussion in recent years of the social responsibilities of enterprises, leading to the introduction of the new concept of *stakeholders*. This term refers to anyone with any interest in the activities of a given enterprise, such as investors, employees, labor unions and civil society as a whole. There is a diversity of opinion on whether a company should pursue the short-term maximization of profits, acting exclusively in the interest of shareholders, or should instead adopt an approach that takes into account the social consequences of its decisions.

As this is not the right place to discuss a problem of such magnitude, we will confine ourselves to pointing out that analyses based on business intelligence systems are affected by this issue and therefore run the risk of being used to maximize profits even when different considerations should prevail related to the social consequences of the decisions made, according to a logic that we believe should be rejected. For example, is it right to develop an optimization model with the purpose of distributing costs on an international scale in order to circumvent the tax systems of certain countries? Is it legitimate to make a decision on the optimal position of the tank in a vehicle in order to minimize production costs, even if this may cause serious harm to the passengers in the event of a collision? As proven by these examples, analysts developing a mathematical model and those who make the decisions cannot remain neutral, but have the moral obligation to take an ethical stance.

1.6 Notes and readings

As observed above, business intelligence methodologies are interdisciplinary by nature and only recently has the scientific community begun to treat them as a separate subject. As a consequence, most publications in recent years have been released in the form of press or promotional reports, with a few exceptions. The following are some suggested readings: Moss and Atre (2003), offering a description of the guidelines to follow in the development of business intelligence systems; Simon and Shaffer (2001) on business intelligence applications for e-commerce; Kudyba and Hoptroff (2001) for a general introduction to the subject; and finally, Giovinazzo (2002) and Marshall *et al.* (2004) focus on business intelligence applications over the Internet. The strategic role of analytical methods, in the form of predictive and optimization mathematical models, has been pointed out recently by a number of authors, among them Davenport and Harris (2007) and Ayres (2007).

The integration of business intelligence architectures, decision support systems and knowledge management is examined by Bolloju *et al.* (2002), Nemati *et al.* (2002) and Malone *et al.* (2003). The volume by Rasmussen *et al.* (2002) describes the role of business intelligence methodologies for financial applications, which are not covered in this text. For considerations of a general nature on the ethical implications of corporate decisions, see Bakan (2005). Snapper (1998) examines the ethical aspects involved in the application of business intelligence methodologies in the medical sector.

2

Decision support systems

A *decision support system* (DSS) is an interactive computer-based application that combines data and mathematical models to help decision makers solve complex problems faced in managing the public and private enterprises and organizations. As described in Chapter 1, the analysis tools provided by a business intelligence architecture can be regarded as DSSs capable of transforming data into information and knowledge helpful to decision makers. In this respect, DSSs are a basic component in the development of a business intelligence architecture.

In this chapter we will first discuss the structure of the decision-making process. Further on, the evolution of information systems will be briefly sketched. We will then define DSSs, outlining the major advantages and pointing out the critical success factors relative to their introduction. Finally, the development phases of a DSS project will be described, addressing the most relevant issues concerning its implementation.

2.1 Definition of system

The term *system* is often used in everyday language: for instance, we refer to the solar system, the nervous system or the justice system. The entities that we intuitively denominate *systems* share a common characteristic, which we will adopt as an abstract definition of the notion of system: each of them is made up of a set of components that are in some way connected to each other so as to provide a single collective result and a common purpose.

Every system is characterized by boundaries that separate its internal components from the external environment. A system is said to be *open* if its boundaries can be crossed in both directions by flows of materials and information.

When such flows are lacking, the system is said to be *closed*. In general terms, any given system receives specific input flows, carries out an internal transformation process and generates observable output flows.

As can be imagined, this abstract definition of system can be used to describe a broad class of real-world phenomena. For example, the logistic structure of an enterprise is a system that receives as input a set of materials, services and information and returns as output a set of products, services and information. More generally, even an enterprise, taken as a whole or in part, may be represented in its turn as a system, provided the boundaries as well as input and output flows are clearly defined.

Figure 2.1 shows the structure that we will use as a reference to describe the concept of system. A system receives a set of *input* flows and returns a set of *output* flows through a *transformation* process regulated by *internal conditions* and *external conditions*. The effectiveness and efficiency of a system are assessed using measurable performance indicators that can be classified into different categories. The figure shows the main types of metrics used to evaluate systems embedded within the enterprises and the public administration.

A system will often incorporate a *feedback* mechanism. Feedback occurs when a system component generates an output flow that is fed back into the system itself as an input flow, possibly as a result of a further transformation. Systems that are able to modify their own output flows based on feedback are called *closed cycle systems*. For example, the closed cycle system outlined in Figure 2.2 describes the development of a sequence of marketing campaigns.

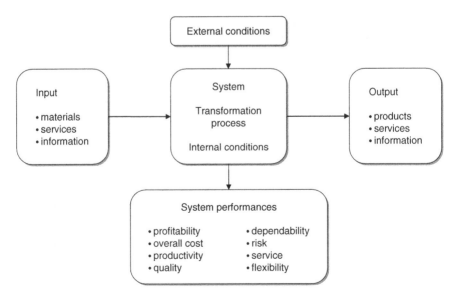

Figure 2.1   Abstract representation of a system

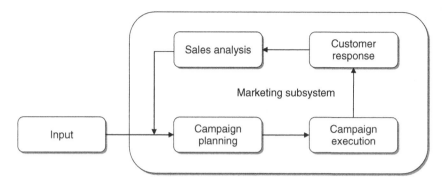

Figure 2.2   A closed cycle marketing system with feedback effects

The sales results for each campaign are gathered and become available as feedback input so as to design subsequent marketing promotions.

In connection with a decision-making process, whose structure will be described in the next section, it is often necessary to assess the performance of a system. For this purpose, it is appropriate to categorize the evaluation metrics into two main classes: *effectiveness* and *efficiency*.

Effectiveness. Effectiveness measurements express the level of conformity of a given system to the objectives for which it was designed. The associated performance indicators are therefore linked to the system output flows, such as production volumes, weekly sales and yield per share.

Efficiency. Efficiency measurements highlight the relationship between input flows used by the system and the corresponding output flows. Efficiency measurements are therefore associated with the quality of the transformation process. For example, they might express the amount of resources needed to achieve a given sales volume.

Generally speaking, effectiveness metrics indicate whether the *right* action is being carried out or not, while efficiency metrics show whether the action is being carried out in the *best* possible way or not.

2.2   Representation of the decision-making process

In order to build effective DSSs, we first need to describe in general terms how a decision-making process is articulated. In particular, we wish to understand the steps that lead individuals to make decisions and the extent of the influence exerted on them by the subjective attitudes of the decision makers and the specific context within which decisions are taken.

2.2.1 Rationality and problem solving

A *decision* is a choice from multiple alternatives, usually made with a fair degree of rationality. Each individual faces on a continual basis decisions that can be more or less important, both in their personal and professional life. In this section, we will focus on decisions made by knowledge workers in public and private enterprises and organizations. These decisions may concern the development of a strategic plan and imply therefore substantial investment choices, the definition of marketing initiatives and related sales predictions, and the design of a production plan that allows the available human and technological resources to be employed in an effective and efficient way.

The decision-making process is part of a broader subject usually referred to as *problem solving*, which refers to the process through which individuals try to bridge the gap between the current operating conditions of a system (*as is*) and the supposedly better conditions to be achieved in the future (*to be*). In general, the transition of a system toward the desired state implies overcoming certain obstacles and is not easy to attain. This forces decision makers to devise a set of alternative feasible options to achieve the desired goal, and then choose a decision based on a comparison between the advantages and disadvantages of each option. Hence, the decision selected must be put into practice and then verified to determine if it has enabled the planned objectives to be achieved. When this fails to happen, the problem is reconsidered, according to a recursive logic.

Figure 2.3 outlines the structure of the problem-solving process. The *alternatives* represent the possible actions aimed at solving the given problem and helping to achieve the planned objective. In some instances, the number of alternatives being considered may be small. In the case of a credit agency that has to decide whether or not to grant a loan to an applicant, only two options exist, namely acceptance and rejection of the request. In other instances, the number of alternatives can be very large or even infinite. For example, the development of the annual logistic plan of a manufacturing company requires a choice to be made from an infinite number of alternative options.

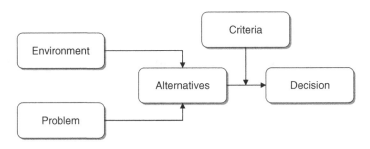

Figure 2.3 Logical flow of a problem-solving process

Criteria are the measurements of effectiveness of the various alternatives and correspond to the different kinds of system performance shown in Figure 2.1. A *rational* approach to decision making implies that the option fulfilling the best performance criteria is selected out of all possible alternatives. Besides economic criteria, which tend to prevail in the decision-making process within companies, it is however possible to identify other factors influencing a rational choice.

Economic. Economic factors are the most influential in decision-making processes, and are often aimed at the minimization of costs or the maximization of profits. For example, an annual logistic plan may be preferred over alternative plans if it achieves a reduction in total costs.

Technical. Options that are not technically feasible must be discarded. For instance, a production plan that exceeds the maximum capacity of a plant cannot be regarded as a feasible option.

Legal. Legal rationality implies that before adopting any choice the decision makers should verify whether it is compatible with the legislation in force within the application domain.

Ethical. Besides being compliant with the law, a decision should abide by the ethical principles and social rules of the community to which the system belongs.

Procedural. A decision may be considered ideal from an economic, legal and social standpoint, but it may be unworkable due to cultural limitations of the organization in terms of prevailing procedures and common practice.

Political. The decision maker must also assess the political consequences of a specific decision among individuals, departments and organizations.

The process of evaluating the alternatives may be divided into two main stages, shown in Figure 2.4: *exclusion* and *evaluation*. During the exclusion stage, compatibility rules and restrictions are applied to the alternative actions that were originally identified. Within this assessment process, some alternatives will be dropped from consideration, while the rest represent feasible options that will be promoted to evaluation. In the evaluation phase, feasible alternatives are compared to one another on the basis of the performance criteria, in order to identify the preferred decision as the best opportunity.

2.2.2 The decision-making process

A compelling representation of the decision-making process was proposed in the early 1960s, and still remains today a major methodological reference. The

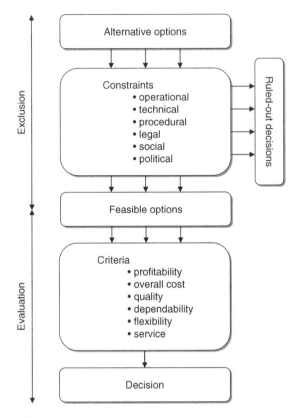

Figure 2.4   Logical structure of the decision-making process

model includes three phases, termed *intelligence*, *design* and *choice*. Figure 2.5 shows an extended version of the original scheme, which results from the inclusion of two additional phases, namely *implementation* and *control*.

Intelligence. In the *intelligence* phase the task of the decision maker is to identify, circumscribe and explicitly define the problem that emerges in the system under study. The analysis of the context and all the available information may allow decision makers to quickly grasp the signals and symptoms pointing to a corrective action to improve the system performance. For example, during the execution of a project the intelligence phase may consist of a comparison between the current progress of the activities and the original development plan. In general, it is important not to confuse the problem with the symptoms. For example, suppose that an e-commerce bookseller receives a complaint concerning late delivery of a book order placed on-line. Such inconvenience may be interpreted as the problem and be tackled by arranging a second delivery by priority shipping to circumvent the dissatisfaction of the customer. On the other

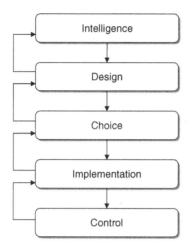

Figure 2.5   Phases of the decision-making process

hand, this may be the symptom of a broader problem, due to an understaffed shipping department where human errors are likely to arise under pressure.

Design. In the *design* phase actions aimed at solving the identified problem should be developed and planned. At this level, the experience and creativity of the decision makers play a critical role, as they are asked to devise viable solutions that ultimately allow the intended purpose to be achieved. Where the number of available actions is small, decision makers can make an explicit enumeration of the alternatives to identify the best solution. If, on the other hand, the number of alternatives is very large, or even unlimited, their identification occurs in an implicit way, usually through a description of the rules that feasible actions should satisfy. For example, these rules may directly translate into the constraints of an optimization model.

Choice. Once the alternative actions have been identified, it is necessary to evaluate them on the basis of the performance criteria deemed significant. Mathematical models and the corresponding solution methods usually play a valuable role during the *choice* phase. For example, optimization models and methods allow the best solution to be found in very complex situations involving countless or even infinite feasible solutions. On the other hand, decision trees can be used to handle decision-making processes influenced by stochastic events.

Implementation. When the best alternative has been selected by the decision maker, it is transformed into actions by means of an *implementation* plan. This involves assigning responsibilities and roles to all those involved into the action plan.

Control. Once the action has been implemented, it is finally necessary to verify and check that the original expectations have been satisfied and the effects of the action match the original intentions. In particular, the differences between the values of the performance indicators identified in the choice phase and the values actually observed at the end of the implementation plan should be measured. In an adequately planned DSS, the results of these evaluations translate into experience and information, which are then transferred into the data warehouse to be used during subsequent decision-making processes.

The most relevant aspects characterizing a decision-making process can be briefly summarized as follows.

- Decisions are often devised by a group of individuals instead of a single decision maker.

- The number of alternative actions may be very high, and sometimes unlimited.

- The effects of a given decision usually appear later, not immediately.

- The decisions made within a public or private enterprise or organization are often interconnected and determine broad effects. Each decision has consequences for many individuals and several parts of the organization.

- During the decision-making process knowledge workers are asked to access data and information, and work on them based on a conceptual and analytical framework.

- Feedback plays an important role in providing information and knowledge for future decision-making processes within a given organization.

- In most instances, the decision-making process has multiple goals, with different performance indicators, that might also be in conflict with one another.

- Many decisions are made in a fuzzy context and entail risk factors. The level of propensity or aversion to risk varies significantly among different individuals.

- Experiments carried out in a real-world system, according to a *trial-and-error* scheme, are too costly and risky to be of practical use for decision making.

- The dynamics in which an enterprise operates, strongly affected by the pressure of a competitive environment, imply that knowledge workers need to address situations and make decisions quickly and in a timely fashion.

2.2.3 Types of decisions

Defining a taxonomy of decisions may prove useful during the design of a DSS, since it is likely that decision-making processes with similar characteristics may be supported by the same set of methodologies. Decisions can be classified in terms of two main dimensions, according to their *nature* and *scope*. Each dimension will be subdivided into three classes, giving a total of nine possible combinations, as shown in Figure 2.6.

According to their nature, decisions can be classified as *structured*, *unstructured* or *semi-structured*.

Structured decisions. A decision is structured if it is based on a well-defined and recurring decision-making procedure. In most cases structured decisions can be traced back to an algorithm, which may be more or less explicit for decision makers, and are therefore better suited for automation. More specifically, we have a structured decision if input flows, output flows and the transformations performed by the system can be clearly described in the three phases of intelligence, design and choice. In this case, we will also say that each component phase is structured in its turn. Actually, even decisions that appear fully structured require in most cases the direct intervention of decision makers to cope with unexpected events, caused for example by unusual values of some input flows.

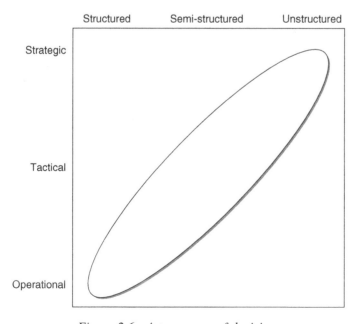

Figure 2.6 A taxonomy of decisions

Unstructured decisions. A decision is said to be unstructured if the three phases of intelligence, design and choice are also unstructured. This means that for each phase there is at least one element in the system (input flows, output flows and the transformation processes) that cannot be described in detail and reduced to a predefined sequence of steps. Such an event may occur when a decision-making process is faced for the first time or if it happens very seldom. In this type of decisions the role of knowledge workers is fundamental, and business intelligence systems may provide support to decision makers through timely and versatile access to information.

Semi-structured decisions. A decision is semi-structured when some phases are structured and others are not. Most decisions faced by knowledge workers in managing public or private enterprises or organizations are semi-structured. Hence, they can take advantage of DSSs and a business intelligence environment primarily in two ways. For the unstructured phases of the decision-making process, business intelligence tools may offer a passive type of support which translates into timely and versatile access to information. For the structured phases it is possible to provide an active form of support through mathematical models and algorithms that allow significant parts of the decision-making process to be automated.

Sometimes situations may arise where the nature of a decision cannot be easily identified unambiguously. When facing the same problem, such as establishing the sale price of a product, different decision makers operating in different organizations may come up with dissimilar choices. For example, a first decision maker may believe that the best sale price can be obtained by comparing cost and price–demand elasticity curves. As a consequence, such decision maker may consider the choice phase of the decision-making process as structured. By contrast, a second decision maker may believe that the elasticity curve does not reflect all the factors influencing the response of the market to price variations since some of these elements cannot be quantified. For this individual the choice phase turns out to be unstructured or at most semi-structured. Examples 2.1, 2.2 and 2.3 describe structured, semi-structured and unstructured decisions, respectively.

Example 2.1 – A structured decision. A paper mill produces for the company warehouse paper sheets in different standard sizes that are subsequently cut to size for customers. Specifically, customers submit orders in terms of type of paper, quantity and size. The sizes specified in the orders are usually smaller than standard sizes and must be cut out of these.

The paper mill is therefore forced to consider how the sizes required to fulfill orders should best be combined and cut from standard sizes so as to minimize paper waste. This decision is common to many industries (paper, aluminum, wood, steel, glass, fabric) and can be very well supported by optimization models. However, even in connection with such structured decisions, particular circumstances and specific input values may require intervention by the decision maker to modify the plans obtained by means of optimization models. For example, the company may wish to favor a specific request of a customer considered strategic, introducing a *fast-processing lane* in the cutting plan, even if this may involve more wasted material during the cutting stage.

Example 2.2 – A semi-structured decision. The logistics manager of a manufacturing company needs to develop an annual plan. The logistic plan determines the allocation to each plant of the production volumes forecasted for the different market areas, the purchase of materials from each supplier with the related volumes and delivery times, the production lots for each manufacturing stage, the stock levels of sub-assemblies and end items, and the distribution of end items to the market areas. These decisions have a great economic and organizational impact that might greatly benefit from the adoption of a DSS based on large-scale optimization models. However, it is likely that in a real situation some elements are left to discretion of the decision makers, who may prefer a given logistic plan over another, even if it implies moderately higher costs compared to the optimal plan proposed by the model. For example, it might be appropriate to maintain unaltered the supply of parts purchased from a given supplier who is considered strategic for the future even though this supplier is less competitive than others, that are instead preferred by the optimization model in terms of minimum cost.

Example 2.3 – An unstructured decision. Consider an enterprise that is the target of a hostile takeover by a public offer made by a direct competitor. There are various possible defensive decisions and actions that are strongly dependent on the context in which the enterprise operates

> and the offer is made. It is difficult to envisage a systematic description
> of the decision process that might be later reproduced in other similar
> cases.

From the above examples it emerges that the nature of a decision process
depends on many factors, including:

- the characteristics of the organization within which the system is placed;

- the subjective attitudes of the decision makers;

- the availability of appropriate problem-solving methodologies;

- the availability of effective decision support tools.

Depending on their scope, decisions can be classified as *strategic*, *tactical* and
operational.

Strategic decisions. Decisions are strategic when they affect the entire organization or at least a substantial part of it for a long period of time. Strategic
decisions strongly influence the general objectives and policies of an enterprise.
As a consequence, strategic decisions are taken at a higher organizational level,
usually by the company top management.

Tactical decisions. Tactical decisions affect only parts of an enterprise and
are usually restricted to a single department. The time span is limited to a
medium-term horizon, typically up to a year. Tactical decisions place themselves within the context determined by strategic decisions. In a company
hierarchy, tactical decisions are made by middle managers, such as the heads
of the company departments.

Operational decisions. Operational decisions refer to specific activities carried
out within an organization and have a modest impact on the future. Operational decisions are framed within the elements and conditions determined by
strategic and tactical decisions. Therefore, they are usually made at a lower
organizational level, by knowledge workers responsible for a single activity or
task such as sub-department heads, workshop foremen, back-office heads.

The characteristics of the information required in a decision-making process will change depending on the scope of the decisions to be supported, and
consequently also the orientation of a DSS will vary accordingly. Figure 2.7
shows variations in the characteristics of the information as the scope of the
decisions changes. The scheme can be used as an assessment tool when designing a DSS: once the scope of the decisions for which the system is intended

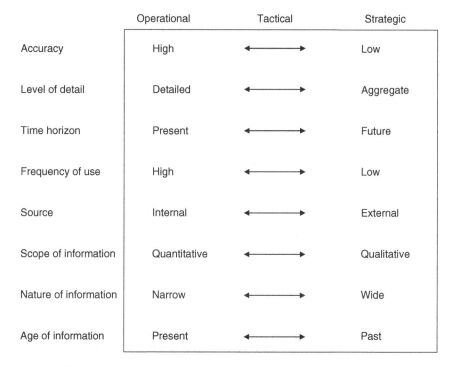

	Operational		Strategic
Accuracy	High	←——→	Low
Level of detail	Detailed	←——→	Aggregate
Time horizon	Present	←——→	Future
Frequency of use	High	←——→	Low
Source	Internal	←——→	External
Scope of information	Quantitative	←——→	Qualitative
Nature of information	Narrow	←——→	Wide
Age of information	Present	←——→	Past

Figure 2.7   Characteristics of the information in terms of the scope of decisions

has been set, the scheme can be used to establish whether the decision-making process is adequately supported by the right information.

Although nature and scope are not perfectly correlated, most real-world decisions fall within the ellipse shown in Figure 2.6: most strategic decisions are unstructured, while most operational decisions are structured and most tactical decisions are semi-structured. This empirical remark is useful when defining in advance the characteristics of a DSS to facilitate a decision-making process of specific nature and scope.

2.2.4   Approaches to the decision-making process

As observed above, the subjective orientation of decision makers across an organization strongly influences the structure of the decision-making process. In this section we will review the major approaches that prevail in the management of complex organizations, and examine the implications when designing a DSS.

A preliminary distinction should be made between a *rational* approach and a *political-organizational* approach.

Rational approach. When a rational approach is followed, a decision maker considers major factors, such as economic, technical, legal, ethical, procedural and political, also establishing the criteria of evaluation so as to assess different options and then select the best decision. In this context, a DSS may help both in a passive way, through timely and versatile access to information, and in an active way, through the use of mathematical models for decision making.

Political-organizational approach. When a political-organizational approach is pursued, a decision maker proceeds in a more instinctual and less systematic way. Decisions are not based on clearly defined alternatives and selection criteria. As a consequence, a DSS can only help in a passive way, providing timely and versatile access to information. It might also be useful during discussions and negotiations in those decision-making processes that involve multiple actors, such as managers operating in different departments.

Within the rational approach we can further distinguish between two alternative ways in which the actual decision-making process influences decisions: *absolute rationality* and *bounded rationality*.

Absolute rationality. The term 'absolute rationality' refers to a decision-making process for which multiple performance indicators can be reduced to a single criterion, which therefore naturally lends itself to an optimization model. For example, a production manager who has to put together a medium-term logistic plan may be able to convert all performance indicators into monetary units, and therefore subsequently derive the solution with the minimum cost. This implies that non-monetary indicators, such as stock volumes or the number of days of delay in handling a given order, should be transformed into monetary measurement units. From a methodological perspective, this implies that a multi-objective optimization problem is transformed into a single-objective problem by expressing all the relevant factors in a common measurement unit that allows the heterogeneous objectives to be added together.

Bounded rationality. Bounded rationality occurs whenever it is not possible to meaningfully reduce multiple criteria into a single objective, so that the decision maker considers an option to be satisfactory when the corresponding performance indicators fall above or below prefixed threshold values. For instance, a production plan is acceptable if its cost is sufficiently low, the stock quantities are within a given threshold, and the service time is below customers expectations. Therefore, the concept of bounded rationality captures the rational choices that are constrained by the limits of knowledge and cognitive capability.

Most decision-making processes occurring within the enterprises and the public administration are aimed at making a decision that appears acceptable with respect to multiple evaluation criteria, and therefore decision processes based on bounded rationality are more likely to occur in practice.

2.3 Evolution of information systems

Decision support systems combine data and mathematical models to help decision makers in their work. To some extent they show connections with the information systems of an organization. Hence, it is worth describing in this section a time-line for the evolution of information systems, since this highlights how data processing has developed and has been used within companies.

Digital computers made their appearance in the late 1940s, and soon began to be applied in the business environment. The first decades saw a rush toward information technology development, usually under the mantra of *data processing*. They were characterized by the widespread diffusion of applications that achieved an increase in efficiency by automating routine operations within companies, especially in administration, production, research and development.

In the 1970s there began to arise within enterprises increasingly complex needs to devise software applications, called *management information systems* (MIS), in order to ease access to useful and timely information for decision makers. However, attempts to develop such systems were hampered by the state of information technologies at the time. The mainframe computers of those days lacked graphic visualization capabilities, and communicated with users through character-based computer terminals and dot printers. A further difficulty lay in the organizational structure of companies, based on a highly centralized information systems department, usually resulting in very long and frustrating time delays in implementing changes or extensions to the available applications.

From the late 1980s the introduction of personal computers with operating systems featuring graphic interfaces and pointing devices, such as a mouse or an optical pen, had two major consequences. On the one hand, it became possible to implement applications capable of sophisticated interactions and graphic presentation of results, a prerequisite for providing decision makers with really useful support tools. On the other hand, knowledge workers could rely on autonomous processing tools which made them substantially independent of the company information systems department, thus avoiding the above-mentioned time lag in data access. This led the most proactive knowledge workers to create local databases and develop simulation models, for example by means of spreadsheets, which can be regarded as true ancestors of today's business intelligence architectures. The increase in independent processing capabilities held by users, usually referred to as *end user computing*, was a critical enabling factor for future developments, as it helped circumscribe the importance of information systems departments.

Meanwhile, the initial concept of *decision support system* was also introduced, whose exact meaning will be described in the next section. Later developments brought to light new types of applications and architectures: *executive information systems* and *strategic information systems* were first introduced toward the late 1980s to support executives in the decision-making process. Such systems were intended for unstructured decision-making processes and therefore represented passive support systems oriented toward timely and easy access to information.

From the early 1990s, network architectures and distributed information systems based on *client–server* computing models began to be widely adopted. Moreover, there arose the need to logically and physically separate the databases intended for DSSs from operational information systems. This brought about the concepts of *data warehouses* and *data marts*, which will be described in Chapter 3.

Finally, toward the end of the 1990s, the term *business intelligence* began to be used to generally address the architecture containing DSSs, analytical methodologies and models used to transform data into useful information and knowledge for decision makers, as discussed in Chapter 1.

2.4   Definition of decision support system

Since the late 1980s, a decision support system has been defined as an interactive computer system helping decision makers to combine data and models to solve semi-structured and unstructured problems. This definition entails the three main elements of a DSS shown in Figure 2.8: a database, a repository of mathematical models and a module for handling the dialogue between the

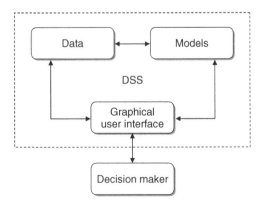

Figure 2.8   Structure of a decision support system

system and the users. It thus highlights the role of DSSs as the focal point of evolution trends in two distinct areas: on the one hand, data processing and information technologies; and on the other hand, the disciplines addressing the study of mathematical models and methods, such as operations research and statistics.

Indeed, despite significant improvements achieved in both areas, the actual implementation of applications that could be used by knowledge workers has been troublesome. Information technology in the early 1970s, as discussed in the previous section, mainly consisted of applications for accounting and administration, their capabilities restricted to the processing of large amounts of transactions and the production of summary reports. The partial failure of management information systems indicated that even the essential objective of attaining timely and versatile access to information was not within easy reach. On the other hand, although mathematical models were made more flexible by resort to innovative user interaction techniques, they were still mostly based on regulatory methods, suitable for operational decisions characterized by unique objectives rather than tactical and strategic decisions, usually less structured. As we have observed, semi-structured and unstructured decision processes are heavily influenced by subjective opinions, so that knowledge workers have to directly analyze multiple performance indicators which it would be difficult to reduce to a single decision-making criterion.

It is worth highlighting the relevant features of a DSS in order to circum-scribe the definition given above and to better understand its role.

Effectiveness. Decision support systems should help knowledge workers to reach more effective decisions. In this respect they are a fundamental compo-nent of business intelligence architectures. Note that this does not necessarily imply an increased efficiency in the decision-making process. In fact, the adop-tion of a DSS may entail a more accurate analysis and therefore require a greater time investment by decision makers. However, the greater effort required will usually result in better decisions.

Mathematical models. In order to achieve more effective decisions, a DSS makes use of mathematical models, borrowed from disciplines such as oper-ations research and statistics, which are applied to the data contained in data warehouses and data marts. The use of analytical models to transform data into knowledge and provide active support is the main characteristic that sets apart a DSS from a simple information system.

Integration in the decision-making process. A DSS should provide help for different kinds of knowledge workers, within the same application domain, particularly in respect of semi-structured and unstructured decision processes, both of an individual and a collective nature. Further, a DSS is intended for decision-making processes that are strategic, tactical and operational in scope. Moreover, decision makers should have the opportunity to integrate in a DSS their preferences and competencies, adapting it to their needs rather than passively accepting what comes out of it. In this way, a DSS may progressively take the role of a key component in the problem-solving methodology adopted by decision makers, enabling a *proactive* and *perceptive* decision-making style, instead of a *reactive* and *by-exception* attitude, in order to anticipate any rapidly evolving dynamic phenomenon.

Organizational role. In many situations the users of a DSS operate at different hierarchical levels within an enterprise, and a DSS tends to encourage communication between the various parts of an organization. By providing support for sequential and interdependent decision processes, a DSS can keep track of the analysis and the information that led to a specific decision.

Flexibility. A DSS must be flexible and adaptable in order to incorporate the changes required to reflect modifications in the environment or in the decision-making process. Moreover, it should be easy to use, with user-friendly and intuitive interaction methods and high-quality graphics for presenting the information extracted or generated. It is becoming increasingly common for DSSs to feature a web-browser interface to communicate with users.

The structure of the DSS shown in Figure 2.8 is extended in Figure 2.9 to include some new components.

Data management. The data management module includes a database designed to contain the data required by the decision-making processes to which the DSS is addressed. In most applications the database is a data mart, as we will see

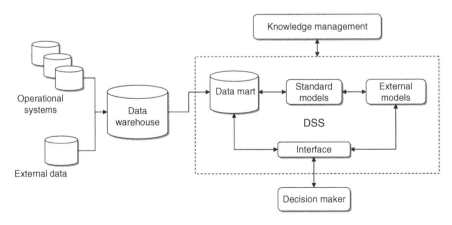

Figure 2.9   Extended structure of a decision support system

in Chapter 3. Keeping in mind current business intelligence architectures, the data management module of a DSS is usually connected with a company data warehouse, also described in Chapter 3, which represents the main repository of the data available to develop a business intelligence analysis.

Model management. The model management module provides end users with a collection of mathematical models derived from operations research, statistics and financial analysis. These are usually relatively simple models that allow analytical investigations to be carried out that are very helpful during the decision-making process. To illustrate the role played by the model management module one can think of the analytical functions offered by current spreadsheets. These include simple optimization models, financial and actuarial analysis models and statistical functionalities. Moreover, the model management module helps the activities of knowledge workers by means of high-level languages for the development of *ad hoc* models. In certain applications such a module may be integrated with more complex models, referred to as *external models* in Figure 2.9, created to carry out specific analysis tasks. For example, a large-scale optimization model formulated to develop the annual logistic plan of a manufacturing company falls in this category.

Interactions. In most applications, knowledge workers use a DSS interactively to carry out their analyses. The module responsible for these interactions is expected to receive input data from users in the easiest and most intuitive way, usually through the graphic interface of a web browser, and then to return the extracted information and the knowledge generated by the system in an appropriate graphical form.

Knowledge management. The knowledge management module is also interconnected with the company knowledge management integrated system. It allows decision makers to draw on the various forms of collective knowledge, usually unstructured, that represents the *corporate culture*.

This section concludes with a summary of the major potential advantages deriving from the adoption of a DSS:

- an increase in the number of alternatives or options considered;

- an increase in the number of effective decisions devised;

- a greater awareness and a deeper understanding of the domain analyzed and the problems investigated;

- the possibility of executing scenario and what-if analyses by varying the hypotheses and parameters of the mathematical models;

- an improved ability to react promptly to unexpected events and unforeseen situations;

- a value-added exploitation of the available data;

- an improved communication and coordination among the individuals and the organizational departments;

- more effective development of teamwork;

- a greater reliability of the control mechanisms, due to the increased intelligibility of the decision process.

2.5   Development of a decision support system

In this section we will describe the development phases of a DSS. Unlike other software applications, such as information systems and office automation tools, DSSs are usually not available as standard programs. Multidimensional analysis environments have facilitated and standardized the access to passive business intelligence functions. However, in order to develop most DSSs a specific project is still required.

Figure 2.10 shows the major steps in the development of a DSS. The logical flow of the activities is shown by the solid arrows. The dotted arrows in the opposite direction indicate revisions of one or more phases that might become necessary during the development of the system, through a feedback mechanism. We describe now in detail how each phase is carried out.

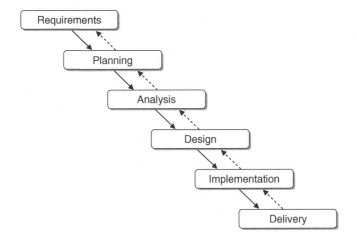

Figure 2.10   Phases in the development of a decision support system

Planning. The main purpose of the planning phase is to understand the needs and opportunities, sometimes characterized by *weak signals*, and to translate them into a project and later into a successful DSS. Planning usually involves a *feasibility study* to address the question: *Why do we wish to develop a DSS?* During the feasibility analysis, general and specific objectives of the system, recipients, possible benefits, execution times and costs are laid down. It is not easy to identify the benefits of a DSS. As already observed, the main advantage for most DSSs is not so much an increased effectiveness as an improvement in decision quality, which is difficult to predict a priori. The elements required to decide whether or not the system should be implemented become available at the end of the feasibility study. A negative decision will call a halt to the development of the project, although of course this may be reconsidered at a later time. If one decides to proceed with the system, the planning phase should be followed by the definition of the activities, tasks, responsibilities and development phases, for which classical project management methodologies should be used.

Analysis. In the analysis phase, it is necessary to define in detail the functions of the DSS to be developed, by further developing and elaborating the preliminary conclusions achieved during the feasibility study. A response should therefore be given to the following question: *What should the DSS accomplish, and who will use it, when and how?* To provide an answer, it is necessary to analyze the decision processes to be supported, to try to thoroughly understand all interrelations existing between the problems addressed and the surrounding environment. The organizational implications determined by a DSS should be assessed. The analysis also involves mapping out the actual decision processes and imagining what the new processes will look like once the DSS is in place. Finally, it is necessary to explore the data in order to understand how much and what type of information already exists and what information can be retrieved from external sources.

Design. During the design phase the main question is: *How will the DSS work?* The entire architecture of the system is therefore defined, through the identification of the hardware technology platforms, the network structure, the software tools to develop the applications and the specific database to be used. It is also necessary to define in detail the interactions with the users, by means of input masks, graphic visualizations on the screen and printed reports. In recent years the web browser has become an important interaction tool, and has certainly contributed to the harmonization of, and to the simplification of the problems related to, communication between knowledge workers and computers. A further aspect that should be clarified during the design phase is the *make-or-buy* choice – whether to subcontract the implementation of the DSS to third parties, in whole or in part.

Implementation. Once the specifications have been laid down, it is time for implementation, testing and the actual installation, when the DSS is rolled out and put to work. Any problems faced in this last phase can be traced back to project management methods. A further aspect of the implementation phase, which is often overlooked, relates to the overall impact on the organization determined by the new system. Such effects should be monitored using *change management* techniques, making sure that no one feels excluded from the organizational innovation process and rejects the DSS.

Sometimes a project may not come to a successful conclusion, may not succeed in fulfilling expectations, or may even turn out to be a complete failure. However, there are ways to reduce the risk of failure. The most significant of these is based on the use of *rapid prototyping development* where, instead of implementing the system as a whole, the approach is to identify a sequence of autonomous subsystems, of limited capabilities, and develop these subsystems step by step until the final stage is reached corresponding to the fully developed DSS.

Rapid prototyping development offers clear advantages. Each subsystem can be actually developed more quickly and therefore is more readily available. Moreover, when a subsystem is released to users it is possible to verify its conformity with the intended purpose and test its functions even if these are still not fully developed. Hence, the evolutionary development of a DSS allows the risk of failure to be minimized. Furthermore, intermediate tests make it possible to promptly correct most design errors. In a situation where there is a clear discrepancy between the features of the prototype developed and the users' expectations it is even possible to come to an early interruption of the project.

Other development methods can be effectively adopted in order to speed up the implementation of the software. These include *agile development techniques* and *extreme programming techniques*.

A further aspect that should not be overlooked is the periodic administration and revision of the DSS. Since no software application should be considered completely finished and unchangeable over time, this is even more critical for a tool intended to support knowledge workers in the decision-making process in rapidly changing dynamic situations. As a consequence, it is necessary to develop a DSS by making provision for future changes and adjustments.

In this respect it is possible to identify two clear advantages offered by the development methods described so far. First, if a DSS has been developed using rapid prototyping techniques, it is likely to be more amenable to periodic extensions and revisions. Furthermore, the mathematical models used to develop a

DSS are suitable for incorporating evolving changes and updates which often reduce to simple modifications in the expressions that describe the model.

We conclude this section by considering the major critical factors that may affect the degree of success of a DSS.

Integration. The design and development of a DSS require a significant number of methodologies, tools, models, individuals and organizational processes to work in harmony. This results in a highly complex project requiring diverse competencies. The role of a *system integrator* is thus essential. A system integrator is an expert in all the aspects involved in the development of a DSS, such as information system architectures, decision-making processes, mathematical models and solution methods. This role is usually performed by a third party who may also exert a positive influence as an *agent of innovation* capable of overcoming most of the resistance to change that often arises in every organization.

Involvement. The exclusion or marginalization from the project team of knowledge workers who will actually use the system once it is implemented is a mistake that is sometimes made during the design and development of a DSS. In many cases this happens because a DSS is mistakenly considered a mere computer application, the development of which is assigned solely or primarily to the information systems department. Conversely, the involvement of decision makers and users during the development process is of primary importance to avert their inclination to reject a tool that they perceive as alien. It is also necessary to promote informal communication processes, especially during the design and implementation phases of the first prototypes.

Uncertainty. In general, costs are not a major concern in the implementation of a DSS, and the advantage of devising more effective decisions largely offsets the development costs incurred. Of course, it is appropriate to reduce the project uncertainty through prototyping, user friendliness, system tests during the preliminary stages and an evolutionary implementation.

2.6   Notes and readings

Among the many volumes devoted to decision support systems, we wish to suggest the following readings: Mallach (2000), Turban *et al.* (2005), Klein and Methlie (1995) and Dhar and Stein (1997). For a recent review, see Shim *et al.* (2002). Moss and Atre (2003) analyze the process of developing a DSS. Among the major contributions dating back to the foundations of DSSs, see Simon (1969, 1977), Gorry and Scott Morton (1971), Keen and Scott Morton (1978) and Sprague (1980).

3

Data warehousing

As observed in Chapter 2, from the mid-1990s the need was felt for a logical and material separation between the databases feeding input data into decision support systems and business intelligence architectures on the one hand, and operational information systems on the other.

In this chapter we will describe the features of *data warehouses* and *data marts*, illustrating the factors that led to their conception, and highlighting the major differences between them and operational systems, and discussing the requirements concerning data quality. Then we will examine the architecture of a data warehouse, pointing out the role of ETL tools and metadata. The last part of the chapter will be devoted to on-line analytical processing operations and analyses that can be performed by using multidimensional cubes and hierarchies of concepts.

We will focus our discussion on the goals and functions of a data warehouse, deliberately avoiding technical issues relating to their development. For these latter, readers may refer to more specific books fully devoted to the subject, indicated in the last section of the chapter.

3.1 Definition of data warehouse

As its name suggests, a data warehouse is the foremost repository for the data available for developing business intelligence architectures and decision support systems. The term *data warehousing* indicates the whole set of interrelated activities involved in designing, implementing and using a data warehouse.

It is possible to identify three main categories of data feeding into a data warehouse: *internal data*, *external data* and *personal data*.

Business Intelligence: Data Mining and Optimization for Decision Making   C. Vercellis
© 2009 John Wiley & Sons, Ltd

Internal data. Internal data are stored for the most part in the databases, referred to as *transactional systems* or *operational systems*, that are the backbone of an enterprise information system. Internal data are gathered through transactional applications that routinely preside over the operations of a company, such as administration, accounting, production and logistics. This collection of transactional software applications is termed *enterprise resource planning* (ERP). The data stored in the operational systems usually deal with the main entities involved in a company processes, namely customers, products, sales, employees and suppliers. These data usually come from different components of the information system:

- *back-office systems*, that collect basic transactional records such as orders, invoices, inventories, production and logistics data;

- *front-office systems*, that contain data originating from call-center activities, customer assistance, execution of marketing campaigns;

- *web-based systems*, that gather sales transactions on e-commerce websites, visits to websites, data available on forms filled out by existing and prospective customers.

External data. There are several sources of external data that may be used to extend the wealth of information stored in the internal databases. For example, some agencies gather and make available data relative to sales, market share and future trend predictions for specific business industries, as well as economic and financial indicators. Other agencies provide data market surveys and consumer opinions collected through questionnaires.

A further significant source of external data is provided by *geographic information systems* (GIS), which represent a set of applications for acquiring, organizing, storing and presenting territorial data. These contain information relative to entities having a specific geographic position. Each entity is therefore associated with latitude and longitude coordinates, along with some other attributes, usually originating from relational databases and actually depending on the application domain. Hence, these data allow to subject-specific analyses to be carried out on the data associated with geographic elements and the results to be graphically visualized.

Personal data. In most cases, decision makers performing a business intelligence analysis also rely on information and personal assessments stored inside worksheets or local databases located in their computers. The retrieval of such information and its integration with structured data from internal and external sources is one of the objectives of knowledge management systems.

Software applications that are at the heart of operational systems are referred to as *on-line transaction processing* (OLTP). On the other hand, the whole set of tools aimed at performing business intelligence analyses and supporting decision-making processes go by the name of *on-line analytical processing* (OLAP). We can therefore assume that the function of a data warehouse is to provide input data to OLAP applications.

There are several reasons for implementing a data warehouse separately from the databases supporting OLTP applications in an enterprise. Among them, we recall here the most relevant.

Integration. In many instances, decision support systems must access information originating from several data sources, distributed across different parts of an organization or deriving from external sources. A data warehouse integrating multiple and often heterogeneous sources is then required to promote and facilitate the access to information. Data integration may be achieved by means of different techniques – for example, by using uniform encoding methods, converting to standard measurement units and achieving a semantic homogeneity of information.

Quality. The data transferred from operational systems into the data warehouse are examined and corrected in order to obtain reliable and error-free information, as much as possible. Needless to say, this increases the practical value of business intelligence systems developed starting from the data contained in a data warehouse.

Efficiency. Queries aimed at extracting information for a business intelligence analysis may turn out to be burdensome in terms of computing resources and processing time. As a consequence, if a 'killer' query were directed to the transactional systems it would risk severely compromising the efficiency required by enterprise resource planning applications, with negative consequences on the routine activities of a company. A better solution is then to direct complex queries for OLAP analyses to the data warehouse, physically separated from the operational systems.

Extendability. The data stored in transactional systems stretch over a limited time span in the past. Indeed, due to limitations on memory capacity, data relative to past periods are regularly removed from OLTP systems and permanently archived in off-line mass-storage devices, such as DVDs or magnetic tapes. On the other hand, business intelligence systems and prediction models need to access all available past data to be able to grasp trends and detect recurrent patterns. This is possible due to the ability of data warehouses to retain historical information.

In light of the previous remarks, we can define a data warehouse as a collection of data supporting decision-making processes and business intelligence systems having the following characteristics.

Entity-oriented. The data contained in a data warehouse are primarily concerned with the main entities of interest for the analysis, such as products, customers, orders and sales. On the other hand, transactional systems are more oriented toward operational activities and are based on each single transaction recorded by enterprise resource planning applications. During a business intelligence analysis, orientation toward the entities allows the performance of a company to be more easily evaluated and any potential source of inefficiencies to be detected.

Integrated. The data originating from the different sources are integrated and homogenized as they are loaded into a data warehouse. For example, measurement units and encodings are harmonized and made consistent.

Time-variant. All data entered in a data warehouse are labeled with the time period to which they refer. We can fairly relate the data stored in a data warehouse to a sequence of nonvolatile snapshot pictures, taken at successive times and bearing the label of the reference period. As a consequence, the temporal dimension in any data warehouse is a critical element that plays a predominant role. In this way decision support applications may develop historical trend analysis.

Persistent. Once they have been loaded into a data warehouse, data are usually not modified further and are held permanently. This feature makes it easier to organize read-only access by users and simplifies the updating process, avoiding concurrency which is of critical importance for operational systems.

Consolidated. Usually some data stored in a data warehouse are obtained as partial summaries of primary data belonging to the operational systems from which they originate. For example, a mobile phone company may store in a data warehouse the total cost of the calls placed by each customer in a week, subdivided by traffic routes and by type of service selected, instead of storing the individual calls recorded by the operational systems. The reason for such consolidation is twofold: on one hand, the reduction in the space required to store in the data warehouse the data accumulated over the years; on the other hand, consolidated information may be able to better meet the needs of business intelligence systems.

Denormalized. Unlike operational databases, the data stored in a data warehouse are not structured in normal form but can instead make provision for redundancies, to allow shorter response time to complex queries.

Granularity represents the highest level of detail expressed by the primary data contained in a data warehouse, also referred to as *atomic data*. Obviously, the granularity of a data warehouse cannot exceed that of the original data

Table 3.1   Differences between OLTP and OLAP systems

Characteristic	OLTP	OLAP
volatility	dynamic data	static data
timeliness	current data only	current and historical data
time dimension	implicit and current	explicit and variant
granularity	detailed data	aggregated and consolidated data
updating	continuous and irregular	periodic and regular
activities	repetitive	unpredictable
flexibility	low	high
performance	high, few seconds per query	may be low for complex queries
users	employees	knowledge workers
functions	operational	analytical
purpose of use	transactions	complex queries and decision support
priority	high performance	high flexibility
metrics	transaction rate	effective response
size	megabytes to gigabytes	gigabytes to terabytes

sources. In general, it is strictly lower due to consolidation aimed at reducing storage occupancy, as described above.

The design philosophy behind data warehouses is quite different from that adopted for operational databases. Table 3.1 summarizes the main differences between OLTP and OLAP systems.

3.1.1   Data marts

Data marts are systems that gather all the data required by a specific company department, such as marketing or logistics, for the purpose of performing business intelligence analyses and executing decision support applications specific to the function itself. Therefore, a data mart can be considered as a functional or departmental data warehouse of a smaller size and a more specific type than the overall company data warehouse.

A data mart therefore contains a subset of the data stored in the company data warehouse, which are usually integrated with other data that the company department responsible for the data mart owns and deems of interest. For example, a marketing data mart will contain data extracted from the central data warehouse, such as information on customers and sales transactions, but also additional data pertaining to the marketing function, such as the results of marketing campaigns run in the past.

Data warehouses and data marts thus share the same technological framework. In order to implement business intelligence applications, some companies prefer to design and develop in an incremental way a series of integrated data marts rather than a central data warehouse, in order to reduce the implementation time and uncertainties connected with the project.

3.1.2 Data quality

The need to verify, preserve and improve the quality of data is a constant concern of those responsible for the design and updating of a data warehouse. The main problems that might compromise the validity and integrity of the data are shown in Table 3.2.

More generally, we can identify the following major factors that may affect data quality.

Accuracy. To be useful for subsequent analyses, data must be highly accurate. For instance, it is necessary to verify that names and encodings are correctly represented and values are within admissible ranges.

Completeness. In order to avoid compromising the accuracy of business intelligence analyses, data should not include a large number of missing values. However, one should keep in mind that most learning and data mining techniques are capable of minimizing in a robust way the effects of partial incompleteness in the data.

Consistency. The form and content of the data must be consistent across the different data sources after the integration procedures, with respect to currency and measurement units.

Table 3.2   Data integrity: problems, causes and remedies

Problem	Cause	Remedy
incorrect data	data collected without due care	systematic checking of input data
	data entered incorrectly	data entry automation
	uncontrolled modification of data	implementation of a safety program for access and modifications
data not updated	data collection does not match user needs	timely updating and collection of data
		retrieval of updated data from the web
missing data	failure to collect the required data	identification of data needed via preliminary analysis and estimation of missing data

Timeliness. Data must be frequently updated, based on the objectives of the analysis. It is customary to arrange an update of the data warehouse regularly on a daily or at most weekly basis.

Non-redundancy. Data repetition and redundancy should be avoided in order to prevent waste of memory and possible inconsistencies. However, data can be replicated when the denormalization of a data warehouse may result in reduced response times to complex queries.

Relevance. Data must be relevant to the needs of the business intelligence system in order to add real value to the analyses that will be subsequently performed.

Interpretability. The meaning of the data should be well understood and correctly interpreted by the analysts, also based on the documentation available in the metadata describing a data warehouse, as illustrated in Section 3.2.2.

Accessibility. Data must be easily accessible by analysts and decision support applications.

3.2   Data warehouse architecture

The reference architecture of a data warehouse, shown in Figure 3.1, includes the following major functional components.

- The data warehouse itself, together with additional data marts, that contains the data and the functions that allow the data to be accessed, visualized and perhaps modified.

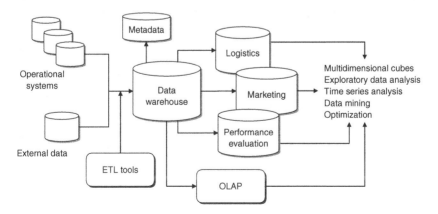

Figure 3.1   Architecture and functions of a data warehouse

- Data acquisition applications, also known as *extract, transform and load* (ETL) or *back-end* tools, which allow the data to be extracted, transformed and loaded into the data warehouse.

- Business intelligence and decision support applications, which represent the *front-end* and allow the knowledge workers to carry out the analyses and visualize the results.

The three-level distinction applies to the architecture shown in Figure 3.1 even from a technological perspective.

- The level of the data sources and the related ETL tools that are usually installed on one or more servers.

- The level of the data warehouse and any data mart, possibly available on one or more servers as well, and separated from those containing the data sources. This second level also includes the metadata documenting the origin and meaning of the records stored in the data warehouse.

- The level of the analyses that increase the value of the information contained in a data warehouse through query, reporting and possibly sophisticated decision support tools. The applications for business intelligence and decision support analysis are usually found on separate servers or directly on the client PC used by analysts and knowledge workers.

The same *database management system* platforms utilized to develop transactional systems are also adopted to implement data warehouses and data marts. Due to the response requirements raised by the complex queries addressed to a data warehouse, database management system platforms used for data warehousing are subject to different structuring and parameterizations with respect to transactional systems.

A data warehouse may be implemented according to different design approaches: *top-down*, *bottom-up* and *mixed*.

Top-down. The top-down methodology is based on the overall design of the data warehouse, and is therefore more systematic. However, it implies longer development times and higher risks of not being completed within schedule since the whole data warehouse is actually being developed.

Bottom-up. The bottom-up method is based on the use of prototypes and therefore system extensions are made according to a step-by-step scheme. This approach is usually quicker, provides more tangible results but lacks an overall vision of the entire system to be developed.

Mixed. The mixed methodology is based on the overall design of the data warehouse, but then proceeds with a prototyping approach, by sequentially implementing different parts of the entire system. This approach is highly practical and usually preferable, since it allows small and controlled steps to be taken while bearing in mind the whole picture.

The steps in the development of a data warehouse or a data mart can be summarized as follows.

- One or more processes within the organization to be represented in the data warehouse are identified, such as sales, logistics or accounting.

- The appropriate granularity to represent the selected processes is identified and the atomic level of the data is defined.

- The relevant measures to be expressed in the fact tables for multidimensional analysis are then chosen, as described in Section 3.3.

- Finally, the dimensions of the fact tables are determined.

3.2.1 ETL tools

ETL refers to the software tools that are devoted to performing in an automatic way three main functions: *extraction*, *transformation* and *loading* of data into the data warehouse.

Extraction. During the first phase, data are extracted from the available internal and external sources. A logical distinction can be made between the initial extraction, where the available data relative to all past periods are fed into the empty data warehouse, and the subsequent incremental extractions that update the data warehouse using new data that become available over time. The selection of data to be imported is based upon the data warehouse design, which in turn depends on the information needed by business intelligence analyses and decision support systems operating in a specific application domain.

Transformation. The goal of the cleaning and transformation phase is to improve the quality of the data extracted from the different sources, through the correction of inconsistencies, inaccuracies and missing values. Some of the major shortcomings that are removed during the data cleansing stage are:

- inconsistencies between values recorded in different attributes having the same meaning;
- data duplication;
- missing data;
- existence of inadmissible values.

During the cleaning phase, preset automatic rules are applied to correct most recurrent mistakes. In many instances, dictionaries with valid terms are used to substitute the supposedly incorrect terms, based upon the level of similarity. Moreover, during the transformation phase, additional data conversions occur in order to guarantee homogeneity and integration with respect to the different data sources. Furthermore, data aggregation and consolidation are performed in order to obtain the summaries that will reduce the response time required by subsequent queries and analyses for which the data warehouse is intended.

Loading. Finally, after being extracted and transformed, data are loaded into the tables of the data warehouse to make them available to analysts and decision support applications.

3.2.2   Metadata

In order to document the meaning of the data contained in a data warehouse, it is recommended to set up a specific information structure, known as *metadata*, i.e. data describing data. The metadata indicate for each attribute of a data warehouse the original source of the data, their meaning and the transformations to which they have been subjected. The documentation provided by metadata should be constantly kept up to date, in order to reflect any modification in the data warehouse structure. The documentation should be directly accessible to the data warehouse users, ideally through a web browser, according to the access rights pertaining to the roles of each analyst.

In particular, metadata should perform the following informative tasks:

- a documentation of the data warehouse structure: layout, logical views, dimensions, hierarchies, derived data, localization of any data mart;

- a documentation of the data genealogy, obtained by tagging the data sources from which data were extracted and by describing any transformation performed on the data themselves;

- a list keeping the usage statistics of the data warehouse, by indicating how many accesses to a field or to a logical view have been performed;

- a documentation of the general meaning of the data warehouse with respect to the application domain, by providing the definition of the terms utilized, and fully describing data properties, data ownership and loading policies.

3.3 Cubes and multidimensional analysis

The design of data warehouses and data marts is based on a multidimensional paradigm for data representation that provides at least two major advantages: on the functional side, it can guarantee fast response times even to complex queries, while on the logical side the dimensions naturally match the criteria followed by knowledge workers to perform their analyses.

The multidimensional representation is based on a *star schema* which contains two types of data tables: *dimension tables* and *fact tables*.

Dimension tables. In general, *dimensions* are associated with the entities around which the processes of an organization revolve. Dimension tables then correspond to primary entities contained in the data warehouse, and in most cases they directly derive from *master tables* stored in OLTP systems, such as customers, products, sales, locations and time. Each dimension table is often internally structured according to *hierarchical* relationships. For example, the temporal dimension is usually based upon two major hierarchies: {day, week, year} and {day, month, quarter, year}. Similarly, the location dimension may be hierarchically organized as {street, zip code, city, province, region, country, area}. Products in their turn have hierarchical structures such as {item, family, type} in the manufacturing industry and {item, category, department} in the retail industry. In a way, dimensions predetermine the main paths along which OLAP analyses will presumably be developed.

Fact tables. Fact tables usually refer to transactions and contain two types of data:

- links to dimension tables, that are required to properly reference the information contained in each fact table;

- numerical values of the attributes that characterize the corresponding transactions and that represent the actual target of the subsequent OLAP analyses.

For example, a fact table may contain sales transactions and make reference to several dimension tables, such as customers, points of sale, products, suppliers, time. The corresponding measures of interest are attributes such as quantity of items sold, unit price and discount. In this example the fact table allows analysts to evaluate the trends of sales over time, either total, or referred to a single customer, or referred to a group of customers, that can be identified through any hierarchy induced by the dimension table associated with the customers. The analyst may also evaluate the trend over time of sales percentages relative to customers located in a specific region.

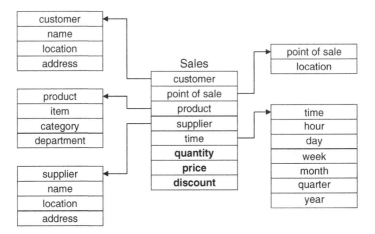

Figure 3.2   Example of a star schema

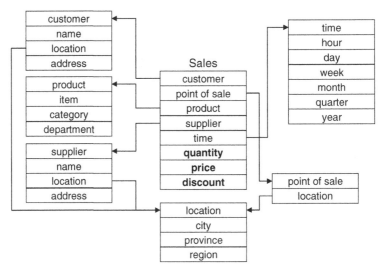

Figure 3.3   Example of a snowflake schema

Figure 3.2 shows the star schema associated with the fact table representing sales transactions. The fact table is placed in the middle of the schema and is linked to the dimension tables through appropriate references. The measures in the fact table appear in bold type.

Sometimes dimension tables are connected in their turn to other dimension tables, as shown in Figure 3.3, through a process of partial data standardization, in order to reduce memory use. In the given example the dimension table

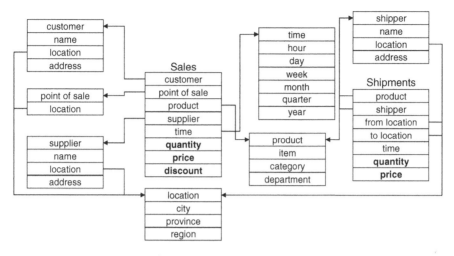

Figure 3.4   Example of a galaxy schema

referring to the location is in turn hierarchically connected with the dimension table containing geographical information. This brings about a *snowflake schema*.

A data warehouse includes several fact tables, interconnected with dimension tables, linked in their turn with other dimension tables. The latter type of schema, shown in Figure 3.4, is termed a *galaxy schema*.

A fact table connected with n dimension tables may be represented by an n-dimensional *data cube* where each axis corresponds to a dimension. Multidimensional cubes are a natural extension of the popular two-dimensional spreadsheets, which can be interpreted as two-dimensional cubes. For instance, consider a sales fact table developed along the three dimensions of {time, product, region}. Suppose we select only two dimensions for the analysis, such as {time, product}, having preset the region attribute along the three values {USA, Asia, Europa}. In this way we obtain the three two-dimensional tables in which the rows correspond to quarters of a year and the columns to products (see Tables 3.3–3.5). The cube shown in Figure 3.5 is a three-dimensional illustration of the same sales fact table. Atomic data are represented by 36 cells that can be obtained by crossing all possible values along the three dimensions: time {Q1, Q2, Q3, Q4}, region {USA, Asia, Europa} and product {TV, PC, DVD}. These atomic cells can be supplemented by 44 cells corresponding to the summary values obtained through consolidation along one or more dimensions, as shown by the cube in the figure.

Suppose that the sales fact table also contains a fourth dimension represented by the suppliers. The corresponding data cube constitutes a structure in

Table 3.3   Two-dimensional view of sales data in the USA

	region = USA		
	product		
time	TV	PC	DVD
Q1	980	546	165
Q2	765	456	231
Q3	879	481	192
Q4	986	643	203

Table 3.4   Two-dimensional view of sales data in Asia

	region = Asia		
	product		
time	TV	PC	DVD
Q1	789	456	187
Q2	654	732	157
Q3	623	354	129
Q4	756	876	231

Table 3.5   Two-dimensional view of sales data in Europe

	region = Europe		
	product		
time	TV	PC	DVD
Q1	638	576	192
Q2	876	723	165
Q3	798	675	154
Q4	921	754	201

four-dimensional space and therefore cannot be represented graphically. However, we can obtain four logical views composed of three-dimensional cubes, called *cuboids*, inside the four-dimensional cube, by fixing the values of one dimension.

More generally, starting from a fact table linked to n dimension tables, it is possible to obtain a lattice of cuboids, each of them corresponding to a different level of consolidation along one or more dimensions. This type of aggregation is equivalent in *structured query language* (SQL) to a query *sum* derived from a *group-by* condition. Figure 3.6 illustrates the lattice composed by the cuboids obtained from the data cube defined along the four dimensions {time, product, region, supplier}.

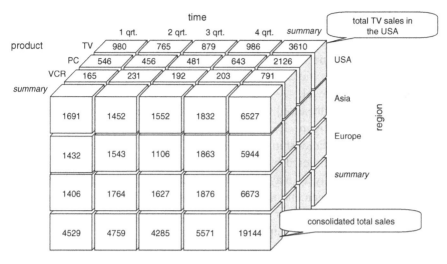

Figure 3.5   Example of a three-dimensional cube

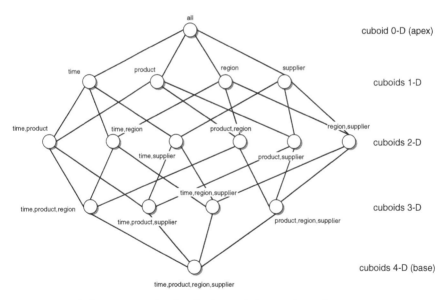

Figure 3.6   Lattice of cuboids derived from a four-dimensional cube

The cuboid associated with the atomic data, which therefore does not imply any type of consolidation, is called a *base cuboid*. At the other extreme, the *apex cuboid* is defined as the cuboid corresponding to the consolidation along all dimensions, therefore associated with the grand total of the measure of interest.

3.3.1 Hierarchies of concepts and OLAP operations

In many instances, OLAP analyses are based on *hierarchies of concepts* to consolidate the data and to create *logical views* along the dimensions of a data warehouse. A concept hierarchy defines a set of maps from a lower level of concepts to a higher level.

For example, the {location} dimension may originate a totally ordered hierarchy, as shown in Figure 3.7, developing along the {address, municipality, province, country} relationship. The temporal dimension, on the other hand, originates a partially ordered hierarchy, also shown in Figure 3.7.

Specific hierarchy types may be predefined in the software platform used for the creation and management of a data warehouse, as in the case of the dimensions shown in Figure 3.7. For other hierarchies it is necessary for analysts to explicitly define the relationships among concepts.

Hierarchies of concepts are also used to perform several visualization operations dealing with data cubes in a data warehouse.

Roll-up. A *roll-up* operation, also termed *drill-up*, consists of an aggregation of data in the cube, which can be obtained alternatively in the following two ways.

- Proceeding upwards to a higher level along a single dimension defined over a concepts hierarchy. For example, for the {location} dimension it is possible to move upwards from the {city} level to the {province} level and to consolidate the measures of interest through a *group-by* conditioned sum over all records whereby the city belongs to the same province.

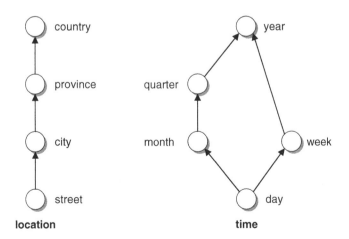

Figure 3.7   Hierarchies of concepts

• Reducing by one dimension. For example, the removal of the {time} dimension leads to consolidated measures through the sum over all time periods existing in the data cube.

Roll-down. A *roll-down* operation, also referred to as *drill-down*, is the opposite operation to roll-up. It allows navigation through a data cube from aggregated and consolidated information to more detailed information. The effect is to reverse the result achieved through a roll-up operation. A drill-down operation can therefore be carried out in two ways.

• Shifting down to a lower level along a single dimension hierarchy. For example, in the case of the {location} dimension, it is possible to shift from the {province} level to the {city} level and to disaggregate the measures of interest over all records whereby the city belongs to the same province.

• Adding one dimension. For example, the introduction of the {time} dimension leads to disaggregate the measures of interest over all time periods existing in a data cube.

Slice and dice. Through the *slice* operation the value of an attribute is selected and fixed along one dimension. For example, Table 3.3 has been obtained by fixing the region at the {Usa} value. The *dice* operation obtains a cube in a subspace by selecting several dimensions simultaneously.

Pivot. The *pivot* operation, also referred to as *rotation*, produces a rotation of the axes, swapping some dimensions to obtain a different view of a data cube.

3.3.2   Materialization of cubes of data

OLAP analyses developed by knowledge workers may need to access the information associated with several cuboids, based on the specific queries and analyses being carried out. In order to guarantee adequate response time, it might be useful to design a data warehouse where all (or at least a large portion of) values of the measures of interest associated with all possible cuboids are pre-calculated. This approach is termed *full materialization* of the information relative to the data cubes.

Observe that where hierarchies of concepts are missing, it is possible to form 2^n distinct cuboids from all possible combinations of n dimensions. The existence of hierarchies along different dimensions makes the number of distinct cuboids even greater. If L_i denotes the number of hierarchical levels associated with the ith dimension, for an n-dimensional data cube it is possible to calculate the full number of cuboids, given by

$$T = \prod_{i=1}^{n}(L_i + 1). \tag{3.1}$$

For example, if a data cube includes 5 dimensions, and if each of these dimensions includes 3 hierarchical levels, the number of cuboids is equal to $4^5 = 2^{10} \approx 10^3$. It is clear that the full materialization of the cuboids for all the cubes associated with the fact tables of a data warehouse would impose storage requirements that could be hardly sustained over time, considering the rate at which new records are gathered.

For all of the above reasons, it is necessary to strike a balance between the need for fast access to information, which would suggest the full materialization of the cuboids, and the need to keep memory use within reasonable limits. As a consequence, preventive materialization should be carried out only for those cuboids that are most frequently accessed, while for the others the computation should be carried out on demand only when actual queries requesting the associated information are performed. This latter approach is referred to as *partial materialization* of the information relative to the data cubes.

3.4   Notes and readings

More in-depth studies of data warehouses, including technical aspects which have deliberately been omitted from this chapter, are provided by Kimball (1996), Kimball *et al.* (1998), Dyche (2000), Inmon (2002), Kimball and Ross (2002) and Nemati *et al.* (2002).

Part II

Mathematical models and methods

4

Mathematical models for decision making

In the previous chapters we have emphasized the critical role played by mathematical models in the development of business intelligence environments and decision support systems aimed at providing *active* support for knowledge workers. In this chapter we will focus on the main characteristics shared by different mathematical models embedded into business intelligence systems. We will also develop a taxonomy of the most common classes of models, identifying for each of them the prevailing application domain.

4.1 Structure of mathematical models

Mathematical models have been developed and used in many application domains, ranging from physics to architecture, from engineering to economics. The models adopted in the various contexts differ substantially in terms of their mathematical structure. However, it is possible to identify a few fundamental features shared by most models.

Generally speaking, a model is a selective abstraction of a real system. In other words, a model is designed to analyze and understand from an abstract point of view the operating behavior of a real system, regarding which it only includes those elements deemed relevant for the purpose of the investigation carried out. In this respect it is worth quoting the words of Einstein on the development of a model: 'Everything should be made as simple as possible, but not simpler.' Figure 4.1 expresses in graphical terms the definition of a model.

Business Intelligence: Data Mining and Optimization for Decision Making   C. Vercellis
© 2009 John Wiley & Sons, Ltd

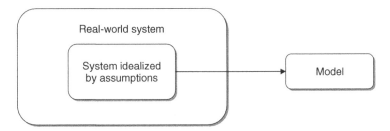

Figure 4.1   A model is a selective abstraction of reality

Scientific and technological development has turned to mathematical models of various types for the abstract representation of real systems. As an example, consider the *thought experiment* (*Gendankenexperiment*) popularized in physics at the beginning of the twentieth century, which involved building a mental model of a given phenomenon and verifying its validity by imagining the consequences caused by hypothetical modifications in the model itself. The analogy is well apparent between this conceptual paradigm and *what-if analyses* that can be easily performed using a simple spreadsheet to find an answer to questions such as: given a model for calculating the budget of a company, how are cash flows affected by a change in the payment terms, such as 90 days vs. 60 days, of invoices issued in favor of the main customers?

According to their characteristics, models can be divided into *iconic*, *analogical* and *symbolic*.

Iconic. An *iconic* model is a material representation of a real system, whose behavior is imitated for the purpose of the analysis. A miniaturized model of a new city neighborhood is an example of iconic model.

Analogical. An *analogical* model is also a material representation, although it imitates the real behavior by analogy rather than by replication. A wind tunnel built to investigate the aerodynamic properties of a motor vehicle is an example of an analogical model intended to represent the actual progression of a vehicle on the road.

Symbolic. A *symbolic* model, such as a mathematical model, is an abstract representation of a real system. It is intended to describe the behavior of the system through a series of symbolic variables, numerical parameters and mathematical relationships.

Business intelligence systems, and consequently the models presented in this book, are exclusively based on symbolic models.

A further relevant distinction concerns the probabilistic nature of models, which can be either *stochastic* or *deterministic*.

Stochastic. In a *stochastic* model some input information represents random events and is therefore characterized by a probability distribution, which in turn can be assigned or unknown. Predictive models, which will be thoroughly described in the following chapters, as well as waiting line models, briefly mentioned below in this chapter, are examples of stochastic models.

Deterministic. A model is called *deterministic* when all input data are supposed to be known a priori and with certainty. Since this assumption is rarely fulfilled in real systems, one resorts to deterministic models when the problem at hand is sufficiently complex and any stochastic elements are of limited relevance. Notice, however, that even for deterministic models the hypothesis of knowing the data with certainty may be relaxed. Sensitivity and scenario analyses, as well as what-if analysis, allow one to assess the robustness of optimal decisions to variations in the input parameters.

A further distinction concerns the temporal dimension in a mathematical model, which can be either *static* or *dynamic*.

Static. *Static* models consider a given system and the related decision-making process within one single temporal stage. For instance, the optimization model described in Section 14.2.7 determines an optimal plan for the distribution of goods in a specific time frame.

Dynamic. *Dynamic* models consider a given system through several temporal stages, corresponding to a sequence of decisions. In many instances the temporal dimension is subdivided into discrete intervals of a previously fixed span: minutes, hours, days, weeks, months and years are examples of discrete subdivisions of the time axis. *Discrete-time* dynamic models, which largely prevail in business intelligence applications, observe the status of a system only at the beginning or at the end of discrete intervals. *Continuous-time* dynamic models consider a continuous sequence of periods on the time axis.

4.2   Development of a model

It is possible to break down the development of a mathematical model for decision making into four primary phases, shown in Figure 4.2. The figure also includes a *feedback* mechanism which takes into account the possibility of changes and revisions of the model.

Problem identification

First of all, the problem at hand must be correctly identified. The observed critical symptoms must be analyzed and interpreted in order to formulate hypotheses

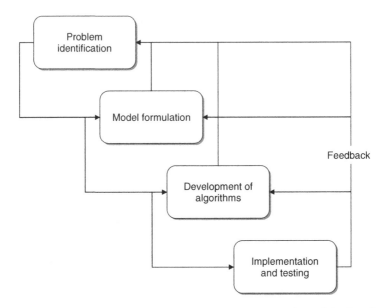

Figure 4.2   Phases in the development of mathematical models for decision making

for investigation. For example, too high a stock level, corresponding to an excessive stock turnover rate, may possibly represent a symptom for a company manufacturing consumable goods. It is therefore necessary to understand what caused the problem, based on the opinion of the production managers. In this case, an ineffective production plan may be the cause of the stock accumulation.

Model formulation

Once the problem to be analyzed has been properly identified, effort should be directed toward defining an appropriate mathematical model to represent the system. A number of factors affect and influence the choice of model, such as the *time horizon*, the *decision variables*, the *evaluation criteria*, the *numerical parameters* and the *mathematical relationships*.

Time horizon. Usually a model includes a temporal dimension. For example, to formulate a tactical production plan over the medium term it is necessary to specify the production rate for each week in a year, whereas to derive an operational schedule it is required to assign the tasks to each production line for each day of the week. As we can see, the time span considered in a model, as well as the length of the base intervals, may vary depending on the specific problem considered.

Evaluation criteria. Appropriate measurable performance indicators should be defined in order to establish a criterion for the evaluation and comparison of the alternative decisions. These indicators may assume various forms in each different application, and may include the following factors:

- monetary costs and payoffs;
- effectiveness and level of service;
- quality of products and services;
- flexibility of the operating conditions;
- reliability in achieving the objectives.

Decision variables. Symbolic variables representing alternative decisions should then be defined. For example, if a problem consists of the formulation of a tactical production plan over the medium term, decision variables should express production volumes for each product, for each process and for each period of the planning horizon.

Numerical parameters. It is also necessary to accurately identify and estimate all numerical parameters required by the model. In the production planning example, the available capacity should be known in advance for each process, as well as the capacity absorption coefficients for each combination of products and processes.

Mathematical relationships. The final step in the formulation of a model is the identification of mathematical relationships among the decision variables, the numerical parameters and the performance indicators defined during the previous phases. Sometimes these relationships may be exclusively deterministic, while in other instances it is necessary to introduce probabilistic relationships. In this phase, the trade-off between the accuracy of the representation achieved through the model and its solution complexity should be carefully considered. It may turn out more helpful at a practical level to adopt a model that sacrifices some marginal aspects of reality in the representation of the system but allows an efficient solution and greater flexibility in view of possible future developments.

Development of algorithms

Once a mathematical model has been defined, one will naturally wish to proceed with its solution to assess decisions and to select the best alternative. In other words, a solution algorithm should be identified and a software tool that incorporates the solution method should be developed or acquired. An analyst in charge of model formulation should possess a thorough knowledge of current solution methods and their characteristics.

Implementation and test

When a model is fully developed, then it is finally implemented, tested and utilized in the application domain. It is also necessary that the correctness of the data and the numerical parameters entered in the model be preliminarily assessed. These data usually come from a data warehouse or a data mart previously set up. Once the first numerical results have been obtained using the solution procedure devised, the model must be validated by submitting its conclusions to the opinion of decision makers and other experts in the application domain. A number of factors should be taken into account at this stage:

- the plausibility and likelihood of the conclusions achieved;

- the consistency of the results at extreme values of the numerical parameters;

- the stability of the results when minor changes in the input parameters are introduced.

4.3   Classes of models

There are several classes of mathematical models for decision making, which in turn can be solved by a number of alternative solution techniques. Each model class is better suited to represent certain types of decision-making processes. In this section we will cover the main categories of mathematical models for decision making, including:

- predictive models;

- pattern recognition and learning models;

- optimization models;

- project management models;

- risk analysis models;

- waiting line models.

Predictive models

A significant proportion of the models used in business intelligence systems, such as optimization models, require input data concerned with future events.

For example, the results of random events determine the future demand for a product or service, the development of new scenarios of technological innovation and the level of prices and costs. As a consequence, *predictive* models play a primary role in business intelligence systems, since they are logically placed upstream with respect to other mathematical models and, more generally, to the whole decision-making process.

Predictions allow input information to be fed into different decision-making processes, arising in strategy, research and development, administration and control, marketing, production and logistics. Basically, all departmental functions of an enterprise make some use of predictive information to develop decision making, even though they pursue different objectives.

Predictive models can be subdivided into two main categories. The purpose of *explanatory* models is to functionally identify a possible relationship between a dependent variable and a set of independent attributes. *Regression* models, described in Chapter 8, belong to this category as well as *classification* models, described in Chapter 10, even though it would be more appropriate to place these latter together with pattern recognition and learning models, covered in the following paragraph. The purpose of *time series* models, described in Chapter 9, is to functionally identify any temporal pattern expressed by a time series of observations referred to the same numerical variable.

Pattern recognition and machine learning models

In a broad sense, the purpose of pattern recognition and learning theory is to understand the mechanisms that regulate the development of intelligence, understood as the ability to extract knowledge from past experience in order to apply it in the future. Mathematical models for learning can be used to develop efficient algorithms that can perform such task. This has led to intelligent machines capable of learning from past observations and deriving new rules for the future, just like the human mind is able to do with great effectiveness due to the sophisticated mechanisms developed and fine-tuned in the course of evolution.

Besides an intrinsic theoretical interest, mathematical methods for learning are applied in several domains, such as recognition of images, sounds and texts; biogenetic and medical diagnosis; relational marketing, for segmenting and profiling customers; manufacturing process control; identification of anomalies and fraud detection.

Mathematical models for learning have two primary objectives. The purpose of *interpretation* models is to identify regular patterns in the data and to express them through easily understandable rules and criteria. *Prediction* models help to forecast the value that a given random variable will assume in the future, based on the values of some variables associated with the entities of a database, as for explanatory models. Based on the existence or not of a *target* attribute, the learning process may be *supervised* or *unsupervised*. In the first case, the target attribute expresses for each record either the membership class or a measurable quantity. *Classification* and *regression* models belong to this category. In the second case, no target attribute exists and consequently the purpose of the analysis is to identify regularities, similarities and differences in the data. It is also possible to derive *association rules*, as described in Chapter 11. Alternatively one can determine groups of records, called *clusters*, characterized by similarity within each cluster and by dissimilarity among the elements of distinct clusters, as described in Chapter 12.

Optimization models

Many decision-making processes faced by companies or complex organizations can be cast according to the following framework: given the problem at hand, the decision maker defines a set of *feasible* decisions and establishes a criterion for the evaluation and comparison of alternative choices, such as monetary costs or payoffs. At this point, the decision maker must identify the *optimal* decision according to the evaluation criterion defined, that is, the choice corresponding to the minimum cost or to the maximum payoff. The conceptual paradigm outlined determines a wide and popular class of mathematical models for decision making, represented by *optimization* models.

In general, optimization models arise naturally in decision-making processes where a set of limited resources must be allocated in the most effective way to different entities. These resources may be personnel, production processes, raw materials, components or financial factors. Among the main application domains requiring an optimal allocation of the resources we find:

- logistics and production planning;

- financial planning;

- work shift planning;

- marketing campaign planning;

- price determination.

Mathematical optimization models represent a fairly substantial class of optimization problems that are derived when the objective of the decision-making process is a function of the decision variables, and the criteria describing feasible decisions can be expressed by a set of mathematical equalities and inequalities in the decision variables. Mathematical optimization models offer ample opportunities for application, as we will see in the following chapters, due to the high flexibility of their formulations and the availability of efficient solution methods.

In light of the structure of the objective function and of the constraints, optimization models may assume different forms:

- linear optimization;

- integer optimization;

- convex optimization;

- network optimization;

- multiple-objective optimization.

Project management models

A *project* is a complex set of interrelated activities carried out in pursuit of a specific goal, which may represent an industrial plant, a building, an information system, a new product or a new organizational structure, depending on the different application domains. The execution of the project requires a planning and control process for the interdependent activities as well as the human, technical and financial resources necessary to achieve the final goal. *Project management* methods are based on the contributions of various disciplines, such as business organization, behavioral psychology and operations research.

Mathematical models for decision making play an important role in project management methods. In particular, network models are used to represent the component activities of a project and the precedence relationships among them. These models allow the overall project execution time to be determined, assuming a deterministic knowledge of the duration of each activity.

Stochastic models, on the other hand, usually referred to as *project evaluation and review techniques* (*PERT*), are used to derive the execution times when stochastic assumptions are made regarding the duration of the activities, represented by random variables.

Finally, different classes of optimization models allow the analysis to be extended to the complex problem of optimally allocating a set of limited resources among the project activities in view of execution costs and times.

Risk analysis models

Some decision problems can be described according to the following conceptual paradigm: the decision maker is required to choose among a number of available alternatives, having uncertain information regarding the effects that these options may have in the future.

For example, assume that senior management wishes to evaluate different alternatives in order to increase the company's production capacity. On the one hand, the company may build a new plant providing a high operating efficiency and requiring a high investment cost. On the other hand, it may expand an existing plant with a lower investment but with higher operating costs. Finally, it may subcontract to external third parties part of its production: in this case, the investment cost is minimized but the operating costs are the highest among the available alternatives.

Clearly, in this situation the effects of the different options are strongly influenced by future stochastic events. In particular, a high level of future demand makes the construction of a new plant advantageous, while low demand levels tend to favor the subcontracting option. At an intermediate level of demand the expansion of an existing plant may be convenient.

However, the decision maker is forced to make a choice *before* knowing with absolute certainty the level of future demand. At best, she may obtain some stochastic information regarding the likelihood of occurrence of future events by carrying out some market research.

In situations of this type, the methodological support offered by risk analysis models, primarily based on Bayesian and utility theories, may prove quite helpful. Indeed, this class of models is used successfully in several application domains, such as technology investment, design of new products, research and development, and financial and real estate investment.

Waiting line models

The purpose of *waiting line* theory is to investigate congestion phenomena occurring when the demand for and provision of a service are stochastic in nature. If the arrival times of the customers and the duration of the service are not known beforehand in a deterministic way, conflicts may arise between customers in the use of limited shared resources. As a consequence, some customers are forced to wait in a line.

Schematically, a waiting line system is made up of three main components: a *source* generating a stochastic process in which entities, also referred to as customers, arrive at a given location to obtain a service; a set of *resources* providing the service; a *waiting area* able to receive the entities whose requests cannot immediately be satisfied.

Waiting line models allow the performance of a system to be evaluated once its structure has been defined, and therefore are mostly useful within the system design phase. Indeed, in order to determine the appropriate values for the parameters that characterize a new system, relevant economic factors are considered, which depend on the service level that the system should guarantee when operating in optimal conditions. More precisely, a model takes into account the cost of meeting the requests, which increases as the service level increases, and the cost generated by customer waiting times, which decreases as the level of service provided decreases. Since customers hope for shorter waiting times while the provider is interested in holding down the cost of service provision, the structure of the system is defined in such a way as to obtain an optimal trade-off between service costs and waiting line costs. In other words, the optimal service level, which in turn determines the ideal structure of the system, can be found at the minimum point of the curve expressing the sum of the two types of cost.

The main components of a waiting line system are the *population*, the *arrivals process*, the *service process*, the *number of stations*, and the *waiting line rules*. The population, which can be finite or infinite, represents the source from which potential customers are drawn and to which they return once the requested service has been received. The arrivals process describes how customers arrive at the system entry point. In general this is a stochastic process described by the probability distribution of the inter-arrival times, that is, the time intervals between the arrival of two consecutive customers. The service process describes how the providers meet the requests of the customers waiting in line. This in turn is a stochastic process, defined by the probability distribution of the service time, that is, the amount of time that customers spend with the resources providing the service. The number of existing stations and the number of providers assigned to each station are additional relevant parameters of the waiting line system. The waiting rules describe the order in which customers are extracted from the line to be admitted to the service. A primary role is finally played by priority schemes in which a level of priority is assigned to each customer. The customer with the highest priority is then served before all the other customers waiting in line.

4.4   Notes and readings

For the topics that will be discussed in the following chapters, such as pattern recognition and predictive models, see the suggested reading section in the respective chapters. For more in-depth information on specific classes of models, see: Winston and Venkataramanan (2002), Bertsekas (2003) and Vercellis (2008) on mathematical optimization, and Nemhauser *et al.* (1994) on optimization in general; Elmaghraby (1978) on network models for project management; Clemen (1997) and Skinner (1999) on risk analysis; Kleinrock (1975) on waiting line theory. Keys (1995) deals with the epistemological aspects involved in operations research.

5

Data mining

As observed in previous chapters, the evolving technologies of information gathering and storage have made available huge amounts of data within most application domains, such as the business world, the scientific and medical community, and public administration. The set of activities involved in the analysis of these large databases, usually with the purpose of extracting useful knowledge to support decision making, has been referred to in different ways, such as *data mining*, *knowledge discovery*, *pattern recognition* and *machine learning*.

In particular, the term *data mining* indicates the process of exploration and analysis of a dataset, usually of large size, in order to find regular patterns, to extract relevant knowledge and to obtain meaningful recurring rules. Data mining plays an ever-growing role in both theoretical studies and applications.

In this chapter we wish to describe and characterize data mining activities with respect to investigation purposes and analysis methodologies. The relevant properties of input data will also be discussed. Finally, we will describe the data mining process and its articulation in distinct phases.

5.1 Definition of data mining

Data mining activities constitute an iterative process aimed at the analysis of large databases, with the purpose of extracting information and knowledge that may prove accurate and potentially useful for knowledge workers engaged in decision making and problem solving.

As described in Section 5.3, the analysis process is iterative in nature since there are distinct phases that might imply feedback and subsequent revisions. Usually such a process represents a cooperative activity between experts in the application domain and data analysts, who use mathematical models for inductive learning. Indeed, experience indicates that a data mining study requires frequent interventions by the analysts across the different investigation phases and therefore cannot easily be automated. It is also necessary that the knowledge extracted be accurate, in the sense that it must be confirmed by data and not lead to misleading conclusions.

The term *data mining* refers therefore to the overall process consisting of data gathering and analysis, development of inductive learning models and adoption of practical decisions and consequent actions based on the knowledge acquired. The term *mathematical learning theory* is reserved for the variety of mathematical models and methods that can be found at the core of each data mining analysis and that are used to generate new knowledge.

The data mining process is based on inductive learning methods, whose main purpose is to derive general rules starting from a set of available examples, consisting of past observations recorded in one or more databases. In other words, the purpose of a data mining analysis is to draw some conclusions starting from a sample of past observations and to generalize these conclusions with reference to the entire population, in such a way that they are as accurate as possible. The models and patterns identified in this way may take on different forms, which will be described in the following chapters, such as linear equations, sets of rules in *if–then–else* form, clusters, charts and trees.

A further characteristic of data mining depends on the procedure for collecting past observations and inserting them into a database. Indeed, these records are usually stored for purposes that are not primarily driven by data mining analysis. For instance, information on purchases from a retail company, or on the usage of each telephone number stored by a mobile phone provider, will basically be recorded for administrative purposes, even if the data may be later used to perform some useful data mining analysis. The data gathering procedure is therefore largely independent and unaware of the data mining objectives, so that it substantially differs from data gathering activities carried out according to predetermined sampling schemes, typical of classical statistics. In this respect, data mining represent a *secondary* form of data analysis.

Data mining activities can be subdivided into two major investigation streams, according to the main purpose of the analysis: *interpretation* and *prediction*.

Interpretation. The purpose of interpretation is to identify regular patterns in the data and to express them through rules and criteria that can be easily understood by experts in the application domain. The rules generated must be original and non-trivial in order to actually increase the level of knowledge and understanding of the system of interest. For example, for a company in the retail industry it might be advantageous to cluster those customers who have taken out loyalty cards according to their purchasing profile. The segments generated in this way might prove useful in identifying new market niches and directing future marketing campaigns.

Prediction. The purpose of prediction is to anticipate the value that a random variable will assume in the future or to estimate the likelihood of future events. For example, a mobile phone provider may develop a data mining analysis to estimate for its customers the probability of churning in favor of some competitor. In a different context, a retail company might predict the sales of a given product during the subsequent weeks. Actually, most data mining techniques derive their predictions from the value of a set of variables associated with the entities in a database. For example, a data mining model may indicate that the likelihood of future churning for a customer depends on features such as age, duration of the contract and percentage of calls to subscribers of other phone providers. There are, however, time series models, described in Chapter 9, which make predictions based only on the past values of the variable of interest.

Sometimes, a model developed for the purpose of prediction may also turn out to be effective for interpretation. In the case of classification trees with parallel axis splitting rules, which will be described in Chapter 10, the models generated for predictive purposes may also prove useful in identifying recurrent explanatory phenomena.

5.1.1   Models and methods for data mining

There are several learning methods that are available to perform the different data mining tasks. A number of techniques originated in the field of computer science, such as classification trees or association rules, and are referred to as *machine learning* or *knowledge discovery in databases*. In most cases an empirically based approach tends to prevail within this class of techniques. Other methods belong to multivariate statistics, such as regression or Bayesian classifiers, and are often parametric in nature but appear more theoretically grounded. More recent developments include mathematical methods for learning, such as *statistical learning theory*, which are based on solid theoretical foundations and place themselves at the crossroads of various disciplines, among which probability theory, optimization theory and statistics.

Example 5.1 – Linear regression.   Linear regression models, described in Chapter 8, are one of the best-known learning and predictive methodologies in classical statistics. In its simplest form, linear regression is used to relate a dependent response variable Y to an independent predictor X

through a linear regression in the form $Y = aX + b$, where a and b are parameters to be determined using the available past observations. For example, Y may represent the sales of a mass consumption product during a week and X the total advertisement cost during the same week. With respect to the development phases of a model, the selection of a linear function determines the type of relationship between the predictor and the response variable. A reasonable evaluation metric is the sum of the squared differences between the values of Y actually observed in the past and the values predicted by the linear model. An appropriate optimization algorithm calculates the value of the parameters a and b in order to minimize the sum of squared errors.

Irrespective of the specific learning method that one wishes to adopt, there are other recurrent steps in the development of a data mining model, as shown in Example 5.1:

- the selection of a class of models to be used for learning from the past and of a specific form for representing patterns in the data;

- the definition of a metric for evaluating the effectiveness and accuracy of the models being generated;

- the design of a computational algorithm in order to generate the models by optimizing the evaluation metric.

In the next chapters we will provide a description of the most popular classes of methods for prediction and pattern recognition. Some methodologies can be applied to perform several data mining tasks, among those described in Section 5.4. Since it is generally possible to relate each class of models to a prevailing activity, in the following chapters each technique will be associated with the most appropriate task.

5.1.2   Data mining, classical statistics and OLAP

Data mining projects differ in many respects from both classical statistics and OLAP analyses. Such differences are shown in Table 5.1, with reference to an example.

The main difference consists of the active orientation offered by inductive learning models, compared with the passive nature of statistical techniques and OLAP. Indeed, in statistical analyses decision makers formulate a hypothesis that then has to be confirmed on the basis of sample evidence. Similarly, in OLAP analyses knowledge workers express some intuition on which they base

Table 5.1   Differences between OLAP, statistics and data mining

OLAP	statistics	data mining
extraction of details and aggregate totals from data	verification of hypotheses formulated by analysts	identification of patterns and recurrences in data
information	validation	knowledge
distribution of incomes of home loan applicants	analysis of variance of incomes of home loan applicants	characterization of home loan applicants and prediction of future applicants

extraction, reporting and visualization criteria. Both methods – on one hand statistical validation techniques and on the other hand information tools to navigate through data cubes – only provide elements to confirm or disprove the hypotheses formulated by the decision maker, according to a *top-down* analysis flow. Conversely, learning models, which represent the core of data mining projects, are capable of playing an active role by generating predictions and interpretations which actually represent new knowledge available to the users. The analysis flow in the latter case has a *bottom-up* structure. In particular, when faced with large amounts of data, the use of models capable of playing an active role becomes a critical success factor, since it is hard for knowledge workers to formulate a priori meaningful and well-founded hypotheses.

5.1.3   Applications of data mining

Data mining methodologies can be applied to a variety of domains, from marketing and manufacturing process control to the study of risk factors in medical diagnosis, from the evaluation of the effectiveness of new drugs to fraud detection.

Relational marketing. Data mining applications in the field of relational marketing, described in Chapter 13, have significantly contributed to the increase in the popularity of these methodologies. Some relevant applications within relational marketing are:

- identification of customer segments that are most likely to respond to targeted marketing campaigns, such as *cross-selling* and *up-selling*;

- identification of target customer segments for retention campaigns;

- prediction of the rate of positive responses to marketing campaigns;

- interpretation and understanding of the buying behavior of the customers;

- analysis of the products jointly purchased by customers, known as *market basket analysis*.

Fraud detection. Fraud detection is another relevant field of application of data mining. Fraud may affect different industries such as telephony, insurance (false claims) and banking (illegal use of credit cards and bank checks; illegal monetary transactions).

Risk evaluation. The purpose of risk analysis is to estimate the risk connected with future decisions, which often assume a dichotomous form. For example, using the past observations available, a bank may develop a predictive model to establish if it is appropriate to grant a monetary loan or a home loan, based on the characteristics of the applicant.

Text mining. Data mining can be applied to different kinds of texts, which represent unstructured data, in order to classify articles, books, documents, emails and web pages. Examples are web search engines or the automatic classification of press releases for storing purposes. Other text mining applications include the generation of filters for email messages and newsgroups.

Image recognition. The treatment and classification of digital images, both static and dynamic, is an exciting subject both for its theoretical interest and the great number of applications it offers. It is useful to recognize written characters, compare and identify human faces, apply correction filters to photographic equipment and detect suspicious behaviors through surveillance video cameras.

Web mining. Web mining applications, which will be briefly considered in section 13.1.9, are intended for the analysis of so-called *clickstreams* – the sequences of pages visited and the choices made by a web surfer. They may prove useful for the analysis of e-commerce sites, in offering flexible and customized pages to surfers, in caching the most popular pages or in evaluating the effectiveness of an e-learning training course.

Medical diagnosis. Learning models are an invaluable tool within the medical field for the early detection of diseases using clinical test results. Image analysis for diagnostic purpose is another field of investigation that is currently burgeoning.

5.2   Representation of input data

In most cases, the input to a data mining analysis takes the form of a two-dimensional table, called a *dataset*, irrespective of the actual logic and material representation adopted to store the information in files, databases, data warehouses and data marts used as data sources. The rows in the dataset correspond to the *observations* recorded in the past and are also called *examples*, *cases*, *instances* or *records*. The columns represent the information available for each observation and are termed *attributes*, *variables*, *characteristics* or *features*.

The attributes contained in a dataset can be categorized as *categorical* or *numerical*, depending on the type of values they take on.

Categorical. Categorical attributes assume a finite number of distinct values, in most cases limited to less than a hundred, representing a qualitative property of an entity to which they refer. Examples of categorical attributes are the province of residence of an individual (which takes as values a series of names, which in turn may be represented by integers) or whether a customer has abandoned her service provider (expressed by the value 1) or remained loyal to it (expressed by the value 0). Arithmetic operations cannot be applied to categorical attributes even when the coding of their values is expressed by integer numbers.

Numerical. Numerical attributes assume a finite or infinite number of values and lend themselves to subtraction or division operations. For example, the amount of outgoing phone calls during a month for a generic customer represents a numerical variable. Regarding two customers A and B making phone calls in a week for €27 and €36 respectively, it makes sense to claim that the difference between the amounts spent by the two customers is equal to €9 and that A has spent three fourths of the amount spent by B.

Sometimes a more refined taxonomy of attributes can prove useful.

Counts. Counts are categorical attributes in relation to which a specific property can be true or false. These attributes can therefore be represented using Boolean variables {true, false} or binary variables {0,1}. For example, a bank's customers may or may not be holders of a credit card issued by the bank.

Nominal. Nominal attributes are categorical attributes without a natural ordering, such as the province of residence.

Ordinal. Ordinal attributes, such as education level, are categorical attributes that lend themselves to a natural ordering but for which it makes no sense to calculate differences or ratios between the values.

Discrete. Discrete attributes are numerical attributes that assume a finite number or a countable infinity of values.[1]

---

[1] If a set A has the same cardinality as the set $\mathbb{N}$ of natural numbers, then we say that A is *countable*. In other words, a set is countable if there is a bijection between that set and $\mathbb{N}$. There exist sets, such as the set $\mathbb{R}$ of real numbers, that are infinite and not countable, and are therefore called *uncountable*.

Continuous. Continuous attributes are numerical attributes that assume an uncountable infinity of values.

To represent a generic dataset $\mathcal{D}$, we will denote by m the number of observations, or rows, in the two-dimensional table containing the data and by n the number of attributes, or columns. Furthermore, we will denote by

$$\mathbf{X} = [x_{ij}], \quad i \in \mathcal{M} = \{1, 2, \ldots, m\}, \quad j \in \mathcal{N} = \{1, 2, \ldots, n\}, \quad (5.1)$$

the matrix of dimensions $m \times n$ that corresponds to the entries in the dataset $\mathcal{D}$. We will write

$$\mathbf{x}_i = (x_{i1}, x_{i2}, \ldots, x_{in}) \quad (5.2)$$

$$\mathbf{a}_j = (x_{1j}, x_{2j}, \ldots, x_{mj}) \quad (5.3)$$

for the n-dimensional row vector associated with the ith record of the dataset and the m-dimensional column vector representing the jth attribute in $\mathcal{D}$, respectively.

5.3 Data mining process

The definition of data mining given at the beginning of Section 5.1 refers to an iterative process, during which learning models and techniques play a key, though non-exhaustive, role. Figure 5.1 shows the main phases of a generic data mining process.

Definition of objectives. Data mining analyses are carried out in specific application domains and are intended to provide decision makers with useful knowledge. As a consequence, intuition and competence are required by the domain experts in order to formulate plausible and well-defined investigation objectives. If the problem at hand is not adequately identified and circumscribed one may run the risk of thwarting any future effort made during data mining activities. The definition of the goals will benefit from close cooperation between experts in the field of application and data mining analysts. With reference to Example 5.2, it is possible to define the problem and the goals of the investigation as the analysis of past data and identification of a model so as to express the propensity of customers to leave the service (churn) based on their characteristics, in order to understand the reasons for such disloyalty and predict the probability of churn.

In the hierarchy of infinities, countable sets, such as the set $\mathbb{Q}$ of rational numbers, are placed on the lowest step and are consequently less 'dense' than others.

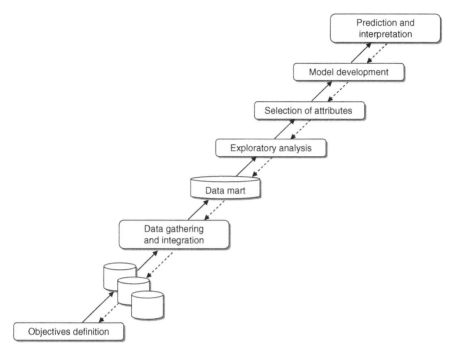

Figure 5.1   Data mining process

Example 5.2 – Retention in the mobile phone industry.   Table 5.2 shows the two-dimensional structure of input data from an example of the analysis of customer loyalty. Suppose that a mobile phone company carries out a data mining analysis with both prediction and interpretation goals. On the one hand, the company wishes to assess the likelihood of future churning by each customer, in order to target marketing actions for retention purposes. On the other hand, the intent is to understand the reasons why customers churn, with the purpose of improving the service level and reducing future churning. Table 5.2 contains 23 observations and 12 attributes, whose meaning is indicated in Table 5.3.
The first 11 attributes represent *explanatory* variables, while the last attribute represents the *target* variable, expressing the class of each record in relation to the objectives of the data mining analysis. The first explanatory variable gives personal demographic information while the rest refer to the use of the service. Observed values are relative to time period of index $t - 2$ for the explanatory attributes, whereas for the target variable they refer to period t. The difference in time placement is required in

Table 5.2   An example of input data for a data mining model

area	numin	timein	numout	Pothers	Pmob	Pland	numsms	numserv	numcall	diropt	churner
3	32	8093	45	0.14	0.75	0.12	18	1	0	0	0
3	277	157842	450	0.26	0.35	0.38	9	3	0	1	0
1	17	15023	20	0.37	0.23	0.40	1	1	0	0	0
1	46	22459	69	0.10	0.39	0.51	33	1	0	0	0
1	19	8640	9	0.00	0.00	1.00	0	0	0	0	0
2	17	7652	66	0.16	0.42	0.43	1	3	0	1	0
3	47	17768	11	0.45	0.00	0.55	0	0	0	0	0
3	19	9492	42	0.18	0.34	0.48	3	1	0	0	1
1	1	84	9	0.09	0.54	0.37	0	0	0	0	0
2	119	87605	126	0.84	0.02	0.14	12	1	0	0	1
4	24	6902	47	0.25	0.26	0.48	4	1	0	0	0
1	32	28072	43	0.28	0.66	0.06	0	2	0	0	0
3	103	112120	24	0.61	0.28	0.11	24	2	0	0	0
3	45	21921	94	0.34	0.47	0.19	45	1	3	1	0
1	8	25117	89	0.02	0.89	0.09	189	0	0	0	1
3	4	945	16	0.00	0.00	1.00	0	0	0	0	1
2	83	44263	83	0.00	0.00	0.67	0	2	0	0	1
2	22	15979	59	0.05	0.53	0.41	5	1	1	1	1
2	0	0	57	0.00	1.00	0.00	15	3	0	1	1
4	162	114108	273	0.18	0.15	0.41	2	1	0	1	1
4	21	4141	70	0.14	0.58	0.28	0	3	0	1	1
4	33	10066	45	0.12	0.21	0.67	0	1	0	0	1
4	5	965	40	0.41	0.27	0.32	64	1	0	0	1

Table 5.3   Meaning of the attributes in Table 5.2

attribute	meaning
area	residence area
numin	number of calls received in period $t - 2$
timein	duration in seconds of calls received in period $t - 2$
numout	number of calls placed in the period $t - 2$
Pothers	percentage of calls placed to other mobile telephone companies in period $t - 2$
Pmob	percentage of calls placed to the same mobile telephone company in period $t - 2$
Pland	percentage of calls placed to land numbers in period $t - 2$
numsms	number of messages sent in period $t - 2$
numserv	number of calls placed to special services in period $t - 2$
numcall	number of calls placed to the call center in period $t - 2$
diropt	binary variable indicating whether the customer corresponding to the record has subscribed to a special rate plan for calls placed to selected numbers
churner	binary variable indicating whether the customer corresponding to the record has left the service in period t

order to use the model for predictive purposes. Indeed, it is necessary to predict during the current period which customers will leave the service within 2 periods, based on the available information, in order to develop timely and effective retention actions; see also the discussion on time latency for predictive models in Section 13.1.4. Of course, in a real application the number of available attributes is much higher, of the order of hundreds or thousands, and the number of rows representing the customers is far greater, of the order of hundreds of thousands or millions of records. The purpose here is to obtain an inductive model that is capable of learning from past available observations and identifying a plausible relationship between the target variable and the explanatory attributes. Once the model has been created based on past records, it is possible to use it to predict the target class of new records or to understand common characteristics of customers who churn compared to those who remain loyal.

Data gathering and integration. Once the objectives of the investigation have been identified, the gathering of data begins. Data may come from different sources and therefore may require integration. Data sources may be internal, external or a combination of the two. The integration of distinct data sources may be suggested by the need to enrich the data with new descriptive dimensions, such as geomarketing variables, or with lists of names of potential customers, termed *prospects*, not yet existing in the company information system. In some instances, data sources are already structured in data warehouses and data marts for OLAP analyses and more generally for decision support activities. These are favorable situations where it is sufficient to select the attributes deemed relevant for the purpose of a data mining analysis. There is a risk, however, that, in order to limit memory uptake, the information stored in a data warehouse has been aggregated and consolidated to such an extent as to render useless any subsequent analysis. For example, if a company in the retail industry stores for each customer the total amount of every receipt, without keeping track of each individual purchased item, a future data mining analysis aimed at investigating the actual purchasing behavior may be compromised. In other situations, the original data have a heterogeneous format with no predefined structure. In this case, the process of data gathering and integration becomes more arduous and therefore more prone to errors. Regardless of the original structure, input datasets of data mining analyses almost always take the form of two-dimensional tables, as observed above. Unlike many standard sampling procedures of classical statistics, datasets for data mining represent samples extracted in accordance with an unknown distribution, with the analysts not being able to influence and affect the data gathering process. Chapter 6 will discuss data preparation issues in more detail.

Exploratory analysis. In the third phase of the data mining process, a preliminary analysis of the data is carried out with the purpose of getting acquainted with the available information and carrying out *data cleansing*. Usually, the data stored in a data warehouse are processed at loading time in such a way as to remove any *syntactical* inconsistencies. For example, dates of birth that fall outside admissible ranges and negative sales charges are detected and corrected. In the data mining process, data cleansing occurs at a *semantic* level. First of all, the distribution of the values for each attribute is studied, using histograms for categorical attributes and basic summary statistics for numerical variables. In this way, any abnormal values (*outliers*) and missing values are also highlighted. These are studied by experts in the application domain who may consider excluding the corresponding records from the investigation. Chapter 7 will discuss the techniques used to develop exploratory data analysis.

Attribute Selection. In the subsequent phase, the relevance of the different attributes is evaluated in relation to the goals of the analysis. Attributes that prove to be of little use are removed, in order to cleanse irrelevant information from the dataset. Furthermore, new attributes obtained from the original variables through appropriate transformations are included into the dataset. For example, in most cases it is helpful to introduce new attributes that reflect the trends inherent in the data through the calculation of ratios and differences between original variables. Exploratory analysis and attribute selection are critical and often challenging stages of the data mining process and may influence to a great extent the level of success of the subsequent stages. The methods described in Chapters 6 and 7 may be useful for transforming and selecting the attributes.

Model development and validation. Once a high quality dataset has been assembled and possibly enriched with newly defined attributes, pattern recognition and predictive models can be developed. Usually the *training* of the models is carried out using a sample of records extracted from the original dataset. Then, the predictive accuracy of each model generated can be assessed using the rest of the data. More precisely, the available dataset is split into two subsets. The first constitutes the *training set* and is used to identify a specific learning model within the selected class of models. Usually the sample size of the training set is chosen to be relatively small, although significant from a statistical standpoint – say, a few thousands observations. The second subset is the *test set* and is used to assess the accuracy of the alternative models generated during the training phase, in order to identify the best model for actual future predictions. The most popular classes of learning models will be discussed in detail in the following chapters.

Prediction and interpretation. Upon conclusion of the data mining process, the model selected among those generated during the development phase should be implemented and used to achieve the goals that were originally identified. Moreover, it should be incorporated into the procedures supporting decision-making processes so that knowledge workers may be able to use it to draw predictions and acquire a more in-depth knowledge of the phenomenon of interest.

The data mining process includes feedback cycles, represented by the dotted arrows in Figure 5.1, which may indicate a return to some previous phase, depending on the outcome of the subsequent phases.

Finally, we should emphasize the importance of the involvement and interaction of several professional roles in order to achieve an effective data mining process:

- an expert in the application domain, expected to define the original objectives of the analysis, to provide appropriate understanding during the subsequent data mining activities and to contribute to the selection of the most effective and accurate model;

- an expert in the company information systems, expected to supervise the access to the information sources;

- an expert in the mathematical theory of learning and statistics, for exploratory data analysis and for the generation of predictive models.

Figure 5.2 illustrates the competencies and the involvement in the different activities for each actor in the data mining process.

5.4   Analysis methodologies

Data mining activities can be subdivided into a few major categories, based on the tasks and the objectives of the analysis. Depending on the possible existence of a target variable, one can draw a first fundamental distinction between *supervised* and *unsupervised* learning processes.

Supervised learning. In a supervised (or *direct*) learning analysis, a target attribute either represents the class to which each record belongs, as shown in Example 5.2 on loyalty in the mobile phone industry, or expresses a measurable quantity, such as the total value of calls that will be placed by a customer in a future period. As a second example of the supervised perspective, consider an investment management company wishing to predict the balance sheet of its customers based on their demographic characteristics and past investment transactions. Supervised learning processes are therefore oriented toward prediction and interpretation with respect to a target attribute.

Unsupervised learning. Unsupervised (or *indirect*) learning analyses are not guided by a target attribute. Therefore, data mining tasks in this case are aimed at discovering recurring patterns and affinities in the dataset. As an example, consider an investment management company wishing to identify clusters of customers who exhibit homogeneous investment behavior, based on data on past transactions. In most unsupervised learning analyses, one is interested in identifying *clusters* of records that are similar within each cluster and different from members of other clusters.

Taking the distinction even further, seven basic data mining tasks can be identified:

- characterization and discrimination;

- classification;

- regression;

- time series analysis;

- association rules;

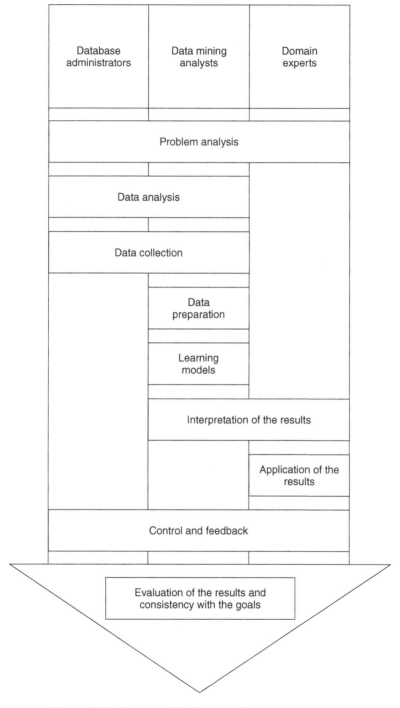

Figure 5.2   Actors and roles in a data mining process

- clustering;

- description and visualization.

The first four tasks correspond to supervised data mining analyses, since a specific target variable exists that must be explained based on the available attributes or throughout its evolution over time. The remaining three tasks represent unsupervised analyses whose purpose is the development of models capable of expressing the interrelationship among the available attributes.

Characterization and discrimination. Where a categorical target attribute exists, before starting to develop a classification model, it is often useful to carry out an exploratory analysis whose purpose is twofold. On the one hand, the aim is to achieve a characterization by comparing the distribution of the values of the attributes for the records belonging to the same class. On the other hand, the purpose is to detect a difference, through a comparison between the distribution of the values of the attributes for the records of a given class and the records of a different class, or between the records of a given class and all remaining records. This data mining task is primarily conducted by means of exploratory data analysis and therefore it is based on queries and counts that do not require the development of specific learning models. The information so acquired is usually presented to users in the form of histograms and other types of charts, as described in Chapter 7. The value of the information generated is, however, remarkable and may often direct the subsequent phase of attribute selection.

Classification. In a classification problem a set of observations is available, usually represented by the records of a dataset, whose target class is known. Observations may correspond, for instance, to mobile phone customers and the binary class may indicate whether a given customer is still active or has churned. Each observation is described by a given number of attributes whose value is known; in the previous example, the attributes may correspond to age, customer seniority and outgoing telephone traffic distinguished by destination. A classification algorithm can therefore use the available observations relative to the past in order to identify a model that can predict the target class of future observations whose attributes values are known. It is worth noting that the target attribute, whose value is to be predicted, is categorical in classification problems and therefore takes on a finite and usually rather small number of values. In most applications the target is even represented by a binary variable. The categorical nature of the target determines the distinction between classification and regression.

Regression. Unlike classification, which is intended for discrete targets, regression is used when the target variable takes on continuous values. Based on the available explanatory attributes, the goal is to predict the value of the target variable for each observation. If one wishes to predict the sales of a product based on the promotional campaigns mounted and the sale price, the target variable may take on a very high number of discrete values and can be treated as a continuous variable. A classification problem may be turned into a regression problem, and vice versa. To see this, a mobile phone company interested in the classification of customers based on their loyalty, may come up with a regression problem by predicting the probability of each customer remaining loyal.

Time series. Sometimes the target attribute evolves over time and is therefore associated with adjacent periods on the time axis. In this case, the sequence of values of the target variable is said to represent a *time series*. For instance, the weekly sales of a given product observed over 2 years represent a time series containing 104 observations. Models for time series analysis investigate data characterized by a temporal dynamics and are aimed at predicting the value of the target variable for one or more future periods.

Association rules. Association rules, also known as *affinity groupings*, are used to identify interesting and recurring associations between groups of records of a dataset. For example, it is possible to determine which products are purchased together in a single transaction and how frequently. Companies in the retail industry resort to association rules to design the arrangement of products on shelves or in catalogs. Groupings by related elements are also used to promote *cross-selling* or to devise and promote combinations of products and services.

Clustering. The term *cluster* refers to a homogeneous subgroup existing within a population. Clustering techniques are therefore aimed at segmenting a heterogeneous population into a given number of subgroups composed of observations that share similar characteristics; observations included in different clusters have distinctive features. Unlike classification, in clustering there are no predefined classes or reference examples indicating the target class, so that the objects are grouped together based on their mutual homogeneity. Sometimes, the identification of clusters represents a preliminary stage in the data mining process, within exploratory data analysis. It may allow homogeneous data to be processed with the most appropriate rules and techniques and the size of the original dataset to be reduced, since the subsequent data mining activities can be developed autonomously on each cluster identified.

Description and visualization. The purpose of a data mining process is sometimes to provide a simple and concise representation of the information stored in a large dataset. Although, in contrast to clustering and association rules, descriptive analysis does not pursue any particular grouping or partition of the records in the dataset, an effective and concise description of information is very helpful, since it may suggest possible explanations of hidden patterns in the data and lead to a better understanding the phenomena to which the data refer. Notice that it is not always easy to obtain a meaningful visualization of the data. However, the effort of representation is justified by the remarkable conciseness of the information achieved through a well-designed chart.

5.5   Notes and readings

There are several books devoted to data mining, such as Hand *et al.* (2001), Han and Kamber (2005), Hastie *et al.* (2001), Witten and Frank (2005) and Pyle (2003). The volume by Miller and Han (2000) focuses on the analysis of geographical data, whereas Berry and Linoff (2002) discusses the extraction of information from the web. A survey is presented in Powell (2001). Also worth mentioning are the following books which present relevant applications of data mining: Berry and Linoff (1999, 2004), Berson and Smith (1997), Berson *et al.* (1999), Parr Rud (2000) and Giudici (2003). A survey of optimization methods applied in the field of data mining is presented in Bradley *et al.* (1999).

6

Data preparation

Business intelligence systems and mathematical models for decision making can achieve accurate and effective results only when the input data are highly reliable. However, the data extracted from the available primary sources and gathered into a data mart may have several anomalies which analysts must identify and correct.

This chapter deals with the activities involved in the creation of a high quality dataset for subsequent use for business intelligence and data mining analysis. Several techniques can be employed to reach this goal: data validation, to identify and remove anomalies and inconsistencies; data integration and transformation, to improve the accuracy and efficiency of learning algorithms; data size reduction and discretization, to obtain a dataset with a lower number of attributes and records but which is as informative as the original dataset. For further readings on the subject and for basic concepts of descriptive statistics, see the notes at the end of Chapter 7.

6.1 Data validation

The quality of input data may prove unsatisfactory due to *incompleteness*, *noise* and *inconsistency*.

Incompleteness. Some records may contain missing values corresponding to one or more attributes, and there may be a variety of reasons for this. It may be that some data were not recorded at the source in a systematic way, or that they were not available when the transactions associated with a record took place. In other instances, data may be missing because of malfunctioning recording devices. It is also possible that some data were deliberately removed during previous stages of the gathering process because they were deemed

Business Intelligence: Data Mining and Optimization for Decision Making   C. Vercellis
© 2009 John Wiley & Sons, Ltd

incorrect. Incompleteness may also derive from a failure to transfer data from the operational databases to a data mart used for a specific business intelligence analysis.

Noise. Data may contain erroneous or anomalous values, which are usually referred to as *outliers*. Other possible causes of noise are to be sought in malfunctioning devices for data measurement, recording and transmission. The presence of data expressed in heterogeneous measurement units, which therefore require conversion, may in turn cause anomalies and inaccuracies.

Inconsistency. Sometimes data contain discrepancies due to changes in the coding system used for their representation, and therefore may appear inconsistent. For example, the coding of the products manufactured by a company may be subject to a revision taking effect on a given date, without the data recorded in previous periods being subject to the necessary transformations in order to adapt them to the revised encoding scheme.

The purpose of data validation techniques is to identify and implement corrective actions in case of incomplete and inconsistent data or data affected by noise.

6.1.1 Incomplete data

To partially correct incomplete data one may adopt several techniques.

Elimination. It is possible to discard all records for which the values of one or more attributes are missing. In the case of a supervised data mining analysis, it is essential to eliminate a record if the value of the target attribute is missing. A policy based on systematic elimination of records may be ineffective when the distribution of missing values varies in an irregular way across the different attributes, since one may run the risk of incurring a substantial loss of information.

Inspection. Alternatively, one may opt for an inspection of each missing value, carried out by experts in the application domain, in order to obtain recommendations on possible substitute values. Obviously, this approach suffers from a high degree of arbitrariness and subjectivity, and is rather burdensome and time-consuming for large datasets. On the other hand, experience indicates that it is one of the most accurate corrective actions if skilfully exercised.

Identification. As a third possibility, a conventional value might be used to encode and identify missing values, making it unnecessary to remove entire records from the given dataset. For example, for a continuous attribute that assumes only positive values it is possible to assign the value $\{-1\}$ to all missing data. By the same token, for a categorical attribute one might replace missing values with a new value that differs from all those assumed by the attribute.

Substitution. Several criteria exist for the automatic replacement of missing data, although most of them appear somehow arbitrary. For instance, missing values of an attribute may be replaced with the mean of the attribute calculated for the remaining observations. This technique can only be applied to numerical attributes, but it will clearly be ineffective in the case of an asymmetric distribution of values. In a supervised analysis it is also possible to replace missing values by calculating the mean of the attribute only for those records having the same target class. Finally, the maximum likelihood value, estimated using regression models or Bayesian methods, can be used as a replacement for missing values. However, estimate procedures can become rather complex and time-consuming for a large dataset with a high percentage of missing data.

6.1.2 Data affected by noise

The term *noise* refers to a random perturbation within the values of a numerical attribute, usually resulting in noticeable anomalies. First, the outliers in a dataset need to be identified, so that subsequently either they can be corrected and regularized or entire records containing them are eliminated. In this section we will describe a few simple techniques for identifying and regularizing data affected by noise, while in Chapter 7 we will describe in greater detail the tools from exploratory data analysis used to detect outliers.

The easiest way to identify outliers is based on the statistical concept of *dispersion*. The sample mean $\bar{\mu}_j$ and the sample variance $\bar{\sigma}_j^2$ of the numerical attribute $\mathbf{a}_j$ are calculated. If the attribute follows a distribution that is not too far from normal, the values falling outside an appropriate interval centered around the mean value $\bar{\mu}_j$ are identified as outliers, by virtue of the central limit theorem. More precisely, with a confidence of $100(1 - \alpha)\%$ (approximately 96% for $\alpha = 0.05$) it is possible to consider as outliers those values that fall outside the interval

$$(\bar{\mu}_j - z_{\alpha/2}\bar{\sigma}_j, \bar{\mu}_j + z_{\alpha/2}\bar{\sigma}_j), \tag{6.1}$$

where $z_{\alpha/2}$ is the $\alpha/2$ quantile of the standard normal distribution. This technique is simple to use, although it has the drawback of relying on the critical assumption that the distribution of the values of the attribute is bell-shaped and roughly normal. However, by applying Chebyshev's theorem, described in Chapter 7, it is possible to obtain analogous bounds independent of the distribution, with intervals that are only slightly less stringent. Once the outliers have been identified, it is possible to correct them with values that are deemed more plausible, or to remove an entire record containing them.

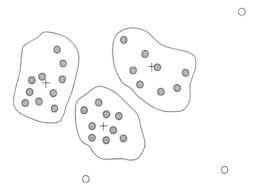

Figure 6.1   Identification of outliers using cluster analysis

An alternative technique, illustrated in Figure 6.1, is based on the distance between observations and the use of clustering methods. Once the clusters have been identified, representing sets of records having a mutual distance that is less than the distance from the records included in other groups, the observations that are not placed in any of the clusters are identified as outliers. Clustering techniques offer the advantage of simultaneously considering several attributes, while methods based on dispersion can only take into account each single attribute separately.

A variant of clustering methods, also based on the distances between the observations, detects the outliers through two parametric values, p and d, to be assigned by the user. An observation $\mathbf{x}_i$ is identified as an outlier if at least a percentage p of the observations in the dataset are found at a distance greater than d from $\mathbf{x}_i$.

The above techniques can be combined with the opinion of experts in order to identify actual outliers with respect to regular observations, even though these fall outside the intervals where regular records are expected to lie. In marketing applications, in particular, it is appropriate to consult with experts before adopting corrective measures in the case of anomalous observations.

Unlike the above methods, aimed at identifying and correcting each single anomaly, there exist also regularization techniques which automatically correct anomalous data. For example, simple or multiple regression models predict the value of the attribute $\mathbf{a}_j$ that one wishes to regularize based on other variables existing in the dataset. Once the regression model has been developed, and the corresponding confidence interval around the prediction curve has been calculated, it is possible to substitute the value computed along the prediction curve for the values of the attribute $\mathbf{a}_j$ that fall outside the interval.

A further automatic regularization technique, described in Section 6.3.4, relies on data discretization and grouping based on the proximity of the values of the attribute $\mathbf{a}_j$.

6.2   Data transformation

In most data mining analyses it is appropriate to apply a few transformations to the dataset in order to improve the accuracy of the learning models subsequently developed. Indeed, outlier correction techniques are examples of transformations of the original data that facilitate subsequent learning phases. The principal component method, described in Section 6.3.3, can also be regarded as a data transformation process.

6.2.1   Standardization

Most learning models benefit from a preventive *standardization* of the data, also called *normalization*. The most popular standardization techniques include the *decimal scaling* method, the *min-max* method and the *z-index* method.

Decimal scaling. Decimal scaling is based on the transformation

$$x'_{ij} = \frac{x_{ij}}{10^h},\qquad(6.2)$$

where h is a given parameter which determines the scaling intensity. In practice, decimal scaling corresponds to shifting the decimal point by h positions toward the left. In general, h is fixed at a value that gives transformed values in the range $[-1, 1]$.

Min-max. Min-max standardization is achieved through the transformation

$$x'_{ij} = \frac{x_{ij} - x_{\min, j}}{x_{\max, j} - x_{\min, j}}(x'_{\max, j} - x'_{\min, j}) + x'_{\min, j},\qquad(6.3)$$

where

$$x_{\min, j} = \min_i x_{ij}, \quad x_{\max, j} = \max_i x_{ij},\qquad(6.4)$$

are the minimum and maximum values of the attribute $\mathbf{a}_j$ before transformation, while $x'_{\min, j}$ and $x'_{\max, j}$ are the minimum and maximum values that we wish to obtain after transformation. In general, the extreme values of the range are defined so that $x'_{\min, j} = -1$ and $x'_{\max, j} = 1$ or $x'_{\min, j} = 0$ and $x'_{\max, j} = 1$.

z-index. z-index based standardization uses the transformation

$$x'_{ij} = \frac{x_{ij} - \bar{\mu}_j}{\bar{\sigma}_j}, \tag{6.5}$$

where $\bar{\mu}_j$ and $\bar{\sigma}_j$ are respectively the sample mean and sample standard deviation of the attribute $\mathbf{a}_j$. If the distribution of values of the attribute $\mathbf{a}_j$ is roughly normal, the z-index based transformation generates values that are almost certainly within the range $(-3, 3)$.

6.2.2 Feature extraction

The aim of standardization techniques is to replace the values of an attribute with values obtained through an appropriate transformation. However, there are situations in which more complex transformations are used to generate new attributes that represent a set of additional columns in the matrix $\mathbf{X}$ representing the dataset $\mathcal{D}$. Transformations of this kind are usually referred to as *feature extraction*. For example, suppose that a set of attributes indicate the spending of each customer over consecutive time intervals. It is then possible to define new variables capable of capturing the trends in the data through differences or ratios between spending amounts of contiguous periods.

In other instances, the transformations may take even more complex forms, such as *Fourier transforms*, *wavelets* and *kernel* functions. The use of such methods will be explained within the classification methods called *support vector machines* in Chapter 10.

Attribute extraction may also consist of the creation of new variables that summarize within themselves the relevant information contained in a subset of the original attributes. For example, in the context of image recognition one is often interested in identifying the existence of a face within a digitalized photograph. There are different indicators intended for the synthesis of each piece of information contained in a group of adjacent pixels, which make it easier for classification algorithms to detect faces.

6.3   Data reduction

When dealing with a small dataset, the transformations described above are usually adequate to prepare input data for a data mining analysis. However, when facing a large dataset it is also appropriate to reduce its size, in order to make learning algorithms more efficient, without sacrificing the quality of the results obtained.

There are three main criteria to determine whether a data reduction technique should be used: *efficiency*, *accuracy* and *simplicity* of the models generated.

Efficiency. The application of learning algorithms to a dataset smaller than the original one usually means a shorter computation time. If the complexity of the algorithm is a superlinear function, as is the case for most known methods, the improvement in efficiency resulting from a reduction in the dataset size may be dramatic. As described in Chapter 5, within the data mining process it is customary to run several alternative learning algorithms in order to identify the most accurate model. Therefore, a reduction in processing times allows the analyses to be carried out more quickly.

Accuracy. In most applications, the accuracy of the models generated represents a critical success factor, and it is therefore the main criterion followed in order to select one class of learning methods over another. As a consequence, data reduction techniques should not significantly compromise the accuracy of the model generated. As shown below, it may also be the case that some data reduction techniques, based on attribute selection, will lead to models with a higher generalization capability on future records.

Simplicity. In some data mining applications, concerned more with interpretation than with prediction, it is important that the models generated be easily translated into simple rules that can be understood by experts in the application domain. As a trade-off for achieving simpler rules, decision makers are sometimes willing to allow a slight decrease in accuracy. Data reduction often represents an effective technique for deriving models that are more easily interpretable.

Since it is difficult to develop a data reduction technique that represents the optimal solution for all the criteria described, the analyst will aim for a suitable trade-off among all the requirements outlined.

Data reduction can be pursued in three distinct directions, described below: a reduction in the number of observations through *sampling*, a reduction in the number of attributes through *selection* and *projection*, and a reduction in the number of values through *discretization* and *aggregation*.

6.3.1   Sampling

A further reduction in the size of the original dataset can be achieved by extracting a sample of observations that is significant from a statistical standpoint. This type of reduction is based on classical inferential reasoning. It is therefore necessary to determine the size of the sample that guarantees the level of accuracy required by the subsequent learning algorithms and to define an adequate sampling procedure. Sampling may be *simple* or *stratified* depending on whether one wishes to preserve in the sample the percentages of the original dataset with respect to a categorical attribute that is considered critical.

Generally speaking, a sample comprising a few thousand observations is adequate to train most learning models. It is also useful to set up several independent samples, each of a predetermined size, to which learning algorithms should be applied. In this way, computation times increase linearly with the number of samples determined, and it is possible to compare the different models generated, in order to assess the robustness of each model and the quality of the knowledge extracted from data against the random fluctuations existing in the sample. It is obvious that the conclusions obtained can be regarded as robust when the models and the rules generated remain relatively stable as the sample set used for training varies.

6.3.2   Feature selection

The purpose of *feature selection*, also called *feature reduction*, is to eliminate from the dataset a subset of variables which are not deemed relevant for the purpose of the data mining activities. One of the most critical aspects in a learning process is the choice of the combination of predictive variables more suited to accurately explain the investigated phenomenon.

Feature reduction has several potential advantages. Due to the presence of fewer columns, learning algorithms can be run more quickly on the reduced dataset than on the original one. Moreover, the models generated after the elimination from the dataset of uninfluential attributes are often more accurate and easier to understand.

Feature selection methods can be classified into three main categories: *filter* methods, *wrapper* methods and *embedded* methods.

Filter methods. Filter methods select the relevant attributes before moving on to the subsequent learning phase, and are therefore independent of the specific algorithm being used. The attributes deemed most significant are selected for learning, while the rest are excluded. Several alternative statistical metrics have been proposed to assess the predictive capability and relevance of a group of attributes. Generally, these are monotone metrics in that their value increases or decreases according to the number of attributes considered. The simplest filter method to apply for supervised learning involves the assessment of each single attribute based on its level of correlation with the target. Consequently, this lead to the selection of the attributes that appear mostly correlated with the target.

Wrapper methods. If the purpose of the data mining investigation is classification or regression, and consequently performances are assessed mainly in terms of accuracy, the selection of predictive variables should be based not only on the level of relevance of each single attribute but also on the specific learning algorithm being utilized. Wrapper methods are able to meet this need, since they assess a group of variables using the same classification or regression algorithm used to predict the value of the target variable. Each time, the algorithm uses a different subset of attributes for learning, identified by a search engine that works on the entire set of all possible combinations of variables, and selects the set of attributes that guarantees the best result in terms of accuracy. Wrapper methods are usually burdensome from a computational standpoint, since the assessment of every possible combination identified by the search engine requires one to deal with the entire training phase of the learning algorithm. An example of the use of a wrapper method for attribute selection is given in Section 8.5 in the context of multiple linear regression models.

Embedded methods. For the embedded methods, the attribute selection process lies *inside* the learning algorithm, so that the selection of the optimal set of attributes is directly made during the phase of model generation. Classification trees, described in Chapter 10, are an example of embedded methods. At each tree node, they use an evaluation function that estimates the predictive value of a single attribute or a linear combination of variables. In this way, the relevant attributes are automatically selected and they determine the rule for splitting the records in the corresponding node.

Filter methods are the best choice when dealing with very large datasets, whose observations are described by a large number of attributes. In these cases, the application of wrapper methods is inappropriate due to very long computation times. Moreover, filter methods are flexible and in principle can be associated with any learning algorithm. However, when the size of the problem at hand is moderate, it is preferable to turn to wrapper or embedded methods which afford in most cases accuracy levels that are higher compared to filter methods.

As described above, wrapper methods select the attributes according to a search scheme that inspects in sequence several subsets of attributes and applies the learning algorithm to each subset in order to assess the resulting accuracy of the corresponding model. If a dataset contains n attributes, there are 2^n possible subsets and therefore an exhaustive search procedure would require excessive computation times even for moderate values of n. As a consequence, the procedure for selecting the attributes for wrapper methods is usually of a heuristic nature, based in most cases on a *greedy* logic which evaluates for each attribute a relevance indicator adequately defined and then selects the attributes based on their level of relevance.

In particular, three distinct myopic search schemes can be followed: *forward*, *backward* and *forward–backward* search.

Forward. According to the forward search scheme, also referred to as *bottom-up* search, the exploration starts with an empty set of attributes and subsequently introduces the attributes one at a time based on the ranking induced by the relevance indicator. The algorithm stops when the relevance index of all the attributes still excluded is lower than a prefixed threshold.

Backward. The backward search scheme, also referred to as *top-down* search, begins the exploration by selecting all the attributes and then eliminates them one at a time based on the preferred relevance indicator. The algorithm stops when the relevance index of all the attributes still included in the model is higher than a prefixed threshold.

Forward–backward. The forward–backward method represents a trade-off between the previous schemes, in the sense that at each step the best attribute among those excluded is introduced and the worst attribute among those included is eliminated. Also in this case, threshold values for the included and excluded attributes determine the stopping criterion.

The various wrapper methods differ in the choice of the relevance measure as well as well as the threshold preset values for the stopping rule of the algorithm.

6.3.3   Principal component analysis

Principal component analysis (PCA) is the most widely known technique of attribute reduction by means of projection. Generally speaking, the purpose of this method is to obtain a projective transformation that replaces a subset of the original numerical attributes with a lower number of new attributes obtained as their linear combination, without this change causing a loss of information. Experience shows that a transformation of the attributes may lead in many instances to better accuracy in the learning models subsequently developed.

Before applying the principal component method, it is expedient to standardize the data, so as to obtain for all the attributes the same range of values, usually represented by the interval $[-1, 1]$. Moreover, the mean of each attribute $\mathbf{a}_j$ is made equal to 0 by applying the transformation

$$\tilde{x}_{ij} = x_{ij} - \frac{1}{m} \sum_{i=1}^{m} x_{ij}. \tag{6.6}$$

Let $\mathbf{X}$ denote the matrix resulting from applying the transformation (6.6) to the original data, and let $\mathbf{V} = \mathbf{X}'\mathbf{X}$ be the covariance matrix of the attributes (for a definition of the covariance and variance matrices, see Section 7.3.1). If the correlation matrix is used to develop the principal component analysis method instead of the covariance matrix, the transformation (6.6) is not required.

Starting from the n attributes in the original dataset, represented by the matrix $\mathbf{X}$, the principal component method derives n orthogonal vectors, namely the *principal components*, which constitute a new basis of the space $\mathbb{R}^n$. Principal components are better suited than the original attributes to explain fluctuations in the data, in the sense that usually a subset consisting of q principal components, with $q < n$, has an information content that is almost equivalent to that of the original dataset. As a consequence, the original data are projected into a lower-dimensional space of dimension q having the same explanatory capability.

Principal components are generated in sequence by means of an iterative algorithm. The first component is determined by solving an appropriate optimization problem, in order to explain the highest percentage of variation in the data. At each iteration the next principal component is selected, among those vectors that are orthogonal to all components already determined, as the one which explains the maximum percentage of variance not yet explained by the previously generated components. At the end of the procedure the principal components are ranked in non-increasing order with respect to the amount of variance that they are able to explain.

Let $\mathbf{p}_j$, $j \in \mathcal{N}$, denote the n principal components, each of them being obtained as a linear combination $\mathbf{p}_j = \mathbf{X}\mathbf{w}_j$ of the available attributes, where the weights $\mathbf{w}_j$ have to be determined. The projection of a generic example $\mathbf{x}_i$ in the direction of the weights vector $\mathbf{w}_j$ is given by $\mathbf{w}'_j\mathbf{x}_i$. It can easily be seen that its variance is given by

$$E[\mathbf{w}'_j\mathbf{x}_i - E[\mathbf{w}'_j\mathbf{x}_i]]^2 = E[(\mathbf{w}'_j(\mathbf{x}_i - E[\mathbf{x}_i]))^2]$$
$$= \mathbf{w}'_j\,E[(\mathbf{x}_i - E[\mathbf{x}_i])'(\mathbf{x}_i - E[\mathbf{x}_i])]\mathbf{w}_j$$
$$= \mathbf{w}'_j\mathbf{V}\mathbf{w}_j. \qquad (6.7)$$

The first principal component $\mathbf{p}_1$ represents a vector in the direction of maximum variance in the space of the original attributes and therefore its weights may be obtained by solving the quadratic constrained maximization problem

$$\max_{\mathbf{w}_1}\ \{\mathbf{w}'_1\mathbf{V}\mathbf{w}_1 : \mathbf{w}'_1\mathbf{w}_1 = 1\}, \qquad (6.8)$$

where the unit norm constraint for $\mathbf{w}_1$ is introduced in order to derive a well-posed problem. By introducing the Lagrangian function

$$L(\mathbf{w}_1, \lambda_1) = \mathbf{w}'_1\mathbf{V}\mathbf{w}_1 - \lambda_1(\mathbf{w}'_1\mathbf{w}_1 - 1), \qquad (6.9)$$

and applying the Karush–Kuhn–Tucker conditions, the solution of the maximization problem reduces to the solution of the system

$$\frac{\partial L(\mathbf{w}_1, \lambda_1)}{\partial \mathbf{w}_1} = 2\mathbf{V}\mathbf{w}_1 - 2\lambda_1\mathbf{w}_1 = \mathbf{0},　\qquad (6.10)$$

$$\frac{\partial L(\mathbf{w}_1, \lambda_1)}{\partial \lambda_1} = 1 - \mathbf{w}_1'\mathbf{w}_1 = 0,　\qquad (6.11)$$

which can be rewritten as

$$(\mathbf{V} - \lambda_1\mathbf{I})\mathbf{w}_1 = \mathbf{0},　\qquad (6.12)$$

subject to the unit norm condition $\mathbf{w}_1'\mathbf{w}_1 = 1$, where $\mathbf{I}$ is the identity matrix. The solution of the maximization problem is therefore given by $\mathbf{w}_1 = \mathbf{u}_1$, where $\mathbf{u}_1$ is the eigenvector of unit norm associated with the maximum eigenvalue λ_1 of the covariance matrix $\mathbf{V}$. Since the variance we wish to maximize is given by

$$\mathbf{w}_1'\mathbf{V}\mathbf{w}_1 = \lambda_1\mathbf{w}_1'\mathbf{w}_1 = \lambda_1\mathbf{u}_1'\mathbf{u}_1 = \lambda_1,　\qquad (6.13)$$

the first principal component is obtained by means of the eigenvector $\mathbf{u}_1$, associated with the maximum eigenvalue λ_1 of $\mathbf{V}$, through the relation $\mathbf{p}_j = \mathbf{X}\mathbf{u}_j$.

The second principal component may be determined by solving an optimization problem similar to (6.8), adding the condition of orthogonality to the previously obtained principal component, expressed by the constraint

$$\mathbf{w}_2'\mathbf{u}_1 = 0.　\qquad (6.14)$$

Proceeding in an iterative way, it is possible to derive the n principal components starting from the eigenvectors $\mathbf{u}_j$, $j \in \mathcal{N}$, of $\mathbf{V}$ ordered by non-increasing eigenvalues $\lambda_1 \geq \lambda_2 \geq \ldots \geq \lambda_n$, through the equalities $\mathbf{p}_j = \mathbf{X}\mathbf{u}_j$. The variance of the principal component $\mathbf{p}_j$ is given by $\mathrm{var}(\mathbf{p}_j) = \lambda_j$.

The n principal components constitute a new basis in the space $\mathbb{R}^n$, since the vectors are orthogonal to each other. Therefore, they are also uncorrelated and can be ordered according to a relevance indicator expressed by the corresponding eigenvalue. In particular, the first principal component explains the greatest proportion of variance in the data, the second explains the second greatest proportion of variance, and so on. If $\mathbf{U}$ denotes the $n \times n$ matrix whose columns are the eigenvectors $\mathbf{u}_j$, $j \in \mathcal{N}$, and $\mathbf{P}$ indicates the $n \times n$ matrix whose columns are the principal components $\mathbf{p}_j$, $j \in \mathcal{N}$, then the equality $\mathbf{P} = \mathbf{X}\mathbf{U}$ holds true. The total variance of the principal components is equal to the total variance of the original attributes, that is,

$$\sum_{j=1}^{n} \mathrm{var}(\mathbf{p}_j) = \sum_{j=1}^{n} \lambda_j = \mathrm{tr}(\mathbf{V}) = \sum_{j=1}^{n} \mathrm{var}(\mathbf{a}_j).　\qquad (6.15)$$

The interpretation of the principal components may be obtained from the coefficients of the vector $\mathbf{w}_j = \mathbf{u}_j$ which express their relationship with the original attributes. To this end, notice that the principal component $\mathbf{p}_h$ assumes the form

$$\mathbf{p}_h = u_{h1}\mathbf{a}_1 + u_{h2}\mathbf{a}_2 + \cdots + u_{hn}\mathbf{a}_n. \tag{6.16}$$

The coefficient u_{hj} can be therefore interpreted as the weight of the attribute $\mathbf{a}_j$ in determining the component $\mathbf{p}_h$. The greater the absolute value of u_{hj} is, the more the component $\mathbf{p}_h$ is characterized by the attribute $\mathbf{a}_j$. At the same time, $\mathrm{var}(\mathbf{p}_h) = \lambda_h$ represents a measure of the proportion of total variance explained by the principal component $\mathbf{p}_h$. For this reason, the index

$$I_q = \frac{\lambda_1 + \lambda_2 + \cdots + \lambda_q}{\lambda_1 + \lambda_2 + \cdots + \lambda_n} \tag{6.17}$$

expresses the percentage of total variance explained by the first q principal components and provides an indication of the amount of information preserved by the first q components. In order to determine the number of principal components to be appropriately used, it is possible to go on until the level of overall importance I_q of the considered components exceeds a threshold $I_{\min}$ deemed reasonable, in relation to the properties of the dataset. The number of principal components is therefore determined as the smallest value q such that $I_q > I_{\min}$.

An example of application of principal component analysis

To illustrate the application of the principal component method, consider the *mtcars* dataset, described in Appendix B, which contains 11 attributes representing the main characteristics of different types of cars.

Table 6.1 indicates the overall importance of the principal components. It displays for each component $\mathbf{p}_h$ the standard deviation, given by $\sqrt{\lambda_h}$, the proportion of variance explained by each component, equal to $\lambda_h / \sum_j \lambda_j$, and the cumulative proportion explained by the first components up to $\mathbf{p}_h$ included. The analysis shows that the first two components explain 84% of the total variation in the data while the first 5 explain 94%.

Table 6.2 shows the values of the coefficients $w_{hj} = u_{hj}$; equivalently, it indicates the eigenvectors of the covariance matrix $\mathbf{V}$. The first component, which alone explains 60% of the variance, is negatively correlated with the attributes $\{cyl, disp, wt, carb\}$, whose meaning is explained in Appendix B, while it is positively correlated with all the other attributes.

Finally, Figure 6.2 shows through a *scree plot* the decreasing progression of the variance explained by the principal components. The chart may prove helpful in identifying the relevant components to be included in the transformed dataset.

Table 6.1   Overall importance of principal components in the *mtcars* dataset

	Comp.1	Comp.2	Comp.3	Comp.4
Standard deviation	2.5706809	1.6280258	0.7919578	0.5192277
Proportion of variance	0.6007637	0.2409516	0.0570179	0.0245088
Cumulative proportion	0.6007637	0.8417153	0.8987332	0.9232420
	Comp.5	Comp.6	Comp.7	Comp.8
Standard deviation	0.4727061	0.4599957	0.3677798	0.3505730
Proportion of variance	0.0203137	0.0192360	0.0122965	0.0111728
Cumulative proportion	0.9435558	0.9627918	0.9750883	0.9862612
	Comp.9	Comp.10	Comp.11	
Standard deviation	0.2775727	0.2281127	0.1484735	
Proportion of variance	0.0070042	0.0047304	0.0020040	
Cumulative proportion	0.9932654	0.9979959	1.0000000	

Table 6.2   Principal component coefficients for the *mtcars* dataset

	Comp.1	Comp.2	Comp.3	Comp.4	Comp.5	Comp.6
mpg	0.363	−0.226	−0.103	−0.109	0.368	0.754
cyl	−0.374	−0.175	0.169	0.231	−0.846	
disp	−0.368	0.257	−0.394	−0.336	0.214	0.198
hp	−0.33	−0.249	0.14	−0.54	0.222	−0.576
drat	0.294	−0.275	0.161	0.855	0.244	−0.101
wt	−0.346	0.143	0.342	0.246	−0.465	0.359
qsec	0.2	0.463	0.403	0.165	−0.33	0.232
vs	0.307	0.232	0.429	−0.215	−0.6	0.194
am	0.235	−0.429	−0.206	−0.571	−0.587	−0.178
gear	0.207	−0.462	0.29	−0.265	−0.244	0.605
carb	−0.214	−0.414	0.529	−0.127	0.361	0.184
	Comp.7	Comp.8	Comp.9	Comp.10	Comp.11	
mpg	0.236	0.139			0.125	
cyl					0.141	
disp					−0.661	
hp	0.248				0.256	
drat						
wt					0.567	
qsec	−0.528	−0.271			−0.181	
vs	−0.266	0.359	−0.159			
am						
gear	−0.336	−0.214				
carb	−0.175	0.396	0.171		−0.32	

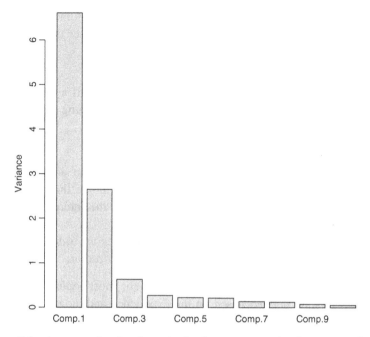

Figure 6.2   A scree plot for the principal components in the mtcars dataset

6.3.4   Data discretization

The general purpose of data reduction methods is to obtain a decrease in the number of distinct values assumed by one or more attributes. Data discretization is the primary reduction method. On the one hand, it reduces continuous attributes to categorical attributes characterized by a limited number of distinct values. On the other hand, its aim is to significantly reduce the number of distinct values assumed by the categorical attributes.

For instance, the weekly spending of a mobile phone customer is a continuous numerical value, which might be discretized into, say, five classes: low, $[0 - 10)$ euros; medium low, $[10 - 20)$ euros; medium, $[20 - 30)$ euros; medium high, $[30 - 40)$ euros; and high, over 40 euros.

As a further example applied to a categorical variable, consider the province of residence of each customer, and suppose it can assume a hundred distinct values. If instead of the province one uses the region of residence, the new attribute might take on twenty distinct values.

In both cases, the discretization process has brought about a reduction in the number of distinct values assumed by each attribute. The models that can be generated on the reduced dataset are likely to be more intuitive and less arbitrary. For instance, using a classification tree, it is possible to generate a rule of the form if spending is in the medium low range, and if a customer resides in region A, then the probability of churning is higher than 0.85.

This is much more interpretable than the rule

if spending is in the [12.21, 14.79] euro range, and if a customer resides in province B, then the probability of churning is higher than 0.85,

which could have been generated for the original dataset.

The examples shown above suggest that discretization and reduction of the number of values taken by each attribute can improve the generalization capability of predictive models, thus making easier the interpretation of the rules obtained.

Among the most popular discretization techniques are *subjective subdivision*, *subdivision into classes* and *hierarchical discretization*.

Subjective subdivision. Subjective subdivision is the most popular and intuitive method. Classes are defined based on the experience and judgment of experts in the application domain.

Subdivision into classes. Subdivision into categorical classes may be achieved in an automated way using the techniques described below. In particular, the subdivision can be based on classes of equal size or equal width.

Hierarchical discretization. The third type of discretization is based on hierarchical relationships between concepts and may be applied to categorical attributes, just as for the hierarchical relationships between provinces and regions. In general, given a hierarchical relationship of the one-to-many kind, it is possible to replace each value of an attribute with the corresponding value found at a higher level in the hierarchy of concepts.

Subdivision into classes

The automated procedure of subdivision into classes consists of ordering in a non-decreasing way the values of the attribute $\mathbf{a}_j$ and grouping them into a predetermined number K of contiguous classes. It is possible to form the classes of either different *size* or different *width*.

In the first case, the m observed values available for the attribute $\mathbf{a}_j$ are distributed by placing $\lfloor m/K \rfloor$ or $\lceil m/K \rceil$ contiguous values in each class, so as to divide the m observed values almost equally among the K classes.

In the second case, the range of total variation between the minimum value and the maximum value taken by the attribute $\mathbf{a}_j$ is subdivided into K contiguous intervals and the observed values are placed in the class corresponding to the interval where they fall. This second procedure is less effective if the distribution of the values significantly moves away from the uniform distribution. Once the K classes have been constructed, each observed value of a_j is replaced by the average value of the corresponding class. As an alternative, instead of using the average value for regularization, it is possible to use the boundary value of the class that is closest to the original value taken by a_j.

The following examples show the different regularization methods based on the subdivision into classes.

Example 6.1 – Subdivision into equal size classes.   Ordered values of a_j: 3, 4, 4, 7, 12, 15, 21, 23, 27.
Class 1: 3, 4, 4
Class 2: 7, 12, 15
Class 3: 21, 23, 27

Example 6.2 – Regularization by the mean value.   Ordered values of a_j: 3, 4, 4, 7, 12, 15, 21, 23, 27.
Class 1: 3.66, 3.66, 3.66
Class 2: 11.33, 11.33, 11.33
Class 3: 23.66, 23.66, 23.66

Example 6.3 – Regularization by boundary values.   Ordered values of a_j: 3, 4, 4, 7, 12, 15, 21, 23, 27.
Class 1: 3, 4, 4
Class 2: 7, 15, 15
Class 3: 21, 21, 27

Example 6.4 – Subdivision into equal width classes.   Ordered values of a_j: 3, 4, 4, 7, 12, 15, 21, 23, 27.
Class 1 - interval [3, 11): 3, 4, 4, 7
Class 2 - interval [11, 19): 12, 15
Class 3 - interval [19, 27]: 21, 23, 27

7

Data exploration

The primary purpose of exploratory data analysis is to highlight the relevant features of each attribute contained in a dataset, using graphical methods and calculating summary statistics, and to identify the intensity of the underlying relationships among the attributes. Exploratory data analysis includes three main phases:

- *univariate* analysis, in which the properties of each single attribute of a dataset are investigated;

- *bivariate* analysis, in which pairs of attributes are considered, to measure the intensity of the relationship existing between them (for supervised learning models, it is of particular interest to analyze the relationships between the explanatory attributes and the target variable);

- *multivariate* analysis, in which the relationships holding within a subset of attributes are investigated.

7.1 Univariate analysis

Univariate analysis is used to study the behavior of each attribute, considered as an entity independent of the other variables of the dataset. It is of interest to assess the tendency of the values of a given attribute to arrange themselves around a specific central value (*location*), to measure the propensity of the variable to assume a more or less wide range of values (*dispersion*) and to extract information on the underlying probability distribution.

Univariate analysis has several objectives. On the one hand, some learning models make specific statistical hypotheses regarding the distribution of the variables being examined, and it is therefore necessary to verify the validity

Business Intelligence: Data Mining and Optimization for Decision Making   C. Vercellis
© 2009 John Wiley & Sons, Ltd

of such assumptions before proceeding with the subsequent investigation. On the other hand, univariate analysis intuitively draws conclusions concerning the information content that each attribute may provide. For example, an attribute that assumes the same value at 95% of the available observations may be considered fairly constant for practical purposes and the information that it provides is therefore of little value. Moreover, univariate analysis plays a key role in pointing out anomalies and non-standard values – that is, in identifying the outliers already mentioned in Section 6.1.2.

Suppose that a given dataset $\mathcal{D}$ contains m observations, and denote by $\mathbf{a}_j$ the generic attribute being analyzed. Since for the purpose of univariate analysis a single attribute at a time is being considered, for the sake of clarity in this section the subscript j will be suppressed from the attribute $\mathbf{a}_j$. Therefore, the vector $(x_{1j}, x_{2j}, \ldots, x_{mj})$ of m observations corresponding to attribute $\mathbf{a}_j$ will be denoted by $\mathbf{a} = (x_1, x_2, \ldots, x_m)$ in the present context.

Graphical representations are often the starting point for exploratory data analysis in order to gain insights into the properties of a single attribute. The eyes and the brain are endowed with an astonishing ability to perceive patterns and recurrences in data presented in a well-designed chart. The example described in Section 7.2.2 will stress the importance of exploratory graphical analysis.

For the purpose of a graphical representation, it is necessary to make a distinction between categorical and numerical attributes.

7.1.1   Graphical analysis of categorical attributes

A categorical attribute may be graphically analyzed by resorting to various representations for the empirical distribution of the observations – that is, the relative frequencies with which the different values occur. Denote by

$$V = \{v_1, v_2, \ldots, v_H\} \tag{7.1}$$

the set of H distinct values that are taken by the categorical attribute $\mathbf{a}$, and let $\mathcal{H} = \{1, 2, \ldots, H\}$. Taking the dataset shown in Example 5.2 and Table 5.2, for the attribute *area* we have $H = 4$ and $V = \{1, 2, 3, 4\}$.

The most natural representation for the graphical analysis of a categorical attribute is a *vertical bar chart*, which indicates along the vertical axis or ordinate the empirical frequencies, that is the number of observations of the dataset corresponding to each of the values assumed by the attribute $\mathbf{a}$, indicated along the horizontal axis or abscissa. Frequencies may be expressed by the relation

$$e_h = \text{card}\{i \in \mathcal{M} : x_i = v_h\}, \quad h \in \mathcal{H}. \tag{7.2}$$

For the attribute *area* in Table 5.2, part (a) of Figure 7.1 shows the corresponding vertical bar chart. Part (b) of the figure depicts the vertical bar chart

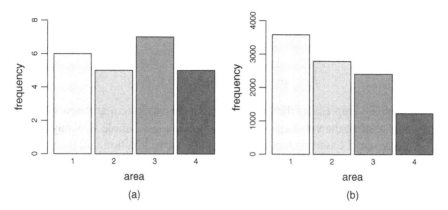

Figure 7.1   Vertical bar charts for a categorical attribute

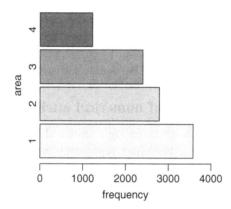

Figure 7.2   Horizontal bar chart for a categorical attribute

obtained for a sample consisting of 10 000 observations. The difference in the profiles of parts (a) and (b) highlights the low robustness of frequencies estimated through a small sample.

Sometimes, a *horizontal bar chart* is preferable in place of a vertical bar chart. Here the positions of the vertical and horizontal axes are swapped, as shown in Figure 7.2.

It is also possible to calculate the *relative empirical frequency*, or *empirical density*,

$$f_h = \frac{e_h}{m} = \frac{\text{card}\{i \in \mathcal{M} : x_i = v_h\}}{m}, \quad h \in \mathcal{H}, \tag{7.3}$$

with which each value v_h is assumed by the attribute **a**. For a sample of adequate size, by virtue of the central limit theorem, the relative empirical frequency represents a good approximation of the probability density of attribute **a**. More precisely, it can be claimed that for a sample of sufficiently large size the approximation

$$f_h \approx p_h = \Pr\{x = v_h\}, \quad h \in \mathcal{H}, \quad (7.4)$$

holds, where p_h represents the probability that the attribute **a** at a new observation x will assume the value v_h. Notice that the smallest sample size may be precisely determined based on the desired level of approximation and the required level of confidence, using fundamental results from inferential statistics.

The empirical density function may be represented by a chart similar to that for frequencies, the only difference being that the heights of the bars are divided by the number m of observations in the sample. As a consequence, it is necessary to introduce a change in the scale along the vertical axis. Sometimes, it is preferable to use a pie chart to represent the empirical density function. In this case, the amplitude of each sector in the circle is equal to the density associated with the corresponding value assumed by the categorical attribute. However, notice that the representation by pie charts is not so effective when one wishes to catch the differences between relative frequencies.

7.1.2 Graphical analysis of numerical attributes

For discrete numerical attributes assuming a finite and limited number of values, it is possible to resort to a bar chart representation, just as in the case of categorical attributes. In the presence of continuous or discrete attributes that might assume infinite distinct values, this type of representation cannot be used, as it would require an infinite number of vertical bars.

We must therefore subdivide the horizontal axis corresponding to the values assumed by the attribute, into a finite and moderate number of intervals, usually of equal width, which in practice are considered as distinct classes. In essence, this is a discretization procedure similar to what was described in Section 6.1.2, which inevitably introduces a degree of approximation, since all the observations that fall within the same interval are considered equivalent and indistinguishable among themselves.

As a consequence, it is appropriate to carry out a partition into R intervals which should be as narrow as possible, while at the same time keeping low the number of distinct classes so generated. Once the partition has been performed, the number of observations $e_r, r = 1, 2, \ldots, R$, falling into each interval is counted, and a chart consisting of contiguous rectangles is generated. Procedure 7.1 describes the generation of the intervals for the calculation of the empirical density.

Procedure 7.1 – Histogram for the empirical density

• The number R of classes loosely depends on the number m of observations in the sample and on the uniformity of the data. Usually the goal is to obtain between 5 and 20 classes, making sure that in each class the frequency is higher than 5.

• The total range and the width l_r of each class is then defined. Usually the total range, given by the difference between the highest value and the lowest value of the attribute, is divided by the number of classes, so as to obtain intervals of equal width.

• The boundaries of each class are properly assigned so as to keep the classes disjoint, making sure that no value falls simultaneously into contiguous classes. For example, each interval should be closed on the left and open on the right, except for the last one which should also be closed on the right.

• Finally, the number of observations in each interval is counted and the corresponding rectangle is assigned a height equal to the empirical density p_r defined as

$$p_r = \frac{e_r}{ml_r}. \tag{7.5}$$

Observe that the total area of the rectangles, included under the empirical density curve, is equal to 1:

$$\sum_{r=1}^{R} p_r l_r = \frac{1}{m} \sum_{r=1}^{R} e_r = 1. \tag{7.6}$$

The main difference between a vertical bar chart and a histogram lies in the order of the classes along the horizontal axis. This is exactly determined by the sequence of adjacent intervals in the case of histograms, while it is arbitrary in bar charts.

The properties already highlighted with regard to the relative frequencies for categorical attributes also apply to empirical densities calculated for numerical attributes. These properties provide two interpretations for the area of the rectangles of the histogram: on the one hand, they express for each interval the percentage of observations of the dataset falling inside the interval itself; on the other hand, by virtue of the central limit theorem, the height of a

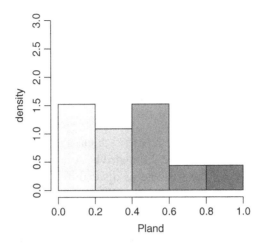

Figure 7.3   Empirical density histogram for a numerical attribute

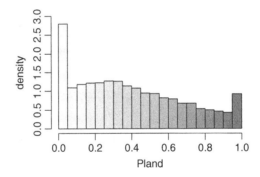

Figure 7.4   Histograms for the empirical density – large sample

rectangle approximates the probability that a new observation extracted from the population will fall within the associated interval.

An example of an empirical density histogram is shown in Figure 7.3 for the numerical attribute *Pland* in Example 5.2. Figure 7.4 illustrates the corresponding histogram for a sample made of 10 000 observations. Also for histograms, as in the case of bar charts for categorical attributes, the estimate of the empirical frequencies obtained from small size samples appears to be of low robustness.

7.1.3   Measures of central tendency for numerical attributes

In this section we will describe the main *location* statistics, also called measures of *central tendency*.

Mean

The best-known measure of location used to describe a numerical attribute is certainly the *sample arithmetic mean*, defined by the expression

$$\bar{\mu} = \frac{x_1 + x_2 + \cdots + x_m}{m} = \frac{1}{m} \sum_{i=1}^{m} x_i. \tag{7.7}$$

Since all the observations of the attribute are used to calculate the sample mean, its value is strongly affected by the extreme values in the sample. As a consequence, the sample mean value reflects the presence of outliers and therefore is not very robust. It may be that the mean significantly differs from each observation existing in the sample, making it appear a somewhat abstract concept.

If the size of the dataset is sufficiently large, the sample arithmetic mean $\bar{\mu}$ approximates the theoretical mean μ of the attribute **a** for the entire population.

It can easily be verified that the sum of the differences between each value and the sample mean, referred to as its *deviation* or *spread*, is equal to zero, as in

$$\sum_{i=1}^{m} (x_i - \bar{\mu}) = 0. \tag{7.8}$$

Moreover, the arithmetic mean is the value that minimizes the sum of squared deviations from a constant reference value:

$$\sum_{i=1}^{m} (x_i - \bar{\mu})^2 = \min_{c} \sum_{i=1}^{m} (x_i - c)^2. \tag{7.9}$$

At times, each value x_i is found to be associated with a numerical coefficient w_i which can be interpreted as a weight of the corresponding value and used to calculate the *weighted sample mean*,

$$\bar{\mu} = \frac{w_1 x_1 + w_2 x_2 + \cdots + w_m x_m}{w_1 + w_2 + \cdots + w_m} = \frac{\sum_{i=1}^{m} w_i x_i}{\sum_{i=1}^{m} w_i}. \tag{7.10}$$

For example, if x_i stands for the unit sale price of the ith shipping lot consisting of w_i product units, the average sale price should be calculated as a weighted sample mean using formula (7.10).

Median

The *median* of m observations can be defined as the central value, assuming that the observations have been ordered in a non-decreasing way. Procedure 7.2 shows how to calculate the median.

Procedure 7.2 – Median calculation

- If m is an odd number, the median is the observation occupying the position $(m + 1)/2$:

$$x^{\text{med}} = x_{(m+1)/2}. \tag{7.11}$$

- If m is an even number, the median is the middle point in the interval between the observations of position $m/2$ and $(m + 2)/2$:

$$x^{\text{med}} = \frac{x_{m/2} + x_{(m+2)/2}}{2}. \tag{7.12}$$

The median is affected by the number of elements in the series but not by the extreme values, and consequently it is more robust than the sample mean. Therefore, the median is suitable for asymmetric distributions and distributions with open extreme classes. However, if the data are not concentrated in the central part of the distribution, the median value looses any statistical significance.

Mode

The third measure of central tendency is the *mode*, defined as the value that corresponds to the peak of the empirical density curve for the attribute **a**. If the empirical density curve has been calculated by a partition into intervals, as shown for graphical methods, each value of the interval that corresponds to the maximum empirical frequency can be assumed as the mode. For example, it is possible to use the central value of the maximum frequency interval. Just like the median, the mode can also be used for distributions with open extreme classes.

With multimodal empirical densities – that is, densities with multiple local maxima – the mode has little significance as a measure of location. Moreover, it has no statistical relevance in connection with a small size sample.

Midrange

Another positioning index is the *midrange*, defined as the mid-point in the interval between the minimum value and the maximum value:

$$x^{\text{midr}} = \frac{x^{\max} + x^{\min}}{2}, \quad x^{\max} = \max_i x_i, \quad x^{\min} = \min_i x_i. \tag{7.13}$$

It should be clear that the midrange, just like the mean, is scarcely robust with respect to extreme values and therefore to the presence of outliers.

Geometric mean

A final location indicator is represented by the *geometric mean*, defined as the mth root of the product of the m observations of the attribute **a**:

$$\bar{\mu}^{\text{geom}} = \sqrt[m]{x_1 x_2 \dots x_m} = \sqrt[m]{\prod_{i=1}^{m} x_i}. \qquad (7.14)$$

7.1.4 Measures of dispersion for numerical attributes

The location measures of the previous subsection gave an indication of the central part of the observed values of a numerical attribute. However, it is necessary to define other indicators that describe the *dispersion* of the data, representing the level of variability expressed by the observations with respect to central values.

By way of illustration, consider the two empirical density curves shown in Figure 7.5. Both density curves identify the center of the observations at $x = 5$, a value where mean, median and mode coincide, and are therefore equivalent from a location standpoint. However, for curve (a) the observations occur between 4 and 6, while for curve (b) they vary in the range $1 - 9$, with a greater dispersion. In most applications, it is desirable that the data dispersion be as small as possible. For example, in a manufacturing process a wide variation in a given critical measure is likely to point to an undesirable level of defective items. Conversely, there are situations in which a higher

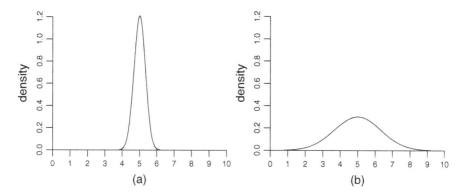

Figure 7.5   Dispersion for a numerical attribute

dispersion may be desired. This is the case for the results of a test intended to discriminate between the abilities of a group of candidates, in relation to which one wishes to obtain a fairly wide spectrum of values.

Range

The simplest measure of dispersion is the *range*, which is defined as the difference between the maximum and the minimum of the observations:

$$x^{\text{range}} = x^{\text{max}} - x^{\text{min}}, \quad x^{\text{max}} = \max_i x_i, \quad x^{\text{min}} = \min_i x_i. \tag{7.15}$$

Although the range is useful in identifying the interval in which the values of the attribute **a** fall, it is unable to catch the actual dispersion of the data in this interval. The densities shown in Figure 7.6 have the same range, but the dispersion for density (b) is greater than that for density (a).

Mean absolute deviation

The *deviation*, or *spread*, of a value is defined as the signed difference from the sample arithmetic mean

$$s_i = x_i - \bar{\mu}, \quad i \in \mathcal{M}. \tag{7.16}$$

As observed when we introduced the sample mean, the equality

$$\sum_{i=1}^{m} s_i = 0 \tag{7.17}$$

always holds. We can express a measure of dispersion of the observations around their sample mean through the sum of the absolute values of the spreads, called *mean absolute deviation* (MAD),

$$\text{MAD} = \frac{1}{m} \sum_{i=1}^{m} |s_i| = \frac{1}{m} \sum_{i=1}^{m} |x_i - \bar{\mu}|. \tag{7.18}$$

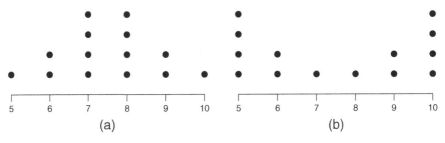

Figure 7.6   Observations with the same range and different dispersions

The lower the measure of the mean absolute deviation is, the more the values fall in proximity of their sample mean and the lower the dispersion is.

Variance

More widely used than the mean absolute deviation is the *sample variance*. There are both theoretical reasons for this, connected with the role played by the variance σ^2 of a random variable in probability theory, and practical reasons, since the mean absolute deviation is the sum of absolute values and is therefore a non-differentiable function. The sample variance is defined as

$$\bar{\sigma}^2 = \frac{1}{m-1} \sum_{i=1}^{m} s_i^2 = \frac{1}{m-1} \sum_{i=1}^{m} (x_i - \bar{\mu})^2. \tag{7.19}$$

As for the mean absolute deviation, a lower sample variance implies a lower dispersion of the values around the sample mean. Table 7.1 shows the calculation of spreads, squared deviations, mean absolute deviation and sample variance for 10 observations of the attribute *numout* in Example 5.2. Notice that $\bar{\sigma}^2 = 1070.23 = 9632.1/9$. As previously observed regarding the relationship between the sample mean $\bar{\mu}$ and the population mean μ, as the size of the sample increases the sample variance $\bar{\sigma}^2$ approximates the variance σ^2 of the distribution from which the values of the attribute **a** are drawn.

The sample variance is obtained as the mean of the squares of the deviations and therefore tends to dilate the most conspicuous errors. To bring the measure of dispersion back to the original scale in which the observations are expressed, the *sample standard deviation* is then introduced:

$$\bar{\sigma} = \sqrt{\bar{\sigma}^2}. \tag{7.20}$$

Table 7.1   An example of calculation of mean, deviations, MAD and variance

x_i	$x_i - \bar{\mu}$	$|x_i - \bar{\mu}|$	$(x_i - \bar{\mu})^2$
45.0	−4.3	4.3	18.5
20.0	−29.3	29.3	858.5
69.0	19.7	19.7	388.1
66.0	16.7	16.7	278.9
11.0	−38.3	38.3	1466.9
42.0	−7.3	7.3	53.3
126.0	76.7	76.7	5882.9
47.0	−2.3	2.3	5.3
43.0	−6.3	6.3	39.7
24.0	−25.3	25.3	640.1
$\sum = 493.0$	$\sum = 0.0$	$\sum = 226.2$	$\sum = 9632.1$
$\bar{\mu} = 49.3$		MAD $= 22.62$	$\bar{\sigma}^2 = 1070.23$

Basically, the variance can be used to delimit the interval around the sample mean where it is reasonable to expect that the sample values will fall. In this way, the values falling outside such interval are identified as anomalous values or outliers. To obtain the appropriate neighborhood of the sample mean, we should draw a distinction between two cases.

Normal distribution. If the distribution of attribute **a** is normal, or at least bell-shaped and approximately normal, we can say that:

- the interval $(\bar{\mu} \pm \bar{\sigma})$ contains approximately 68% of the observed values;

- the interval $(\bar{\mu} \pm 2\bar{\sigma})$ contains approximately 95% of the observed values;

- the interval $(\bar{\mu} \pm 3\bar{\sigma})$ contains approximately 100% of the observed values.

The values of the sample that fall outside the interval $(\bar{\mu} \pm 3\bar{\sigma})$ can therefore be considered suspicious outliers and candidates for removal from the sample. See Section 6.1 for a discussion of the tactics to adopt when outliers are detected.

Arbitrary distribution. If the distribution of the attribute **a** differs significantly from the normal, it is still possible to obtain intervals within which one may reasonably expect the values of the sample to fall. Such intervals apply to any distribution and are inevitably more conservative than the corresponding intervals for (approximately) normal distributions. The theoretical result that is invoked to obtain distribution-free intervals is as follows.

Theorem 7.1 (Chebyshev). *Given a number $\gamma \geq 1$ and a group of m values* **a** *$= (x_1, x_2, \ldots, x_m)$, a proportion at least equal to $(1 - 1/\gamma^2)$ of the values will fall within the interval $(\bar{\mu} \pm \gamma \bar{\sigma})$, namely at no more than γ standard deviations from the sample mean.*

Figure 7.7 provides a graphical illustration of Chebyshev's theorem. If we choose for example $\gamma = 3$, the theorem guarantees that at least $(1 - 1/3^2) = 8/9 \approx 89\%$ of the values will fall within the interval $(\bar{\mu} \pm 3\bar{\sigma})$. This percentage should be compared with the corresponding percentage under the normal curve, roughly equal to 100%. As an example of the application of Chebyshev's theorem, consider the attribute *glucose_tol* expressing the glucose tolerance for the *diabete* dataset, described in Appendix B. Figure 7.8 depicts the corresponding histogram and the intervals determined by Chebyshev's theorem for the values {1,2,3} of the parameter γ.

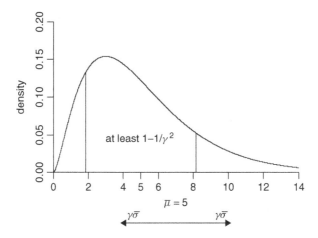

Figure 7.7   Illustration of Chebyshev's theorem

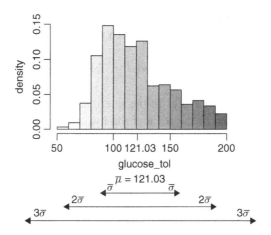

Figure 7.8   An application of Chebyshev's theorem

Coefficient of variation

A further dispersion index is represented by the *coefficient of variation*, which is defined as the ratio between the sample standard deviation and the sample mean, expressed in percentage terms as

$$\mathrm{CV} = 100\frac{\bar{\sigma}}{\bar{\mu}}. \tag{7.21}$$

The coefficient of variation is used to compare two or more groups of data, usually obtained from different distributions.

7.1.5   Measures of relative location for numerical attributes

Measures of relative location for a numerical attribute are used to examine the localization of a value with respect to other values in the sample.

Quantiles

Suppose we arranged the m values $\{x_1, x_2, \ldots, x_m\}$ of an attribute **a** in nondecreasing order. Given any value p, with $0 \le p \le 1$, the *p-order quantile* is the value q_p such that pm observations will fall on the left of q_p and the remaining $(1 - p)m$ on its right. Sometimes, p quantiles are called 100 pth *percentiles*. It should be clear that the 0.5-order quantile coincides with the median, described in Section 7.1.3. Other noteworthy quantiles include the 0.25- and 0.75-order quantiles, respectively called the *lower* and *upper* quartiles and denoted by q_L and q_U. The two quartiles and the median divide the observations into four portions of equal size, except for the small rounding required when m is not divisible by 4. Therefore, in a density histogram the quartiles and the median split the axis of the observations into four parts, such that the area under the density curve for each portion is equal to 0.25, as shown in Figure 7.9.

The quartiles for the attribute *numout* in Table 5.2 assume the values indicated in Table 7.2.

Sometimes it is useful to refer to the *interquartile range*, which is defined as the difference between the upper and the lower quartiles:

$$D_q = q_U - q_L = q_{0.75} - q_{0.25}. \tag{7.22}$$

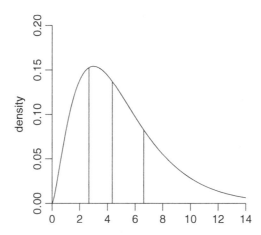

Figure 7.9   Density curve and position of median and quartiles

Table 7.2 Quartiles for the attribute *numout* in Table 5.2

q_0	$q_{0.25}$	$q_{0.5}$	$q_{0.75}$	q_1
9.0	32.0	47.0	76.5	450.0

Measures of central tendency based on quantiles

In order to reduce the dependence of the mean on the extreme values of the sample, one may resort to measures of central tendency based on quantiles, which in general are more robust.

Mid-mean. The *mid-mean* is obtained by calculating the mean of the values of the attribute **a** falling between the lower and the upper quartiles.

Trimmed mean. The *trimmed mean* is a generalization of the mid-mean, since only the values falling between quantiles of order p and $(1 - p)$ are used to calculate the mean. Usually, $p = 0.05$ is the preferred choice, in order to exclude from the calculation the lowest 5% and highest 5% of the values of **a**.

Winsorized mean. Unlike the trimmed mean, the *winsorized mean* does not exclude from the calculation the values falling in the tails; rather, it modifies them according to an intuitive rule: the values lower than the p-order quantile are increased to q_p, while the values higher than the $(1 - p)$-order quantile are decreased to q_{1-p}.

z-index

The *z-index* is another valuable measure of relative location, defined for a generic observation x_i as

$$z_i^{\text{ind}} = \frac{x_i - \bar{\mu}}{\bar{\sigma}}. \tag{7.23}$$

Intuitively, the z-index expresses the signed distance between a value and the sample mean by using the sample standard deviation as a measurement unit. As a consequence, a positive z-index indicates that the corresponding value is greater than the sample mean, while a negative z-index indicates that it is smaller. The z-index is also useful for identifying suspicious values and possible outliers existing in the sample, as we will see in the next section.

7.1.6 Identification of outliers for numerical attributes

To better illustrate the problems connected with the identification of outliers, consider Figure 7.10. Given the data shown in chart (a), suppose that we wish

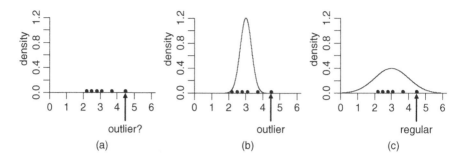

Figure 7.10   Dispersion and outliers for a numerical attribute

to determine if the highest value of the sample, which lies away from the other values, constitutes an outlier. The two possible densities shown in charts (b) and (c) suggest that the answer is to be found in the distribution from which the values of attribute **a** have been drawn. In case (b), characterized by a lower dispersion, the maximum value actually represents an outlier, while in case (c) it appears to be a regular value.

The z-index can be used in most cases to identify, with reasonable confidence, any possible outlier. Based on the intervals obtained using Chebyshev's theorem, possibly reinforced for approximately normal distributions, we can consider as suspicious those values for which $\left|z_i^{\mathrm{ind}}\right| > 3$ and as highly suspicious those values for which $\left|z_i^{\mathrm{ind}}\right| \gg 3$.

Box plots

A second way to identify outliers is based on the use of *box plots*, sometimes called *box-and-whisker* plots, in which the median and the lower and upper quartiles are represented on the axis where the observations are placed.[1]

An observation is identified as an outlier if it falls outside four threshold values, called *edges*, defined as

$$\text{external lower edge} = q_L - 3D_q, \tag{7.24}$$

$$\text{internal lower edge} = q_L - 1.5D_q, \tag{7.25}$$

$$\text{internal upper edge} = q_U + 1.5D_q, \tag{7.26}$$

$$\text{external upper edge} = q_U + 3D_q. \tag{7.27}$$

---

[1]In fact, box plots use two parameters, usually referred to as *hinges*, which represent an approximation of the quartiles. Hinges and quartiles coincide when m is odd, whereas they differ when m is even. Quartiles coincide with observed values only if $m \equiv 1 \pmod 4$, while hinges do so also if $m \equiv 2 \pmod 4$. In all other cases, both quartiles and hinges assume values that bisect the interval between two consecutive observed values. Since the difference between quartiles and hinges is negligible and tends to vanish for high values of m, we prefer to use quartiles for defining box plots.

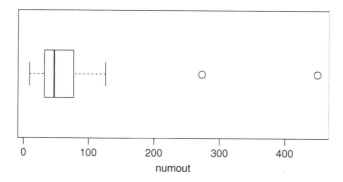

Figure 7.11   A box plot for the attribute numout in Table 5.2

Compared to the z-index method, using a box plot to identify outliers has the advantage of being basically independent of the extreme values of the observations, since both the median and the quartiles enjoy such independence. In contrast, the z-index is strongly affected by extreme values.

The box plot for the attribute *numout* in Table 5.2 is shown in Figure 7.11. The four edges are equal to -101.5, -34.75, 143.25 and 210. We can see that the box lies between the lower and the upper quartiles, respectively equal to 32.0 and 76.5. The figure also shows the median, equal to 47.0.

The horizontal lines departing from the box end in thick marks, called *whiskers*, which correspond respectively to the minimum and the maximum values of the attribute falling inside the internal edges. In our example the whiskers are equal to 9 and 126. In general, the values lying outside the external edges are considered outliers. Therefore, the two values $\{273, 450\}$ are outliers in our example.

7.1.7   Measures of heterogeneity for categorical attributes

For categorical attributes the foregoing measures of central tendency, dispersion and relative location cannot be used. For a categorical attribute **a** it is preferable to define some measures that express the regularity of the arrangement of the data $\{x_1, x_2, \ldots, x_m\}$ within the set of H distinct values taken by the attribute.

In particular, the highest heterogeneity is achieved when the relative empirical frequencies f_h are equal for all classes $h \in \mathcal{H}$. In contrast, the lowest heterogeneity occurs when $f_g = 1$ for a class $g \in \mathcal{H}$ and $f_h = 0$ for the remaining classes $h \in \mathcal{H}, h \neq g$.

We will now describe two measures of heterogeneity for categorical attributes: the *Gini index* and the *entropy index*. As we will show in Chapter 10, these are among the indices most frequently utilized for assessing the local splitting rules in classification trees.

Gini index

The *Gini index* is defined as

$$G = 1 - \sum_{h=1}^{H} f_h^2. \qquad (7.28)$$

Its value is equal to 0 in the case of lowest heterogeneity, that is, when a class is assumed at a frequency equal to 1 and all the other classes are never assumed. In contrast, when all classes have the same relative empirical frequency and the highest heterogeneity is recorded, the Gini index achieves its maximum value $(H-1)/H$.

It is possible to normalize the Gini index so that it assumes values in the interval [0,1], by using the following transformation which leads to the *relative Gini index*:

$$G_{\text{rel}} = \frac{G}{(H-1)/H}. \qquad (7.29)$$

Entropy index

The *entropy index* is defined as

$$E = - \sum_{h=1}^{H} f_h \log_2 f_h. \qquad (7.30)$$

Its value is equal to 0 in the case of lowest heterogeneity, that is, when a class is assumed at a frequency equal to 1 and all the other classes are never assumed. In contrast, when all classes have the same relative empirical frequency and the highest heterogeneity is recorded, the entropy index achieves its maximum value $\log_2 H$.

It is also possible to normalize the entropy index so that it assumes values in the interval [0,1], by using the following transformation which leads to the *relative entropy index*:

$$E_{\text{rel}} = \frac{E}{\log_2 H}. \qquad (7.31)$$

7.1.8   Analysis of the empirical density

As shown at the beginning of the section, the relative empirical frequency histogram is a valuable tool for the graphical analysis of both categorical and numerical attributes. It is therefore convenient to make use of summary indicators in order to investigate the properties of the empirical density curve.

Asymmetry of the density curve

A density curve is called *symmetric* if the mean coincides with the median. From a graphical standpoint, in a symmetric density the two halves of the curve on either side of the mean, and therefore of the median, are mirror images.

A density curve is said to be *asymmetric* when the mean and the median do not coincide. When the mean is greater than the median we say that the curve is *skewed to the right*, whereas when the median is greater than the mean the curve is said *skewed to the left*.

The symmetry or asymmetry of an empirical density will be apparent from graphical analysis of its histogram. Figure 7.12 shows three examples of empirical density curves, namely a curve skewed to the left (a), a symmetric curve (b) and a curve skewed to the right (c). The vertical lines in the charts show the quartiles (solid line) and the mean (dashed).

Box plots reflect the type and level of asymmetry in an empirical density in a particularly intuitive way, as also shown in Figure 7.12. If the density is symmetric (b), the median is equally distant from the lower and upper quartiles. When it is skewed to the left (a), the median is closer to the upper quartile, while when it is skewed to the right (c) the median is closer to the lower quartile.

In addition to graphical analysis, we may also make use of an index of asymmetry, the *sample skewness*, based on the third sample moment. The third sample moment is given by

$$\bar{\mu}_3 = \frac{1}{m} \sum_{i=1}^{m} (x_i - \bar{\mu})^3. \tag{7.32}$$

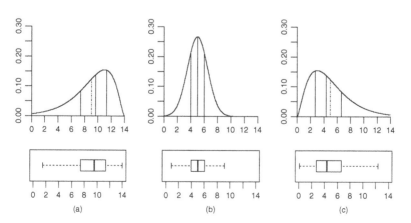

Figure 7.12   Left-skewed, symmetric and right-skewed empirical density curves

The sample skewness is therefore defined as

$$I_{\text{skew}} = \frac{\bar{\mu}_3}{\bar{\sigma}^3}. \tag{7.33}$$

If the density curve is symmetric then $I_{\text{skew}} = 0$. If instead $I_{\text{skew}} > 0$ the density is skewed to the right. Finally, if $I_{\text{skew}} < 0$ the density is skewed to the left.

Kurtosis of the density curve

A further significant problem regarding the density histogram is the identification of the type of theoretical probability distribution, usually unknown beforehand, from which the observations are drawn. We observe that the problem of estimating an unknown distribution based on sampled data is rather complex in its general form. However, in the case of the normal distribution, a continuous distribution which occurs again and again in applications, easy graphical and summary criteria can be used to assess the level of approximation for a given empirical density.

The first criterion is graphical, and is based on a visual comparison between the empirical density histogram and a normal curve having mean $\bar{\mu}$ and standard deviation $\bar{\sigma}$ coinciding with those of the given density. Figure 7.13 shows a histogram with a good approximation to the normal density.

The index of *kurtosis* expresses in a concise way the level of approximation of an empirical density to the normal curve, and it uses the fourth sample moment

$$\bar{\mu}_4 = \frac{1}{m} \sum_{i=1}^{m} (x_i - \bar{\mu})^4. \tag{7.34}$$

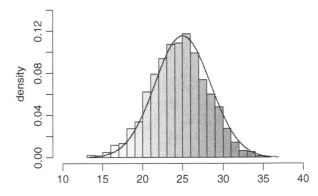

Figure 7.13   An empirical density histogram in comparison to a normal density

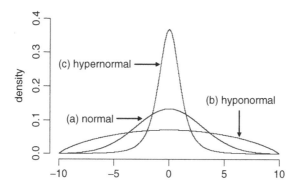

Figure 7.14   Empirical density histograms for hyponormal, normal and hyper-normal distributions

The kurtosis is therefore defined as

$$I_{\text{kurt}} = \frac{\bar{\mu}_4}{\bar{\sigma}^4} - 3. \tag{7.35}$$

If the empirical frequency perfectly fits a normal density, as in curve (a) in Figure 7.14, we have $I_{\text{kurt}} = 0$. If $I_{\text{kurt}} < 0$, the empirical density is said to be *hyponormal*, since it shows a greater dispersion than the normal density and therefore assigns lower frequencies to values close to the mean, as in curve (b) for which $I_{\text{kurt}} = -0.89$. Finally, if $I_{\text{kurt}} > 0$ the empirical density is said to be *hypernormal*, since it shows a dispersion lower than the normal density and therefore assigns higher frequencies to values close to the mean, as in curve (c) for which $I_{\text{kurt}} = 9.36$.

It is possible to use a different type of chart, called a *normal probability plot*, which allows the level of approximation of an empirical density to a normal density to be analyzed. This is a special type of quantile–quantile plot, which will be described in Section 7.2 when graphical methods for bivariate analysis are considered.

In a normal probability plot, the values $\{x_1, x_2, \ldots, x_m\}$ of the attribute of interest are placed in non-decreasing order on the vertical axis. Let q_i be the order of the quantile corresponding to the value x_i. The quantiles N_i of order $q_i, i \in \mathcal{M}$, of the normal distribution are then placed on the horizontal axis and the points with coordinates $(N_i, x_i), i \in \mathcal{M}$, are plotted on the Cartesian plane.[2] It is possible to prove that if the values of the attribute of interest are sampled from a normal distribution, and therefore the empirical density is asymptotically normal as the size of the dataset increases, then the points on

---

[2]Readers should be aware that it is also possible to define a normal probability plot for which the axes are swapped over with respect to the definition given here.

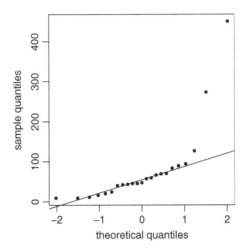

Figure 7.15   Normal probability plot for the attribute numout in Example 5.2

the normal probability plot will fall on a straight line. Conversely, the more the empirical density diverges from normal, the more the points of the normal probability plot diverge from a linear trend.

Figure 7.15 shows the normal probability plot for the attribute *numout* in Example 5.2. The almost concave curve defined by the points in the plot indicates that the empirical density is skewed to the right, and is a hyponormal curve, as shown in the histogram in Figure 7.16. Contrast this with the normal probability plot for the attribute *Pland* in Figure 7.17. The S-shaped curve

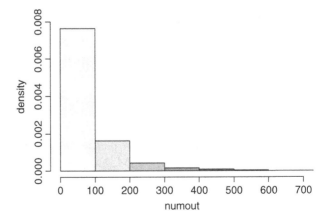

Figure 7.16   Empirical density histogram for the attribute numout in Example 5.2

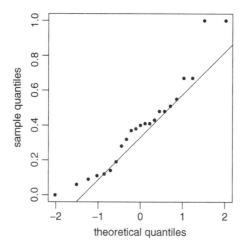

Figure 7.17   Normal probability plot for the attribute Pland in Example 5.2

indicates the occurrence of a hypernormal distribution, and also that the empirical density is significantly far from a normal curve, as clearly shown by the histogram in Figure 7.4.

7.1.9   Summary statistics

In univariate exploratory analysis it is possible to obtain, for each numerical attribute, the various summary indices that we have defined. Table 7.3 shows the main statistics relative to the attribute *numout* in Table 5.2.

Table 7.3   Summary statistics for the attribute *numout* in Table 5.2

index	value	meaning
nobs	23.00	number of observations
NAs	0.00	number of missing values
Minimum	9.00	minimum
Maximum	450.00	maximum
Lower Quartile	32.00	lower quartile
Upper Quartile	76.50	upper quartile
Mean	77.69	mean
Median	47.00	median
Sum	1787.00	sum of values
SE Mean	20.44	mean standard error
LCL Mean	35.30	lower confidence limit for the mean
UCL Mean	120.08	upper confidence limit for the mean
Variance	9611.03	variance
Stdev	98.03	standard deviation
Skewness	2.64	asymmetry index
Kurtosis	6.78	kurtosis index

The only piece of information in Table 7.3 that has not been previously defined is the 95% confidence interval for the sample mean. The standard error of the sample mean (SE mean in the table) is defined as

$$\sigma_{\bar{\mu}} = \frac{\bar{\sigma}}{\sqrt{m}}. \tag{7.36}$$

Consequently, a $100(1 - \alpha)\%$ confidence interval around $\bar{\mu}$ is given by

$$\bar{\mu} \pm t_{\alpha/2}\sigma_{\bar{\mu}}, \tag{7.37}$$

where $t_{\alpha/2}$ represents the $\alpha/2$-order quantile of the Student t-distribution with $m - 1$ degrees of freedom. The LCL mean and UCL mean values give the extremes of the confidence interval for $\alpha = 0.05$.

7.2   Bivariate analysis

Once the characteristics of each attribute contained in a dataset $\mathcal{D}$ have been investigated using the methods for univariate exploratory analysis described in the previous section, it is appropriate to exploit the relationships existing between pairs of attributes through bivariate analysis. We will denote by $\mathbf{a}_j$ and $\mathbf{a}_k$ a generic pair of attributes to be analyzed.

It is useful to distinguish three cases that may occur within bivariate analysis:

- both attributes are numerical;

- one attribute is numerical and the other is categorical;

- both attributes are categorical.

In supervised learning problems it is usually important to investigate the relationship between the target attribute and each explanatory attribute. Hence, it frequently happens that one of the attributes of the pair $\{\mathbf{a}_j, \mathbf{a}_k\}$ represents the target of a supervised learning problem, which is categorical for classification and numerical for regression. In particular, we will denote by $\mathbf{a}_j = (x_{1j}, x_{2j}, \ldots, x_{mj})$ and $\mathbf{a}_k = (x_{1k}, x_{2k}, \ldots, x_{mk})$ the vectors composed of m observations corresponding to the two attributes considered.

7.2.1   Graphical analysis

There are several types of graphical representations that allow the relationship between two attributes to be visualized. For each type of chart, we will indicate the categories of attributes to which it can be applied.

Scatter plots

A *scatter plot* is definitely the most intuitive graphical representation of the relationship between two numerical attributes. This is a two-dimensional Cartesian chart obtained by placing the first attribute $\mathbf{a}_j$ on the horizontal axis and the second attribute $\mathbf{a}_k$ on the vertical axis. The points plotted thus correspond to the pairs of values (x_{ij}, x_{ik}), $i \in \mathcal{M}$, of the dataset $\mathcal{D}$.

For the data in Example 5.2, one might conjecture the existence of a positive dependence between the attributes *numin* and *timein*. Indeed, it is reasonable to expect that a greater number of incoming calls corresponds to a higher total duration. To confirm this hypothesis, the scatter plot in Figure 7.18 shows the intensity of the positive relationship between the two attributes.

As one can easily imagine, scatter plots may prove confusing and ineffective when the number m of observations is very high.

Loess plots

Loess plots are based on scatter plots and can therefore be applied in turn to pairs of numerical attributes. Starting from a scatter plot, it is possible to add a trend curve to express the functional relationship between the attribute $\mathbf{a}_k$ and the attribute $\mathbf{a}_j$. The trend curve can be obtained using local regression techniques. This explains the term *loess*, which stands for *local regression*. To calculate the value of the trend curve at any point, a regression model is applied as described in Chapter 8, but using only the observations lying in a neighborhood of the point being considered.

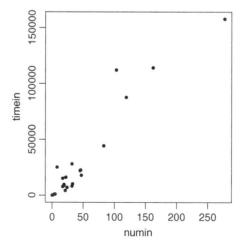

Figure 7.18   Scatter plot for the attributes numin and timein in Example 5.2

The loess curve is regulated by two parameters, described below.

- The degree $\lambda > 0$ of the polynomial that represents the local regression curve; usually the value 1 or 2 is assigned to λ in order to obtain a linear or quadratic local approximation.

- The regularization constant $\alpha > 0$, which regulates the size of the neighborhood. When the value of α is increased, the regularization of the trend line is strengthened, since more observations are used to calculate the regression. Conversely, when α decreases the trend line sticks more closely to the observations, although the generalization capability of the model decreases. In general, values of α ranging between 1/4 and 1 are chosen.

Figure 7.19(a) shows a loess plot with a linear trend ($\lambda = 1$) and an average regularization level ($\alpha = 3/4$) for the attributes *numin* and *numout*. Figure 7.19(b) shows a quadratic trend curve ($\lambda = 2$), while Figure 7.19(c) depicts the effect of an increase in the regularization constant ($\alpha = 1$).

Level curves

Level curves are a further development of scatter plots and can only be used for numerical attributes. They highlight the value of a third numerical attribute $\mathbf{a}_z$ as the attributes $\mathbf{a}_j$ and $\mathbf{a}_k$ placed on the axes of the plot vary.

Connecting to each other the points in the plot that share the value of the third attribute, possibly by using some form of numerical interpolation, curved lines are obtained representing the geometric locus of the points for which the attribute $\mathbf{a}_z$ assumes a given value.

An example of level curves known to everyone are contour lines, which can be obtained by highlighting in a geographical map the points that share the same altitude, which in this case corresponds to the attribute $\mathbf{a}_z$. Figure 7.20 shows an example of contour lines.

Quantile–quantile plots

Quantile–quantile plots (*QQ plots*) are used to compare the distributions of the same attribute for two different characteristics of the population or for samples extracted from two different populations. The analysis is applicable to numerical attributes and is carried out by comparing the quantiles of the two series of observations. It is also possible to obtain a QQ plot even when one of the series is more populated than the other.

Suppose that we have calculated and arranged in non-decreasing order the values of the quantiles $\{q_{lj}\}$ and $\{q_{lk}\}, l \in \mathcal{L}$, for each of the two attributes at

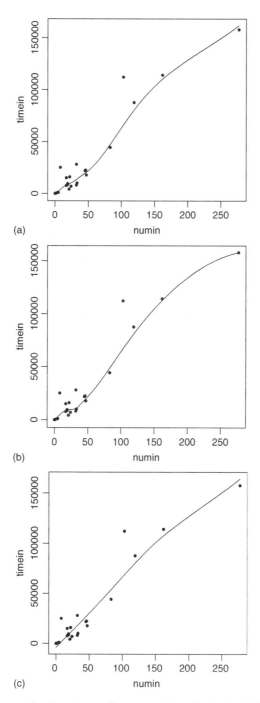

Figure 7.19   Loess plot for the attributes numin and timein in Example 5.2 with (a) λ = 1, *(b)* α = 3/4; *(b)* λ = 2, α = 3/4; *(c)* λ = 1, α = 1

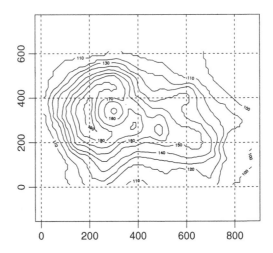

Figure 7.20   Example of contour lines

a predetermined series $\mathcal{L}$ of quantile orders, such as the sequence of deciles. The coordinates of the points in the plot are (x_{lj}, x_{lk}), $l \in \mathcal{L}$. If the two series of data are extracted from the same probability distribution, the points in the chart will fall on a straight line with slope 45 degrees, usually depicted in the QQ plot.

Figure 7.21 shows an example of QQ plot, obtained by comparing the quantiles of 50 observations of the attribute *Pothers* in Example 5.2, corresponding to the records where the target variable is equal to 0 (*churner* = 0) with the

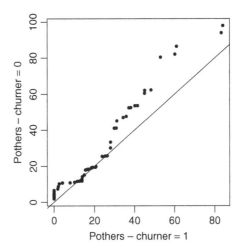

Figure 7.21   Quantile–quantile plot for the attribute Pothers in Example 5.2 at distinct values of the target churner

analogous quantiles evaluated for 50 observations associated with target values equal to 1 (*churner* = 1). As shown, the points are significantly far from the straight line along which they would have to fall if the distributions of the two series of data were coincident. It is therefore apparent that the distribution of the variable *Pothers* varies according to the value of the target attribute. As a consequence, for the corresponding supervised learning problem it can be concluded that the variable *Pothers* is potentially explanatory with respect to the target.

Box plots

In Section 7.1.6 we saw that box plots can be used for the identification of outliers. They are also useful for comparing the distributions of the same variable for observations belonging to distinct groups. This analysis technique is applicable to a pair consisting of a numerical attribute and a categorical attribute. Often, though not necessarily, the categorical attribute represents the target in a supervised learning process.

Figure 7.22 compares the box plots for the variable *Pmob* in Example 5.2, relative to a sample of 10 000 observations, at distinct values of the target variable *churner*. The difference between the two plots is significant and indicates that the variable *Pmob* is potentially relevant in explaining the target value.

Time series

There are situations in which the attribute a_k represents the numerical value of a measurable quantity, for example the sales of a product, while the attribute a_j indicates the sequence of time periods at which the values of a_k have been

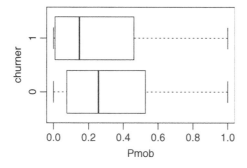

Figure 7.22   Box plots for the attribute Pmob in Example 5.2 at distinct values of the target churner

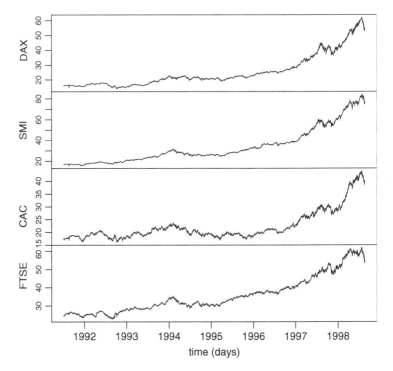

Figure 7.23   Daily closing prices of 4 financial indices (1991–1998)

gathered. Such periods usually correspond to natural time spans, such as hours, days, weeks, months, quarters and years. They are therefore discrete and placed at regular intervals along the horizontal axis.

Without loss of generality, we can denote by $\mathcal{M} = \{1, 2, \ldots, m\}$ the ordered sequence of values assumed by the temporal variable along the horizontal axis, and by $\{x_{ik}\}, i \in \mathcal{M}$, the corresponding sequence of values along the vertical axis. The values $(i, x_{ik}), i \in \mathcal{M}$, assumed by the pair of attributes form a *time series*, a subject that will be covered in Chapter 9.

Figure 7.23 shows four examples of time series consisting of the daily closing values of four financial indices for the German (DAX), Swiss (SMI), French (CAC) and British (FTSE) stock exchanges between 1991 and 1998. Each time series consists of 1860 observations.

7.2.2   Measures of correlation for numerical attributes

As in the case of univariate analysis, besides graphical methods it is useful to introduce summary indicators that express the nature and intensity of the relationship between numerical attributes.

Covariance

Given the pair of attributes $\mathbf{a}_j$ and $\mathbf{a}_k$, let $\bar{\mu}_j$ and $\bar{\mu}_k$ be the corresponding sample means. The *sample covariance* is defined as

$$v_{jk} = \operatorname{cov}(\mathbf{a}_j, \mathbf{a}_k) = \frac{1}{m-2} \sum_{i=1}^{m} (x_{ij} - \bar{\mu}_j)(x_{ik} - \bar{\mu}_k). \qquad (7.38)$$

The sample covariance can be readily interpreted. Indeed, if values of the attribute $\mathbf{a}_j$ greater than the mean $\bar{\mu}_j$ are associated with values of the attribute $\mathbf{a}_k$ also greater than the mean $\bar{\mu}_k$, the two elements of the product in the summation (7.38) will agree in sign and therefore they will provide a positive contribution to the sum. By the same token, if values of the attribute $\mathbf{a}_j$ lower than the mean $\bar{\mu}_j$ are associated with values of the attribute $\mathbf{a}_k$ lower than the mean $\bar{\mu}_k$, the two elements of the product in the summation will still agree in sign and again will provide a positive contribution to the sum. In these cases, the attributes $\mathbf{a}_j$ and $\mathbf{a}_k$ are said to be in *concordance* with one another. Conversely, if values of the attribute $\mathbf{a}_j$ greater than the mean $\bar{\mu}_j$ are associated with values of the attribute $\mathbf{a}_k$ lower than the mean $\bar{\mu}_k$, the two elements of the product in the summation (7.38) will not agree in sign and therefore they will provide a negative contribution to the sum. The same occurs if values of the attribute $\mathbf{a}_j$ lower than the mean $\bar{\mu}_j$ are associated with values of the attribute $\mathbf{a}_k$ greater than the mean $\bar{\mu}_k$. In these cases, the attributes $\mathbf{a}_j$ and $\mathbf{a}_k$ are said to be *discordant* with one another.

We can therefore conclude that positive covariance values indicate that the attributes $\mathbf{a}_j$ and $\mathbf{a}_k$ are concordant, while negative covariance values indicate that the attributes are discordant. For the attributes *numin* and *timein* in Example 5.2 (see also the scatter plot in Figure 7.18), the sample covariance is equal to 2 433 848.0.

Correlation

Although in this specific case the value 2 433 848.0 seems high in absolute terms, the covariance is usually a number ranging on a variable scale and is therefore inadequate to assess the intensity of the relationship between the two attributes. For this reason, the *linear correlation coefficient* between two attributes, also termed the *Pearson coefficient*, is more useful. It is defined as

$$r_{jk} = \operatorname{corr}(\mathbf{a}_j, \mathbf{a}_k) = \frac{v_{jk}}{\bar{\sigma}_j \bar{\sigma}_k}, \qquad (7.39)$$

where $\bar{\sigma}_j$ and $\bar{\sigma}_k$ are the sample standard deviations of $\mathbf{a}_j$ and $\mathbf{a}_k$, respectively. It can be proven that the maximum value achievable by $\operatorname{cov}(\mathbf{a}_j, \mathbf{a}_k)$ is equal to the product $\bar{\sigma}_j \bar{\sigma}_k$ of the sample standard deviations, while the minimum value is equal to $-\bar{\sigma}_j\bar{\sigma}_k$. As a result, the linear correlation coefficient r_{jk} always lies in the interval $[-1, 1]$, and represents a relative index expressing the intensity of a possible linear relationship between the attributes $\mathbf{a}_j$ and $\mathbf{a}_k$.

The main properties of the linear correlation coefficient r_{jk} can be summarized as follows.

- If $r_{jk} > 0$ the attributes are concordant. This means that if the pairs of observations are represented on a scatter plot, they will show a trend consisting of a straight line with a positive slope. The approximation to the line increases as r_{jk} gets closer to 1. If $r_{jk} = 1$ the points will lie exactly on a straight line.

- If $r_{jk} < 0$ the attributes are discordant. In this case the pairs of observations represented on a scatter plot will tend to lie on a line with a negative slope. The approximation to the line increases as r_{jk} gets closer to -1. If $r_{jk} = -1$ the points will lie exactly on a straight line.

- Finally, if $r_{jk} = 0$, or at least $r_{jk} \approx 0$, no linear relationship exists between the two attributes. In this case, the pairs of values on a scatter plot either are placed in a random way or tend to lie on a nonlinear curve.

Figure 7.24 shows the linear correlation coefficients for different sets of data. In particular, notice that the linear regression coefficient can be close to zero when a strong nonlinear relationship exists between the two attributes.

For the attributes *numin* and *timein* in Example 5.2 (see Figure 7.18), the linear correlation coefficient is equal to 0.89 and confirms that the relationship is concordant and corresponds to a linear trend.

The linear correlation coefficient plays a role in the assessment of linear regression models, as shown in Chapter 8.

Before closing this section, we would like to consider four datasets, called *anscombe* and described in Appendix B, whose characteristics illustrate the importance of graphical insights in exploratory data analysis. Each of the four datasets, which will be denoted by A, B, C, D, includes two attributes $\mathbf{a}_1$ and $\mathbf{a}_2$, and 11 observations, such that the sample mean of $\mathbf{a}_1$ is $\bar{\mu}_1 = 9.0$ and the sample mean of $\mathbf{a}_2$ is $\bar{\mu}_2 = 7.5$. The correlation coefficient between the two attributes is $r_{12} = 0.82$ for each dataset, and also the regression line $\mathbf{a}_2 = 3 + 0.5\mathbf{a}_1$ is the same. As a result, based on summary indicators, we are led to believe that the four datasets express similar relationships between the pairs of attributes. However, the graphical analysis of the corresponding scatter plots, shown in Figure 7.25, indicates strong differences. Dataset A shows a moderate linear relationship between the attributes, while for dataset B the relationship follows a curved line. In both datasets C and D there is an outlier, whose

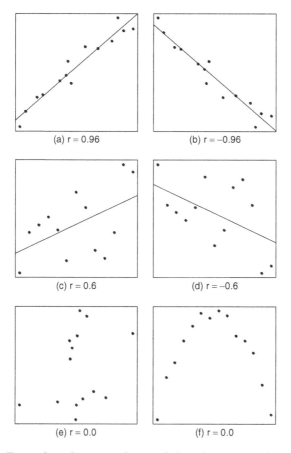

Figure 7.24   Examples of scatter plots and their linear correlation coefficients

removal reveals a perfectly linear relationship in the case of C and the absence
of any relationship in the case of D.

7.2.3   Contingency tables for categorical attributes

When dealing with a pair of categorical attributes $\mathbf{a}_j$ and $\mathbf{a}_k$, let

$$V = \{v_1, v_2, \ldots, v_J\}, \quad U = \{u_1, u_2, \ldots, u_K\} \qquad (7.40)$$

denote the sets of distinct values respectively assumed by each of them. A
contingency table is defined as a matrix $\mathbf{T}$ whose generic element t_{rs} indicates
the frequency with which the pair of values $\{x_{ij} = v_r\}$ and $\{x_{ik} = u_s\}$ appears
in the records of the dataset $\mathcal{D}$. Table 7.4 shows the contingency values for
attributes *area* and *diropt* in Example 5.2.

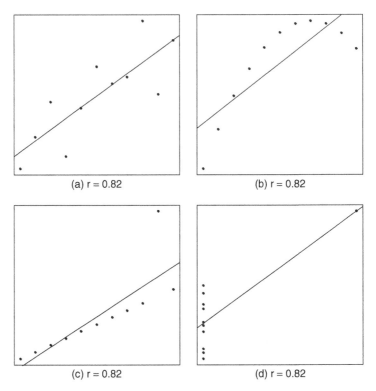

(a) r = 0.82   (b) r = 0.82

(c) r = 0.82   (d) r = 0.82

Figure 7.25   Scatter plots for the four anscombe datasets

Table 7.4   Contingency table for the attributes *area* and *diropt* in Example 5.2

| | diropt | | |
area	0	1	total
1	5	1	6
2	2	3	5
3	5	2	7
4	1	4	5
total	13	10	23

It is also possible to compute the sum of the values for each row and for each column of the contingency table, obtaining the *marginal frequencies*

$$f_r = \sum_{s=1}^{K} t_{rs}, \quad g_s = \sum_{r=1}^{J} t_{rs}. \tag{7.41}$$

Two attributes $\mathbf{a}_j$ and $\mathbf{a}_k$ are said to be *statistically independent* if the following conditions occur:

$$\frac{t_{r1}}{g_1} = \frac{t_{r2}}{g_2} = \cdots = \frac{t_{rK}}{g_K}, \quad r = 1, 2, \ldots, J. \tag{7.42}$$

It can be shown that equalities (7.42) are equivalent to the conditions

$$\frac{t_{1s}}{f_1} = \frac{t_{2s}}{f_2} = \cdots = \frac{t_{Js}}{f_J}, \quad s = 1, 2, \ldots, K, \tag{7.43}$$

which in turn can be used to define the concept of independence between categorical attributes. Intuitively, the two attributes are independent if the analysis of $\mathbf{a}_j$ in relation to the second attribute $\mathbf{a}_k$ is equivalent to the univariate analysis of $\mathbf{a}_j$.

7.3   Multivariate analysis

The purpose of multivariate analysis is to extend the concepts introduced for the bivariate case in order to assess the relationships existing among multiple attributes in a dataset.

7.3.1   Graphical analysis

Notice that all methods for graphical analysis described in this section exclusively apply to numerical attributes.

Scatter plot matrix

Since scatter plots show in an intuitive way the relationships between pairs of numerical attributes, in the case of multivariate analysis it is natural to consider matrices of plots evaluated for every pair of numerical variables. In this way, it is possible to visualize the nature and intensity of the pairwise relationships in a single chart.

Figure 7.26 shows this type of graphical representation for the attributes *numin*, *timein*, *numout*, *Pothers*, *Pmob*, *Pland* and *numsms* in Example 5.2.

Star plots

Star plots belong to the broader class of *icon-based charts*. They show in an intuitive way the differences among values of the attributes for the records of a dataset. To be effective, they should be applied to a limited number of observations, say no more than a few dozen, and the comparison should be based on a small number of attributes.

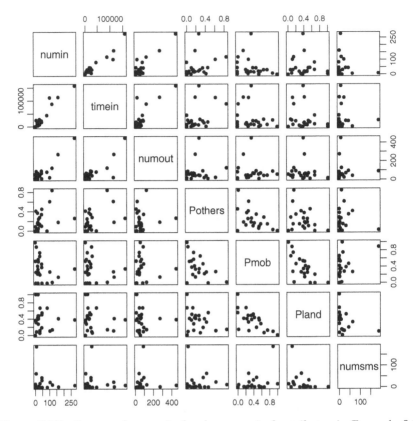

Figure 7.26   Scatter plot matrix for the numerical attributes in Example 5.2

The basic concept involves matching each record with a star-shaped icon, from the center of which depart as many rays as the number of attributes. The length of each ray is equal to the value of the corresponding attribute, normalized to fall in the interval [0,1] so as to give a consistent representation of the various attributes.

Figure 7.27 shows the star plot for the attributes *numin*, *timein*, *numout*, *Pothers*, *Pmob*, *Pland* and *numsms* in Example 5.2. From the figure it can be inferred that records 11, 15 and 23 are of particular interest.

Figure 7.28 is an example of the use of star plots for comparative purposes, using data from the *mtcars* dataset, described in Appendix B. The plot provides a visual comparison of 32 models of vehicles based on their main features.

Spider web chart

Spider web charts are grids where the main rays correspond to the attributes analyzed. For every record the position is calculated on each ray, based on the

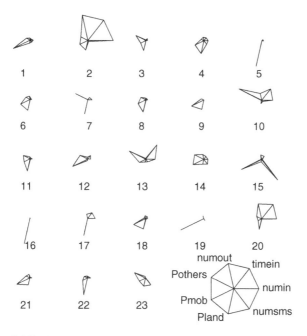

Figure 7.27   Star plot for the numerical attributes in Example 5.2

value of the corresponding attribute. Finally, the points so obtained on the rays for each record are sequentially connected to each other, thus creating a circuit for every record in the dataset.

Figure 7.29 shows a spider web chart corresponding to the same data as in Figure 7.27.

7.3.2   Measures of correlation for numerical attributes

For multivariate analysis of numerical attributes, covariance and correlation matrices are calculated among all pairs of attributes. For notational convenience, we will suppose that all the n attributes of the dataset $\mathcal{D}$ are numerical, having removed for the time being any categorical attribute. Let $\mathbf{V}$ and $\mathbf{R}$ be the two $n \times n$ square matrices whose elements are represented by the covariance values calculated using definition (7.38) and by the correlations calculated using (7.39), respectively. Both matrices $\mathbf{V}$ and $\mathbf{R}$ are symmetric and positive definite. Notice that the covariance matrix $\mathbf{V}$ contains on its main diagonal the sample covariance values of each single attribute, and for this reason it is also called the *variance–covariance* matrix.

Table 7.5 shows the covariance and correlation matrices obtained for the attributes *numin*, *timein*, *numout*, *Pothers*, *Pmob*, *Pland* and *numsms* in Example 5.2.

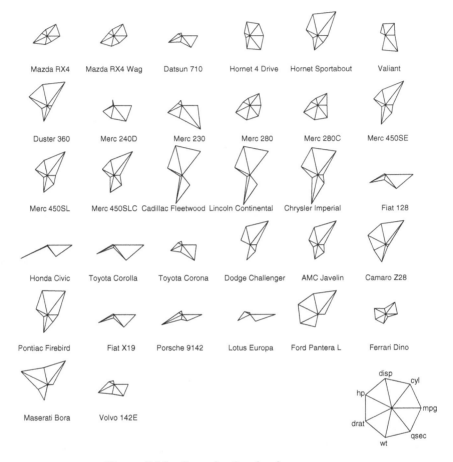

Figure 7.28   Star plot for the dataset mtcars

In order to devise a summary indicator that expresses the total variability of the dataset, and compares different datasets, the *trace* of the matrix $\mathbf{V}$ can be used, defined as the sum of the elements along its main diagonal:

$$\text{tr}(\mathbf{V}) = \sum_{j=1}^{n} v_{jj} = \sum_{j=1}^{n} \bar{\sigma}_{jj}^{2}. \tag{7.44}$$

It can be shown that the trace equals the sum of the eigenvalues λ_j, $j \in \mathcal{N}$, of the covariance matrix $\mathbf{V}$:

$$\text{tr}(\mathbf{V}) = \sum_{j=1}^{n} \lambda_j. \tag{7.45}$$

Table 7.5 Covariance and correlation matrices among the numerical attributes in Example 5.2

covariance	numin	timein	numout	Pothers	Pmob	Pland	numsms
numin	3769.99	2433848.00	5689.04	2.35	-2.50	0.07	-233.25
timein	2433848.00	1997570000.00	3522037.00	1893.14	-2518.33	575.88	-44328.60
numout	5689.04	3522037.00	12525.06	-2.90	0.10	2.76	92.83
Pothers	2.35	1893.14	-2.90	0.04	-0.02	-0.02	0.63
Pmob	-2.50	-2518.33	0.10	-0.02	0.06	-0.04	0.37
Pland	0.07	575.88	2.76	-0.02	-0.04	0.06	-1.00
numsms	-233.25	-44328.60	92.83	0.63	0.37	-1.00	657.63

correlation	numin	timein	numout	Pothers	Pmob	Pland	numsms
numin	1.00	0.89	0.83	0.19	-0.17	0.00	-0.15
timein	0.89	1.00	0.70	0.21	-0.24	0.05	-0.04
numout	0.83	0.70	1.00	-0.13	0.00	0.10	0.03
Pothers	0.19	0.21	-0.13	1.00	-0.41	-0.42	0.12
Pmob	-0.17	-0.24	0.00	-0.41	1.00	-0.65	0.06
Pland	0.00	0.05	0.10	-0.42	-0.65	1.00	-0.16
numsms	-0.15	-0.04	0.03	0.12	0.06	-0.16	1.00

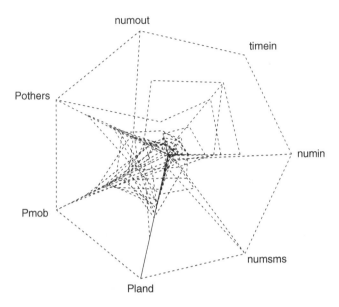

Figure 7.29   Spider web chart for the numerical attributes in Example 5.2

7.4   Notes and readings

There is a vast body of literature on descriptive statistics and data visualization techniques. Confining ourselves to some relevant contributions, the reader may refer to Tukey (1977), Anscombe (1973), Velleman and Hoaglin (1981), Tufte (1983), Cleveland (1993) and Fayyad *et al.* (2002). Data preparation is described in most books on data mining. An accurate discussion is presented in Pyle (1999). For attribute selection methods, see Kohavi and John (1997). For an introduction to basic statistics, readers may refer to Snedecor and Cochran (1989), Mendenhall *et al.* (2000), Weiss (2001), Levine *et al.* (2003) and Hogg *et al.* (2004). Crawley (2005) introduces the fundamental concepts of statistics by making reference to the R software, which we have used for performing the analyses and deriving the charts presented in most chapters of this book; see also Appendix A for a reference to R.

Basic notions of statistics can also be found on the web. For instance, currently active links on the subject are:

http://home.ubalt.edu/ntsbarsh/Business-stat/opre504online.htm,
http://onlinestatbook.com/.

8

Regression

According to the general framework for supervised learning models, *regression* models also deal with a dataset consisting of past observations, for which both the value of the explanatory attributes and the value of the continuous numerical target variable are known. Remember that when dealing with categorical target attributes classification models must be used, as described in Chapter 10.

This chapter is entirely devoted to *regression* models, which are sometimes referred to as *estimation* models.[1] We begin by discussing the general structure of regression models. Then, we will introduce simple linear regression models, for which there is a straightforward geometrical interpretation in the plane. Next, multiple regression models will be described, showing how to derive the regression coefficients in the general case, discussing the main stochastic assumptions concerning the residuals and presenting some extensions of linear regression. We then turn to the validation of regression models, illustrating the main diagnostics, from graphical analysis of the residuals to indicators of significance and accuracy. Finally, we illustrate the development of a multiple linear regression model by means of an example, and also discuss techniques for feature selection.

8.1 Structure of regression models

The purpose of regression models, also known as *explanatory* models, is to identify a functional relationship between the target variable and a subset of the remaining attributes contained in the dataset. Thus, their goal is twofold. On one hand, regression models serve to highlight and interpret the dependency of the

---

[1] Some authors use the term *estimation* to generally address supervised models with a numeric target, reserving the term *regression* for specific modeling approaches based upon regression techniques. However, we prefer to use the latter term to refer to the entire class of models.

target variable on the other variables. On the other hand, they are used to predict the future value of the target attribute, based upon the functional relationship identified and the future value of the explanatory attributes. Therefore, the development of a regression model allows knowledge workers to acquire a deeper understanding of the phenomenon analyzed and to evaluate the effects determined on the target variable by different combinations of values assigned to the remaining attributes. This opportunity is of great interest, particularly for analyzing those attributes that are *control levers* available to decision makers.

Consider, for example, a regression model aimed at interpretating the sales of a product based on investments made in advertising in different media, such as daily newspapers, magazines, TV and radio. Decision makers may use the model to assess the relative importance of the various communication channels, therefore directing future investments toward those media that appear to be more effective. Moreover, they can also use the model to predict the effects on the sales determined by different marketing policies, so as to design a combination of promotional initiatives that appear to be the most advantageous.

Suppose we are given a dataset $\mathcal{D}$ composed of m observations and $n + 1$ attributes, among which we distinguish a *target variable* and n other variables that may play an *explanatory* role with respect to the target. The target attribute is also called the *dependent variable*, *response* or *output*, while the explanatory variables are also termed *independent variables* or *predictors*. The independent variables of each observation may be represented by a vector $\mathbf{x}_i$, $i \in \mathcal{M}$, in the n-dimensional space $\mathbb{R}^n$, while the target attribute is denoted by y_i. For the sake of conciseness, we will write the m vectors of observations as a matrix $\mathbf{X}$ having dimension $m \times n$, as defined in Section 5.2, and the corresponding m-dimensional vector associated with the target variable as $\mathbf{y} = (y_1, y_2, \ldots, y_m)$. Finally, let Y be the random variable that represents the target attribute and X_j, $j \in \mathcal{N}$, the random variables associated with the explanatory attributes.

Regression models conjecture the existence of a function $f : \mathbb{R}^n \to \mathbb{R}$ that expresses the relationship between the dependent variable Y and the n explanatory variables X_j:

$$Y = f(X_1, X_2, \ldots, X_n). \tag{8.1}$$

In general, the process of identifying the function f, called a *hypothesis*, can be divided into two sequential phases. First, a choice is made of an adequate class $\mathcal{F}$ of hypotheses which must fulfill two conflicting requirements clearly and rigorously defined within statistical learning theory. On the one hand, the class must be broad enough to allow the identification of an accurate relationship between the target and the independent attributes, in order to guarantee small errors in explaining past data. On the other hand, it must be narrow enough to guarantee good generalization capability in predicting the target variable of new future observations. Considering these conflicting needs, the most popular classes of hypotheses consist of simple and parametric functional relationships of linear, quadratic, logarithmic and exponential nature. In the second phase, once the class of hypotheses $\mathcal{F}$ has been established, the value of the parameters defining the specific function f within the class $\mathcal{F}$ is determined through the solution of an optimization problem appropriately formulated.

In order to achieve greater robustness in a regression model, it is preferable that the functional relationship between the dependent variable and the independent ones be of a *causal* nature, that is, that it express a cause–effect nexus, where the independent variables clearly play the causal role and the dependent variable the effect role. Indeed, if it is possible to find a logical and causal explanation of the relationship represented in mathematical form by the function f, the interpretation of the model is more convincing and its use for predictive purposes more sound, as long as the causal relationship continues to hold.

A model that is based on a functional relationship between variables devoid of plausible causal nexus is said to be *spurious*. However, the development and use of a spurious model should not be ruled out, provided that the model satisfies the significance requirements discussed in Section 8.4.

In the regression models described in this chapter the class of hypotheses $\mathcal{F}$ will consist of linear functions, which lead to the study of *linear regression* models. In other words, we assume that the functional relationship f between the dependent variable and the independent variables is linear. At first sight, this assumption may appear restrictive, since there are certainly meaningful examples of causal relationships of a nonlinear nature. However, one should bear in mind that most types of nonlinear relationships may be reduced to the linear case by means of appropriate preliminary transformations to the original observations.

For example, a quadratic relationship of the form

$$Y = b + wX + dX^2 \tag{8.2}$$

can be linearized through the transformation $Z = X^2$, which converts it into a linear relationship with two predictors

$$Y = b + wX + dZ. \tag{8.3}$$

As a second example, consider the exponential relationship

$$Y = e^{b+wX} \tag{8.4}$$

which in turn can be linearized through a logarithmic transformation $Z = \log Y$, which converts it into the linear relationship

$$Z = b + wX. \tag{8.5}$$

These examples indicate that linear models are actually much more general than may appear at first sight.

It is appropriate to warn against a possible misunderstanding concerning the nature of the approximation between the target variable and the explanatory variables in a regression model. Indeed, notice that the goal of the analysis is not to identify a function f such that the relation (8.1) is satisfied by all the observations $\mathbf{x}_i$ and the corresponding values of the target y_i. The development of a regression model should not be intended as a problem of interpolation of the available data, for instance through polynomials of a sufficiently high degree or through *spline* functions, according to approaches typical of numerical analysis. Actually, the purpose of an explanatory model is to express a simple and basic relationship between the dependent variable and the independent variables that is able to approximate past data but also to produce a good generalization on future observations. As stated earlier, both theoretical results and empirical evidence suggest that a good trade-off between these opposite requirements is afforded by moderately broad classes of hypotheses.

8.2   Simple linear regression

Linear regression models represent the most widely known family of regression models and are based on a class of hypotheses consisting of linear functions. As a consequence, the functional relation (8.1) reduces to

$$Y = w_1 X_1 + w_2 X_2 + \cdots + w_n X_n + b = \sum_{j=1}^{n} w_j X_j + b. \qquad (8.6)$$

If there is only a single independent variable $X = X_1$ – that is, $n = 1$ – the linear regression model is called *simple*, and we have

$$Y = wX + b. \qquad (8.7)$$

In this section we will describe simple linear regression models, which admit a geometrical interpretation that makes them particularly intuitive. The general case of multiple linear regression will be discussed in Section 8.3.

For simple regression models the dataset actually reduces to m pairs of values (x_i, y_i), $i \in \mathcal{M}$, which are realizations of the random variables X and Y. In a first stage of analysis, the pairs of values can be graphically represented by a scatter plot, as described in Chapter 7, to develop a visual perception of any possible relationship existing between the random variables X and Y.

Example 8.1 – Simple linear regression. A company that manufactures food products wishes to find a relationship between the monthly production cost Y and the corresponding production volume X for one of its manufacturing plants. Table 8.1 gives pairs of values (x_i, y_i), $i \in \mathcal{M}$, recorded over a period of 14 months. The output is expressed in tons of product and the cost in thousands of euros. The regression line, shown in the scatter plot in Figure 8.1, approximates through a linear function the relationship existing between production volume and cost. The coefficients w and b of the line, indicated in the figure, lend themselves to a remarkable economic interpretation potentially useful at a managerial level: the intercept $b = 1.3641$ thousand euro expresses an a posteriori estimate of the fixed production costs, while the slope coefficient $w = 0.2903$ thousand euro per ton is an estimate of the variable costs.

Figure 8.1 shows the scatter plot corresponding to the data of Table 8.1. As shown in the plot, the variable Y seems to depend on the variable X according to a relationship that is well approximated by a linear function.

In general, it is unlikely that m pairs of observations (x_i, y_i), $i \in \mathcal{M}$, will fall exactly on a straight line on the plane. As a consequence, instead of expression (8.7), it is more realistic to suppose that an approximate relationship between

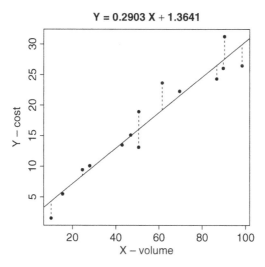

Figure 8.1   Scatter plot, regression line and residuals for Example 8.1

Table 8.1   Production volume and cost for Example 8.1

Volume X (ton)	Cost Y ($€ \times 10^3$)	Volume X (ton)	Cost Y ($€ \times 10^3$)
10.11	1.53	42.87	13.51
50.56	13.14	61.53	23.65
90.28	31.24	24.60	9.43
15.50	5.47	46.85	15.12
69.52	22.27	50.63	18.94
98.40	26.47	89.68	26.06
86.66	24.32	27.91	10.08

Y and X exists, as expressed by the *probabilistic model*

$$Y = wX + b + \varepsilon, \qquad (8.8)$$

in which ε is a random variable, referred to as *error*, which indicates the discrepancy between the response Y and the prediction $f(X) = wX + b$. The variable ε must satisfy some hypotheses of a stochastic nature which will be discussed in Section 8.3.2.

Having established that the relationship of interest is an approximate one, it is necessary to specify a criterion by which it will be possible to identify in the space $\mathcal{F}$ of hypotheses – that is, among all straight lines in the plane, the line that best represents the relationship between Y and X.

8.2.1   Calculating the regression line

Let us suppose that we draw in Figure 8.1 one of the infinite number of lines that can describe the relationship between Y and X. The identification of the regression line then reduces to the specification of its parameters, represented by the slope coefficient w and the intercept b.

In Figure 8.1 the dotted vertical segments highlight the *residuals* e_i, which are the signed lengths of the segments that connect the ordinates y_i of the observed points to the ordinates $f(x_i)$ of the points on the line at abscissas x_i:

$$e_i = y_i - f(x_i) = y_i - wx_i - b, \quad i \in \mathcal{M}. \qquad (8.9)$$

If we assume that the coefficients w and b of the straight line in the figure correspond to the ideal values expressed by the probabilistic model (8.8), the residuals can be thought of as m realizations of the random variable ε. For this reason, in a slight abuse of language, the random variable ε is also referred to as the *residual* variable.

The qualitative analysis of Figure 8.1 indicates that the linear regression model appears quite accurate, since the amount of the residuals is small and therefore the discrepancy between the values on the line and the actual observations is negligible. However, this intuitive perception, based on the analysis of the scatter plot, is insufficient for assessing the plausibility of the model in a rigorous and systematic way. In Section 8.4 we will describe a number of diagnostics aimed at assessing the significance of a linear regression model.

Among the possible criteria for identifying the regression coefficients w and b, the most widely known and intuitive one consists of minimizing the *sum of squared errors*, expressed by the function

$$\text{SSE} = \sum_{i=1}^{m} e_i^2 = \sum_{i=1}^{m} [y_i - f(x_i)]^2 = \sum_{i=1}^{m} [y_i - wx_i - b]^2. \tag{8.10}$$

Notice that SSE is a convex quadratic function of the variables w and b, possessing a single minimum point which can be determined by requiring that the partial derivatives with respect to the regression coefficients be equal to zero:

$$\frac{\partial \, \text{SSE}}{\partial b} = -2 \sum_{i=1}^{m} [y_i - wx_i - b] = 0, \tag{8.11}$$

$$\frac{\partial \, \text{SSE}}{\partial w} = -2 \sum_{i=1}^{m} x_i [y_i - wx_i - b] = 0. \tag{8.12}$$

In order to derive the optimal values of w and b, we have to solve the previous system of two equations in two variables, which can be formulated in the following *normal equation* form:

$$\begin{pmatrix} m & \sum_{i=1}^{m} x_i \\ \sum_{i=1}^{m} x_i & \sum_{i=1}^{m} x_i^2 \end{pmatrix} \begin{pmatrix} b \\ w \end{pmatrix} = \begin{pmatrix} \sum_{i=1}^{m} y_i \\ \sum_{i=1}^{m} x_i y_i \end{pmatrix}. \tag{8.13}$$

It is possibile to obtain the solution of the normal equation in analytic form, by expressing the regression coefficients through the equalities

$$\hat{w} = \frac{\sigma_{xy}}{\sigma_{xx}}, \tag{8.14}$$

$$\hat{b} = \bar{\mu}_y - \hat{w}\bar{\mu}_x, \tag{8.15}$$

where

$$\bar{\mu}_x = \frac{\sum_{i=1}^{m} x_i}{m}, \qquad \bar{\mu}_y = \frac{\sum_{i=1}^{m} y_i}{m} \tag{8.16}$$

represent the sample means of the predictor X and the target Y, respectively, while

$$\sigma_{xy} = \sum_{i=1}^{m} (x_i - \bar{\mu}_x)(y_i - \bar{\mu}_y), \tag{8.17}$$

$$\sigma_{xx} = \sum_{i=1}^{m} (x_i - \bar{\mu}_x)^2, \tag{8.18}$$

$$\sigma_{yy} = \sum_{i=1}^{m} (y_i - \bar{\mu}_y)^2 \tag{8.19}$$

express the sample covariance of x_i and y_i, the sample variance of x_i, and the sample variance of y_i, respectively, except for the multiplicative factors $1/(m-2)$ and $1/(m-1)$.

Notice that the values $\hat{w}$ and $\hat{b}$ of the coefficients, obtained by minimizing the function SSE, actually represent pointwise estimates of the true regression coefficients w and b that appear in the probabilistic model. Such estimators have indeed been derived by using the sample consisting of the m available observations. In Section 8.4 we will address the issue of assessing the accuracy of $\hat{w}$ and $\hat{b}$ as estimates of the regression coefficients.

The simple linear regression model, derived by minimizing the sum of squared errors, turns out to be

$$\hat{Y} = \hat{f}(X) = \hat{b} + \hat{w}X = \bar{\mu}_y + \frac{\sigma_{xy}}{\sigma_{xx}}(X - \bar{\mu}_x). \tag{8.20}$$

By applying the model to the data in Example 8.1 we obtain

$$\sum_{i=1}^{m} x_i = 765.10, \qquad \sum_{i=1}^{m} y_i = 241.25, \tag{8.21}$$

$$\sum_{i=1}^{m} x_i^2 = 52\,883.12, \qquad \sum_{i=1}^{m} x_i y_i = 16\,397.97. \tag{8.22}$$

By substituting these values into the normal equation and then solving it with respect to the regression coefficients, we derive the values $\hat{w} = 0.2903$ and $\hat{b} = 1.3641$, corresponding to the straight line depicted in Figure 8.1.

In those regression problems where the observations x_i, $i \in \mathcal{M}$, assume both positive and negative values, falling therefore on both sides of the value 0 along the x-axis, it may be appropriate to impose the condition $b = 0$, which geometrically corresponds to forcing the straight line to cross the origin. Hence, the expressions for the regression coefficients become

$$\hat{w} = \frac{\sum_{i=1}^{m} x_i y_i}{\sum_{i=1}^{m} x_i^2}, \tag{8.23}$$

$$\hat{b} = b = 0, \tag{8.24}$$

again by minimizing the function SSE, this time only with respect to the coefficient w. However, the condition $b = 0$ compromises one of the two degrees of freedom available in choosing the linear approximation for the data, hence generally worsening the accuracy of the model. In light of this remark, one should resort to this expedient only in those cases in which the condition $b = 0$ is considered mandatory.

8.3   Multiple linear regression

Having completed the description of simple linear regression models, we will now consider the general case, referred to as *multiple linear regression*, whereby the number n of independent variables is greater than one.

Suppose that we are given m observations $\mathbf{x}_i$, comprising n-dimensional vectors which represent realizations of the independent variables X_j, and m values y_i of the dependent variable Y. It is further assumed that a linear probabilistic relationship exists between the response Y and the explanatory variables X_j, $j \in \mathcal{N}$, expressed as

$$Y = w_1 X_1 + w_2 X_2 + \cdots + w_n X_n + b + \varepsilon, \tag{8.25}$$

by analogy with relation (8.8).

It is possible to derive a simple interpretation of the slope coefficients w_j appearing in the multiple regression model (8.25). Indeed, if a single explanatory variable X_j is increased by one, while all the other explanatory attributes remain unchanged, the response variable Y is affected by a variation in value equal to w_j. Therefore, it can be claimed that the regression coefficient w_j expresses the marginal effect of the variable X_j on the target, conditioned on the current value of the remaining predictive variables. Obviously, this information may prove very useful for the interpretation of the regression model and can provide a measure of the relative importance to the response variable of the various predictors.

If we investigate the relationship between a target variable that indicates the sales of a product and the investments made in advertisement through several media communication channels, used as predictive variables, the values of the regression coefficients may provide an indication of the relative advantage afforded by the different channels, and can therefore be used to direct future marketing campaigns.

It is, however, necessary to keep in mind a few conditions that limit the validity of the interpretation illustrated above. First, the value of each coefficient depends on the whole set of explanatory variables. As a consequence, the removal of some variables or the introduction of new predictors implies a change in all the regression coefficients and therefore can also alter their relative ranking. Furthermore, the scale of the values assumed by a predictor, and therefore also the measurement unit in which the values are expressed, influence the value of the corresponding regression coefficient. For this reason, it might be useful to perform a preliminary standardization of all the independent variables before proceeding with the development of a regression model.

8.3.1 Calculating the regression coefficients

For multiple linear regression models, the $n + 1$ parameters w_j and b that identify the hyperplane expressing the linear relationship in the space $\mathbb{R}^n$ can be derived by the least squares principle, through the minimization of the sum of squared errors. If $\mathbf{e} = (e_1, e_2, \ldots, e_m)$ denotes the vector of the residuals associated with the m pairs of observations $(\mathbf{x}_i, y_i)$, the n equalities

$$y_i = w_1 x_{i1} + w_2 x_{i2} + \cdots + w_n x_{in} + b + e_i, \quad i \in \mathcal{M}, \tag{8.26}$$

must hold. Assume that the matrix $\mathbf{X}$ associated with the original dataset has been modified by placing to the left an m-dimensional column vector with all components equal to 1, and denote by $\mathbf{w} = (b, w_1, w_2, \ldots, w_n)$ the vector of the slope regression coefficients, extended in turn to the left with the intercept b. We can therefore reformulate equalities (8.26) in matrix notation as

$$\mathbf{y} = \mathbf{Xw} + \mathbf{e}. \tag{8.27}$$

Hence, the sum of squared errors can be expressed as

$$\begin{aligned} \mathrm{SSE} &= \sum_{i=1}^{m} e_i^2 = \|\mathbf{e}\|^2 = \sum_{i=1}^{m} (y_i - \mathbf{w}'\mathbf{x}_i)^2 \\ &= (\mathbf{y} - \mathbf{Xw})'(\mathbf{y} - \mathbf{Xw}). \end{aligned} \tag{8.28}$$

As in the case of simple regression, SSE is a convex quadratic function possessing a unique minimum point that can be calculated analytically by setting equal to zero the partial derivatives evaluated with respect to the components of the vector $\mathbf{w}$:

$$\frac{\partial \, \mathrm{SSE}}{\partial \mathbf{w}} = -2\mathbf{X}'\mathbf{y} + 2\mathbf{X}'\mathbf{Xw} = \mathbf{0}. \tag{8.29}$$

Again we obtain the normal equation, expressed in this case in matrix form:

$$\mathbf{X'Xw = X'y}. \tag{8.30}$$

The solution of (8.30) uniquely determines the value of the $n + 1$ regression coefficients

$$\hat{\mathbf{w}} = (\mathbf{X'X})^{-1}\mathbf{X'y}, \tag{8.31}$$

provided that the matrix $\mathbf{X'X}$ is invertible.

Notice that for multiple regression, as for simple regression, the solution $\hat{\mathbf{w}}$, obtained through minimization of the function SSE, represents a pointwise estimate of the regression coefficients $\mathbf{w}$, whose accuracy will be discussed in Section 8.4. It can be shown that the vector $\hat{\mathbf{w}}$ coincides with the maximum likelihood estimate of the vector $\mathbf{w}$.

We can easily derive the values of the response variable Y predicted by the model, also termed *fitted values*, as

$$\hat{\mathbf{y}} = \mathbf{X}\hat{\mathbf{w}} = (\mathbf{X}(\mathbf{X'X})^{-1}\mathbf{X'})\mathbf{y} = \mathbf{Hy}, \tag{8.32}$$

where $\mathbf{H} = \mathbf{X}(\mathbf{X'X})^{-1}\mathbf{X'}$ is called the *hat matrix*, which has the properties of being symmetric and idempotent. The matrix $\mathbf{H}$ allows the residuals to be expressed as

$$\mathbf{e = y} - \hat{\mathbf{y}} = (\mathbf{I - H})\mathbf{y}. \tag{8.33}$$

In most applications, the development of a multiple regression model is a complex matter, involving the identification of the most appropriate predictors to be actually included into the model, starting from the whole set of available explanatory variables. Hence, it represents a specific feature selection problem, discussed in general terms in Chapter 6. In Section 8.5 we will describe by means of an example the development of a multiple regression model.

8.3.2   Assumptions on the residuals

The random variable ε, which represents the errors and appears in the probabilistic models (8.8) and (8.25), must satisfy a number of stochastic assumptions, partly dependent on the criterion used for calculating the coefficients of the regression model.

In general, the variable ε should behave as an intrinsically random noise, devoid of effects that can be observed in the response variable Y. Alternatively, ε can also be seen as the result of the action of countless independent variables excluded from the model, each of them having a negligible effect on the target attribute Y.

In particular, if the regression coefficients are determined by minimizing the sum of squared errors SSE, the random variable ε must follow a normal distribution with 0 mean and standard deviation σ. This assumption translates into the following expressions, relative to the expected value and the variance of the residuals conditioned on the value of the observations:

$$E(\varepsilon_i | \mathbf{x}_i) = 0, \tag{8.34}$$

$$\mathrm{var}(\varepsilon_i | \mathbf{x}_i) = \sigma^2. \tag{8.35}$$

Moreover, the residuals ε_i and ε_k, corresponding to two distinct observations $\mathbf{x}_i$ and $\mathbf{x}_k$, should be independent for any choice of i and k.

Notice that the standard deviation σ should ideally remain constant as the values of each single component of the observations vary. This is referred to as *homoscedasticity* in the residuals. Conversely, if the value of the standard deviation σ is not constant – that is, the model is not steady as the variables X_j vary along their axes – we have *heteroscedasticity* in the residuals. This situation compromises the significance of the regression model and must be corrected, usually by resorting to a scaling of the variables.

To better illustrate this technique, consider the numerical example shown in Table 8.2. The dataset shows the relationship between the effectiveness of

Table 8.2   Millions of weekly contacts and TV advertising costs

Company	TV spending (M$)	Milcont (Mil. weekly contacts)
MILLER.LITE	50.1	32.1
PEPSI	74.1	32.5
STROH'S	19.3	11.7
FEDERAL.EXPRESS	22.9	21.9
BURGER.KING	82.4	52.4
COCA-COLA	40.1	47.2
MC.DONALD'S	185.9	41.4
MCI	26.9	43.2
DIET.COLA	20.4	21.4
FORD	166.2	37.3
LEVI'S	123	87.4
BUD.LITE	45.6	20.8
ATT.BELL	154.9	97.9
CALVIN.KLEIN	5	12
WENDY'S	49.7	29.2
POLAROID	26.9	38
SHASTA	5.7	10
MEOW.MIX	7.6	12.3
OSCAR.MEYER	9.2	23.4
CREST	32.4	43.6
KIBBLES.N.BITS	6.1	26.4

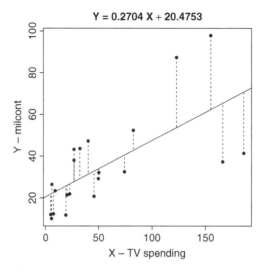

Figure 8.2   Scatter plot, regression line and residuals for Table 8.2

a TV communication, expressed as millions of weekly contacts, and the corresponding TV advertisement costs, expressed in millions of dollars, based on
data collected for 21 companies. The scatter plot and regression line are shown
in Figure 8.2. There is evidently high heteroscedasticity in the residuals, with

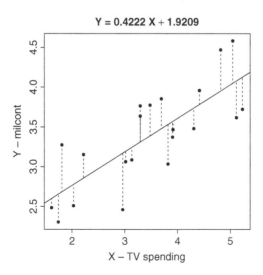

*Figure 8.3   Scatter plot, regression line and residuals on a logarithmic scale
for Table 8.2*

a deviation that increases as the value of the variable X increases. If, however, we use a logarithmic scale on both axes, the new diagram and the corresponding regression line, shown in Figure 8.3, show a substantial homoscedasticity in the residuals.

Notice that a regression model is more accurate, and therefore more useful from an application standpoint, as the deviation σ is close to 0. It is therefore interesting to calculate an unbiased pointwise estimator of the deviation σ^2, given by the expression

$$
\begin{aligned}
\bar{\sigma}^2 &= \frac{\text{SSE}}{m-n-1} = \frac{\sum_{i=1}^{m}(y_i - \mathbf{w}'\mathbf{x}_i)^2}{m-n-1} = \frac{\mathbf{e}'\mathbf{e}}{m-n-1} \\
&= \frac{\mathbf{y}'(\mathbf{I}-\mathbf{H})'(\mathbf{I}-\mathbf{H})\mathbf{y}}{m-n-1} = \frac{\mathbf{y}'(\mathbf{I}-\mathbf{H})^2\mathbf{y}}{m-n-1} = \frac{\mathbf{y}'(\mathbf{I}-\mathbf{H})\mathbf{y}}{m-n-1},
\end{aligned}
\tag{8.36}
$$

in which the denominator takes into account the degrees of freedom of the model, which are equal to the number m of observations less the total number of variables, one of which is dependent and the remaining n are independent. We have also made use of the symmetry and the idempotency of the matrix $\mathbf{H}$, and therefore of the matrix $(\mathbf{I} - \mathbf{H})$. The estimator $\bar{\sigma}$ of the standard deviation σ is referred to as the *standard error* and plays a critical role in the determination of the accuracy of a regression model, since it determines the dispersion of the data around the prediction line.

8.3.3   Treatment of categorical predictive attributes

In applications of regression models, it is common for one or more predictive attributes to be categorical. In this section, we will describe a method that allows us to include also categorical variables among the predictors of the target variable, by means of an *orthonormal* representation based on *dummy* variables.

Suppose that X_j is a categorical variable included in the dataset $\mathcal{D}$, and denote by $V = \{v_1, v_2, \ldots, v_H\}$ the set of H distinct values assumed by the categorical attribute X_j. Of course, it is possible to associate arbitrary numerical values with each of the H distinct levels assumed by the variable X_j, for example the numbers $\{1, 2, \ldots, H\}$. However, if the variable X_j is introduced in a regression model using arbitrary numerical values, the regression coefficients will be affected by the specific scale chosen and the significance of the resulting model will be compromised.

To properly treat the categorical variable, we first need to define $H - 1$ binary variables $D_{j1}, D_{j2}, \ldots, D_{j,H-1}$, called *dummy* variables. The binary variable D_{jh} is associated with level v_h of the predictor X_j and for each observation $\mathbf{x}_i$ the new variable D_{jh} takes the value 1 if $x_{ij} = v_h$ – that is, if the variable X_j assumes the value v_h for the observation $\mathbf{x}_i$. In essence, the binary variable plays the role of an indicator relative to the presence of the corresponding value v_h. Notice that $H - 1$ variables are sufficient since the value of an additional variable D_{jH} would be uniquely determined by the previous $H - 1$ dummy variables, and would consequently have a no informative value.

Suppose that we wish to explain the sales of a product through some predictive variables, among which one indicates for each observation the month in which the observation was recorded. It is then necessary to replace the categorical variable X, which can assume one of the 12 values {Jan, Feb, Mar, Apr, May, Jun, Jul, Aug, Sep, Oct, Nov, Dec}, with the following 11 binary variables

$$\{D_{\text{Jan}}, D_{\text{Feb}}, D_{\text{Mar}}, D_{\text{Apr}}, D_{\text{May}}, D_{\text{Jun}}, D_{\text{Jul}}, D_{\text{Aug}}, D_{\text{Sep}}, D_{\text{Oct}}, D_{\text{Nov}}\}.$$

Each of these takes the value 1 for an observation $\mathbf{x}_i$ if the value of the variable X for $\mathbf{x}_i$ coincides with the month with which the binary variable is associated. Finally, notice that the choice of the level to which the omitted variable corresponds is arbitrary. Actually, we could have included the variable D_{Dec} and excluded any one of the remaining binary variables.

8.3.4   Ridge regression

The computation of the inverse matrix $(\mathbf{X}'\mathbf{X})^{-1}$, usually based on efficient decomposition algorithms, becomes impossible if $\mathbf{X}'\mathbf{X}$ is a singular matrix and turns out to be critical if the number m of observations is insufficient or if some explanatory variables are dependent on each other, thus causing a phenomenon called *multicollinearity*, further discussed in Section 8.4.6.

In such cases, although the matrix $\mathbf{X}'\mathbf{X}$ is in principle invertible, the actual identification of the regression coefficients represents an *ill-conditioned* problem, in the sense that small changes in the input data may cause significant changes in the estimate of the regression parameters. To put it another way, the amount of information contained in the data proves insufficient to create a robust and accurate regression model.

A possible remedy has been proposed within the theory of *regularization* and involves limiting the space $\mathcal{F}$ of hypotheses, that is, those functions f intended to represent the relationship among the response variable and the explanatory variables. In general terms, the restriction of the class $\mathcal{F}$ is obtained by limiting the norm of the candidate functions f, or looking for an ideal trade-off between the norm of f and its accuracy on the training data.

In particular, instead of obtaining the regression coefficients through the minimization of the sum of squared errors SSE, the method of *ridge regression* relies on the optimization model

$$\min_{\mathbf{w}} RR(\mathbf{w}, \mathcal{D}) = \min_{\mathbf{w}} \lambda \|\mathbf{w}\|^2 + \sum_{i=1}^{m} (y_i - \mathbf{w}'\mathbf{x}_i)^2$$

$$= \min_{\mathbf{w}} \lambda \|\mathbf{w}\|^2 + (\mathbf{y} - \mathbf{X}\mathbf{w})'(\mathbf{y} - \mathbf{X}\mathbf{w}),$$

(8.37)

in which λ is a positive parameter controlling the relative importance of the regularization term, expressed by the norm of $\mathbf{w}$, with respect to the sum of squared errors.

Also in this case, the learning problem is brought back to the solution of an appropriate minimization problem, a fact that confirms the key role played by optimization theory within pattern recognition and machine learning models.

8.3.5 Generalized linear regression

It is possible to extend the multiple linear regression model in order to adapt it to a rather broad class of transformations that can be applied to the independent variables. To this end, consider a generalized probabilistic model

$$Y = \sum_{h} w_h g_h(X_1, X_2, \ldots, X_n) + b + \varepsilon,$$

(8.38)

where the functions g_h represent any set of bases, such as polynomials, kernels and other groups of nonlinear functions.

In this case, too, the coefficients w_h and b can be determined through the minimization of the sum of squared errors. However, the corresponding function SSE in this formulation is more complex than in the case of linear regression, and the solution of the minimization problem is therefore more difficult. Usually the solution of problem (8.38) is embodied in the framework of duality theory, through a representation of the regression coefficients $\hat{\mathbf{w}}$ as a linear combination of the observations. For further details see the references at the end of the chapter.

8.4   Validation of regression models

At first sight, one might have the impression that the development of a linear regression model is straightforward, since even the most widely used spreadsheets can calculate regression coefficients. However, such user friendliness may lull one into a false sense of security regarding the real significance and the predictive accuracy of the regression model generated. However, there are different diagnostic criteria, which we will describe in this section, that enable the quality and the predictive accuracy of a linear regression model to be evaluated:

• normality and independence of the residuals;

• significance of the coefficients;

• analysis of variance;

• coefficient of determination;

• linear correlation coefficient;

• multicollinearity of the independent variables;

• confidence and prediction limits.

8.4.1 Normality and independence of the residuals

The aim of the first diagnostic test is to verify the hypotheses of normality and independence of the residuals. This statistical check is usually performed *a posteriori*, after the estimate of the regression coefficients $\hat{\mathbf{w}}$ and the vector **e** of the residuals have been computed.

In order to verify whether the assumption that the residuals follow a normal distribution is correct, one of the alternative goodness-of-fit hypothesis tests can be applied, such as the *chi-square* test or the *Kolmogorov–Smirnov* test.

The statistics on the residuals, as shown in Table 8.3 for Example 8.1, provide further relevant information from which the magnitude of the errors can be evaluated. These are coupled with the standard deviation the residuals, shown in Table 8.6 as the *residual standard error*. For Example 8.1 the deviation $\bar{\sigma} = 2.55$ provides a first insight into the predictive accuracy of the regression model: with a probability of 96% the values of the residuals fall within an interval of semi-width 5.1. Section 8.4.7 will explain how the confidence limits of the predictions can be more accurately evaluated.

Furthermore, it is useful to analyze the distribution of the residuals through a series of graphical representations that we will describe for the regression model of Example 8.1.

The first graph, shown in Figure 8.4, is a scatter plot of the residuals against the fitted values $\hat{\mathbf{y}}$. Such a graph can be used to highlight any abnormality or

Table 8.3   Statistics concerning the residuals in Example 8.1

minimum	lower quartile	median	upper quartile	maximum
-3.46271	-1.98842	-0.07337	0.87332	4.42181

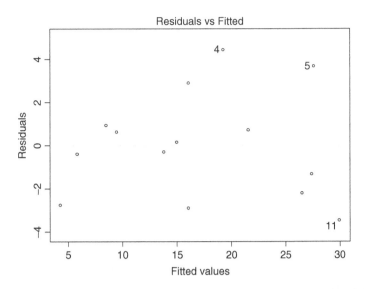

Figure 8.4   Scatter plot of residuals vs. fitted values for Table 8.1

regularity in the profile of residuals. Notice that the plot highlights abnormal residual values by showing the number of the corresponding observation in the dataset. A regular trend in the points on the scatter plot indicates the existence of some explanatory factors not included in the model that might help explain the response variable. Alternatively, a regular pattern in the graph may suggest that the hypothesis f does not adequately express the relationship between the response and the predictors.

The second type of scatter plot (Figure 8.5) shows the square root of the standardized residuals on the vertical axis. These are positive and the differences between them are smaller, as a result of using the square root.

The third graph, shown in Figure 8.6, is a QQ plot of the residuals which provides a way to evaluate visually whether they follow a normal distribution. As described in Chapter 7, for an (approximately) normal distribution the points must fall (approximately) on a straight line. An S-shaped curve indicates that the tails of the distribution of the residuals are shorter compared to the tails of a normal density. On the other hand, for an upside-down 'S' the tails are longer.

Figure 8.7 shows for each observation, placed on the horizontal axis, the value of an indicator, called the *Cook's distance*, which highlights the presence of abnormal and large values of the residuals. The Cook's distance of an observation is a measure of its global influence on all the predicted values, obtained by evaluating the effect of removing the observation from the dataset. In particular, an observation with a Cook's distance greater than 1

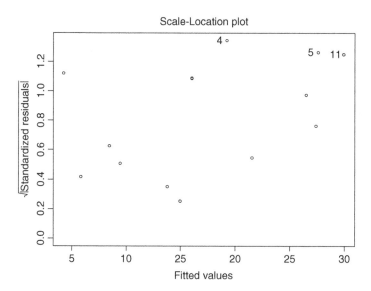

Figure 8.5   Scatter plot of squared standardized residuals vs. fitted values for Table 8.1

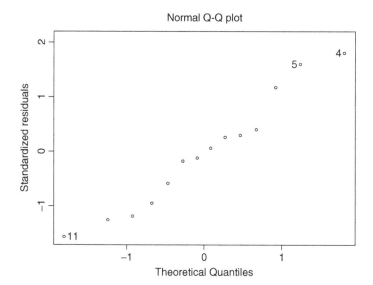

Figure 8.6   QQ plot of the residuals for Table 8.1

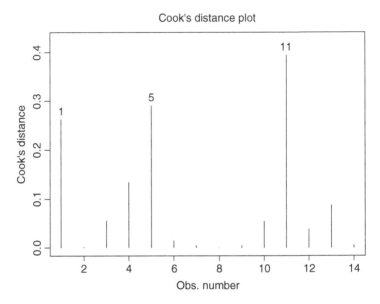

Figure 8.7   Plot of Cook's distances for Table 8.1

is considered abnormal with respect to the regression model, and such as to significantly influence the values of the regression coefficients. As we can see, the Cook's distance is lower than 0.5 for all the observations in Figure 8.7.

8.4.2   Significance of the coefficients

As already observed, the coefficients $\hat{\mathbf{w}}$ obtained through the minimization of the sum of squared errors are pointwise estimates of the regression coefficients $\mathbf{w}$ that appear in the probabilistic model.

The estimation of $\mathbf{w}$ can be regarded as a classical problem of inferential statistics, where the regression coefficients $\mathbf{w}$ are the unknown parameters of the population, the m observations are a sample extracted from the population and the coefficients $\hat{\mathbf{w}}$ are the pointwise estimators with which a confidence interval can be associated.

Observe that the covariance matrix of the estimator $\hat{\mathbf{w}}$ is expressed by

$$\mathbf{V} = \mathrm{cov}(\hat{\mathbf{w}}) = \bar{\sigma}^2(\mathbf{X}'\mathbf{X})^{-1}. \tag{8.39}$$

The elements v_{jj} on the main diagonal of the matrix $\mathbf{V}$ thus provide the variance for each estimated coefficient, leading to a $100(1 - \alpha)\%$ confidence interval of the form

$$\hat{w}_j \pm t_{\alpha/2}\sqrt{v_{jj}}, \tag{8.40}$$

where $t_{\alpha/2}$ is the $\alpha/2$-order quantile of the Student t-distribution with $m - n - 1$ degrees of freedom. For a simple linear regression model, the intervals for the coefficients can be expressed as

$$\hat{w} \pm t_{\alpha/2} \frac{\bar{\sigma}}{\sqrt{\sum_{i=1}^{m}(x_i - \bar{\mu}_x)^2}}, \tag{8.41}$$

$$\hat{b} \pm t_{\alpha/2} \frac{\bar{\sigma}}{\sqrt{n}} \sqrt{1 + \frac{n\bar{\mu}_x}{\sum_{i=1}^{m}(x_i - \bar{\mu}_x)^2}}. \tag{8.42}$$

The confidence intervals for the regression coefficients lead to a third criterion for the validation of a regression model. We can actually conclude that a slope coefficient $\hat{w}_j$ has no significance if its confidence interval contains the value 0. Indeed, if 0 belongs to the confidence interval, the coefficient of the independent variable X_j may be either positive or negative with a non-negligible probability. In other words, the model is unable to establish if the dependent variable Y increases or decreases as the independent variable X_j increases. It is worth noticing that even when the estimator $\hat{w}_j$ appears sufficiently far from 0, it may be the case that the value 0 is included in the confidence interval, provided the corresponding variance v_{jj} of the coefficient $\mathbf{w}_j$ is large enough.

Table 8.4 presents the diagnostics on the significance of the coefficients for the simple regression model in Example 8.1, obtained with respect to the hypothesis test

$$H_0 : w = 0, \quad H_a : w \neq 0. \tag{8.43}$$

The first two columns show the name of the predictive variable and the estimate of the corresponding regression coefficient. The third column provides the value of the standard deviation v_{jj} for each coefficient, using (8.39). The fourth column indicates the t-value, which expresses the z-index of the estimated coefficient value $\hat{w}$ if the null hypothesis H_0 that the actual value w equals 0 is true. Finally, the last column indicates the p-value corresponding to the test, which is the probability that the actual value w is greater in absolute value than the estimated value $\hat{w}$, obtained as the area under the tails for a Student t-density with $m - 2$ degrees of freedom.

A t-value greater than 2 is taken to mean that the interval around $\hat{w}$ does not include the value 0 and that therefore w differs from 0, with 95% confidence.

Table 8.4   Significance of the regression coefficients for Example 8.1

predictor	value	standard error	t-value	$Pr > |t|$
(intercept)	1.36411	1.48944	0.916	0.3780
volume	0.29033	0.02423	11.980	< 0.0001

Analogously, the same conclusion is reached if the p-value is lower than 0.05. For the regression model in Example 8.1, the conclusion can be drawn that w differs significantly from 0, and hence that the positive dependence of Y on X is meaningful. On the other hand, it cannot be concluded that the intercept b is significantly different from 0, since the corresponding t-value is less than 2 and the p-value is greater than 0.05. However, the fact that the intercept has a low significance does not jeopardize the validity of the model and indeed it is rarely taken into account.

8.4.3   Analysis of variance

The analysis of variance, reproduced in Table 8.5 for Example 8.1, is an important diagnostic for the validation of a regression model. The first row in the table, headed *volume*, provides the following information:

df. The number n of degrees of freedom, which in Example 8.1 equals 1.

sum of sq. The sum of squared differences between the predictions and the sample mean of the response variable, defined as

$$\text{RSS}_{\text{reg}} = \sum_{i=1}^{m} (\hat{y}_i - \bar{\mu}_y)^2 = 933.18. \tag{8.44}$$

mean sq. The ratio between the sum of squared prediction differences and the degrees of freedom, given in the two previous columns.

F-value. The value of the F-statistic, defined in equation (8.46) below.

Pr > F. The p-value for the F-statistic.

The second row, headed *residuals*, contains the following information:

df. The number $m - n - 1$ of degrees of freedom.

sum of sq. The sum SSE of squared residuals.

mean sq. The ratio between the sum of squared residuals and the degrees of freedom, given in the two previous columns; this ratio corresponds to the sample variance $\bar{\sigma}^2$ of the residuals.

Table 8.5   Analysis of variance for Example 8.1

predictor	df	sum of sq.	mean sq.	F-value	Pr > F
volume	1	933.18	933.18	143.53	< 0.0001
residuals	12	78.02	6.50		
total	13	1011.20	77.79		

Finally, the last row headed *total* shows the following information:

df. The number $m - 1$ of degrees of freedom.

sum of sq. The sum of squared differences between the actual values and the sample mean for the response variable, given by

$$\text{RSS}_{\text{tot}} = \sum_{i=1}^{m}(y_i - \bar{\mu}_y)^2 = 1011.20. \tag{8.45}$$

mean sq. The ratio between the sum of squared differences and the degrees of freedom, given in the two previous columns, expressing the sample variance of the response variable Y.

The analysis of variance can be interpreted in intuitive terms. To see this, recall that the aim of a regression model is to explain as far as possible through the predictive variables the variance RSS_{tot} inherent in the dependent variable, leaving aside the explanation of purely random fluctuation represented by the residuals. If this goal is achieved, one may expect the sample variance $\bar{\sigma}^2$ of the residuals, calculated using expression (8.36), to be significantly smaller than the sample variance of the response variable Y. Indeed, in Example 8.1 the sample variance of the residuals, equal to $\bar{\sigma}^2 = 6.50$, is much lower than the sample variance of the response variable Y, which is equal to 77.79.

The two quantities above are dependent upon each other so we cannot formulate a meaningful test. However, the sums of the squares in the first and second rows of Table 8.1 are independent. The value 933.18 for the *volume* predictor turns out to be an unbiased estimate of σ^2 if the hypothesis H_0 is true, while it is an excessive estimate if the hypothesis is false. The value 6.50 in the same column for the residuals is an unbiased estimate of σ^2 regardless of whether H_0 is true or false. As a consequence, their ratio is close to 1 if H_0 is true, and therefore the model is not meaningful, while it is greater than 1 if H_0 is false. If the residuals have a normal distribution, the ratio

$$F = \frac{\text{RSS}_{\text{reg}}/n}{\text{SSE}/(m - n - 1)} \tag{8.46}$$

follows an F distribution with n and $m - n - 1$ degrees of freedom. In Example 8.1 this ratio provides an F-value equal to 143.53, much greater than 1, which therefore suggests that hypothesis H_0 is false, as also confirmed by the p-value in the last column which is less than 0.0001.

8.4.4   Coefficient of determination

The coefficient of determination R^2, also known as *multiple R-squared*, expresses the proportion of total variance explained by the predictive variables,

Table 8.6   Summary statistics for Example 8.1

statistics	value
residual standard error	2.5500
multiple R-squared	0.9228
adjusted R-squared	0.9164
regression coeff.	0.9606

and therefore by the regression model. It is defined as

$$R^2 = \frac{\text{RSS}_{\text{reg}}}{\text{RSS}_{\text{tot}}} = \frac{\sum_{i=1}^{m}(\hat{y}_i - \bar{\mu}_y)^2}{\sum_{i=1}^{m}(y_i - \bar{\mu}_y)^2}, \tag{8.47}$$

and it therefore lies in the interval $[0,1]$. If R^2 assumes values close to 1, it may be concluded that the model explains a large portion of the variation inherent in the response variable.

For Example 8.1 the coefficient of determination is $R^2 = 0.9228$, as shown in Table 8.6; thus the model is able to explain approximately 92% of the variation embedded in the dependent variable.

The coefficient of determination tends to increase as the number of observations increases and therefore to overestimate the proportion of variance explained by the model. To avoid this inaccuracy, a corrected coefficient of determination, termed *adjusted R-squared*, is defined as

$$R_{\text{adj}}^2 = 1 - (1 - R^2)\frac{m-1}{m-n-1}. \tag{8.48}$$

In Example 8.1 we have $R_{\text{adj}}^2 = 0.9164$.

8.4.5   Coefficient of linear correlation

In simple regression models it can be observed that the coefficient of determination coincides with the squared coefficient of linear correlation, $R^2 = r^2$, where

$$
\begin{aligned}
r = \text{corr}(\mathbf{y}, \mathbf{x}_i) &= \frac{\sigma_{xy}}{\sqrt{\sigma_{xx}\sigma_{yy}}} \\
&= \frac{\sum_{i=1}^{m}(x_i - \bar{\mu}_x)(y_i - \bar{\mu}_y)}{\sqrt{\sum_{i=1}^{m}(x_i - \bar{\mu}_x)^2 \sum_{i=1}^{m}(y_i - \bar{\mu}_y)^2}}.
\end{aligned}
\tag{8.49}
$$

As observed in Chapter 7, the value of the linear correlation coefficient r lies in the interval $[-1, 1]$. In particular, the linear correlation coefficient can be explained in the following way, as also shown in Figure 7.24:

- If $r > 0$, then X and Y are concordant. This means that the pairs of observations will tend to lie on a straight line with positive slope, the approximation to a straight line increasing as r gets closer to 1; if $r = 1$ the points will lie exactly on a straight line.

- If $r < 0$, then X and Y are discordant. In this case, the pairs of observations will tend to lie on a straight line with negative slope, the approximation to a straight line increasing as r gets closer to -1; if $r = -1$ the points will lie exactly on a straight line.

- Finally, if $r = 0$, or if $r \approx 0$, there is no linear relationship between X and Y; in this case, the pairs of observations may form a totally random pattern, or alternatively may tend to lie on a nonlinear curve.

The linear correlation coefficient for Example 8.6 takes the value $r = 0.9606$, as indicated in Table 8.6.

8.4.6   Multicollinearity of the independent variables

In a multiple linear regression model, predictive variables should not be linearly correlated. If a significant linear correlation exists between two or more independent variables, the model is said to be affected by *multicollinearity*. When there is multicollinearity, the estimate of the regression coefficients is inaccurate and compromises the overall significance of the model. In a situation of multicollinearity it may even occur that the coefficient of determination is close to 1, while the regression coefficients of the predictors are not significantly different from 0. This represents a flaw in the model, which, however, can be overcome either by selecting a subset of non-collinear variables or by eliminating some variables that are collinear with others.

To assess the extent of multicollinearity, the linear correlation coefficients among all pairs of independent variables should first be calculated. For those correlation coefficients that exceed a given threshold, it may be concluded that the corresponding pairs of variables are linearly correlated, while for the remaining pairs one may deduce that no evidence of linear correlation exists.

However, in this way only pairwise correlations can be detected. In order to highlight the presence of multiple linear relationships among independent variables, it is possible to calculate the *variance inflation factor*, which is defined for each predictor X_j as

$$\text{VIF}_j = \frac{1}{1 - R_j^2}, \tag{8.50}$$

where R_j^2 is the coefficient of determination for the linear regression model that explains the variable X_j, treated as a response, through the remaining independent variables. Values of VIF_j greater than 5 point to the existence of multicollinearity.

8.4.7   Confidence and prediction limits

Prediction and *confidence* intervals are two other interesting diagnostics for the validation of a regression model. In particular, the purpose of the prediction interval is to contain the expected value $E[Y]$ of the response variable, whereas the confidence interval includes the value Y of the response variable corresponding to a new vector $\mathbf{x}$ of values assigned to the independent variables.

Let $\hat{y} = \hat{\mathbf{w}}\mathbf{x}$ be the prediction associated with the new observation $\mathbf{x}$. The variance of $\hat{y}$ is given by

$$\text{var}(\hat{y}) = \mathbf{x}'(\mathbf{X}'\mathbf{X})^{-1}\mathbf{x}\bar{\sigma}^2, \tag{8.51}$$

and allows one to derive a confidence interval around $\hat{y}$, within which one can expect to find the future observed value $E[Y]$. In case of a simple linear regression model, an interval at $100(1 - \alpha)\%$ confidence is given by

$$\hat{y} \pm t_{\alpha/2}\bar{\sigma}\sqrt{\frac{1}{m} + \frac{x - \bar{\mu}_x}{\sum_{i=1}^{m}(x_i - \bar{\mu}_x)^2}}, \tag{8.52}$$

where x is the value of the new observation of X and $t_{\alpha/2}$ is the percentile of order $100\alpha/2$ of a Student t-distribution with $m - n - 1 = m - 1 - 1 = m - 2$ degrees of freedom. As we can see, the magnitude varies as x varies, and tends to increase in a nonlinear way when x goes away from its average $\bar{\mu}_x$. Intuitively, this means that the uncertainty relative to future predictions increases near the boundaries of the domain of variation of the independent variable.

For simple regression, a similar prediction interval within which we can expect to find the single value Y associated with the new observation x, with a confidence level $(1 - \alpha)\%$, is given by

$$\hat{y} \pm t_{\alpha/2}\bar{\sigma}\sqrt{1 + \frac{1}{m} + \frac{x - \bar{\mu}_x}{\sum_{i=1}^{m}(x_i - \bar{\mu}_x)^2}}. \tag{8.53}$$

Also in this case the uncertainty increases when x goes away from its average $\bar{\mu}_x$. Obviously, the prediction interval appears wider than the confidence interval, since besides the uncertainty relative to the prediction it also includes the uncertainty caused by the fluctuation of Y around its average.

Figure 8.8 shows for Example 8.1 the confidence interval, which corresponds to the innermost (dashed) lines, and the prediction interval, represented by the outermost (dotted) lines.

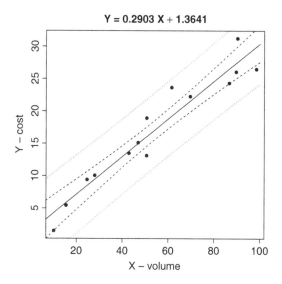

Figure 8.8   Prediction and confidence intervals for Example 8.1

8.5   Selection of predictive variables

The development of a multiple linear regression model requires the selection of a subset of variables that, among all the predictive variables available in the dataset $\mathcal{D}$, are most effective in explaining the response variable. The model obtained using all the predictive variables is rarely significant, since it is often affected by multicollinearity among the independent attributes.

The selection of the predictive variables is a complex activity, particularly when the dataset includes hundreds of explanatory attributes. It is necessary to adopt a wrapper method, as described in Section 6.3.2, whereby a regression model is developed for each subset of predictive variables, then evaluate the significance diagnostics and finally select the best model.

As the number n of attributes increases it clearly becomes impossible to carry out the analysis for all 2^n combinations of the attributes. Therefore, it is usually preferable to automate the procedure for attribute selection by calculating a summary indicator that may direct the algorithms in search for the best subset of predictors. The resulting sequential greedy algorithms are based on myopic criteria of local improvement and do not guarantee the identification of the optimal set of predictors. However, they may prove useful in the preliminary stages of development of a model. Yet, even starting from the suggestions obtained through an automatic search algorithm, a subsequent careful consideration on the part of the analyst is highly recommended in order to identify the most effective predictive variables.

The selection methods are based on three distinct patterns: *forward inclusion*, *backward exclusion* and *exhaustive search*.

Forward inclusion. Forward inclusion involves adding one predictive variable at a time, starting with an empty set of predictors in the initial stage. At each iteration of the algorithm, the variable that affords the greatest increase in predictive power should be chosen out of all the variables still excluded from the model, provided that the increase exceeds a minimum preset threshold. There are several alternative indicators to express the potential for improvement. One of the most widely applied criteria consists of assessing the reduction in the sum of squared residuals brought about by each variable if introduced into the model. A further indicator is based on the increase in the adjusted coefficient of determination R^2_{adj}. A third possible criterion is based on the increase in the value of the F-statistic. It is obviously possible to devise other criteria, perhaps consisting of an appropriate combination of the previous ones, or based on totally different indicators.

Backward exclusion. Backward exclusion is based on an opposite perspective to that of forward inclusion, and involves starting with all the predictive variables in the model, and then removing one variable at a time until a stopping condition is met. At each iteration, the variable whose removal causes the largest increase in the explanatory power of the model is selected for exclusion. In this case too, the applied criteria are analogous to those previously described for forward schemes.

Exhaustive search. Exhaustive search analyzes all subsets of predictors, thus potentially identifying the optimal model, so that it is applicable only when the number of explanatory variables is low, due to the exponential growth in the number of combinations.

8.5.1 Example of development of a regression model

Let us consider the *mtcars* dataset, described in Appendix B, to develop a multiple regression model whose purpose is to explain the fuel economy in automobiles, expressed (in miles per gallon) by the *mpg* response variable, by means of 10 features of a car, represented by the remaining attributes (*cyl, disp, hp, drat, wt, qsec, vs, am, gear, carb*), whose explanation is given in Appendix B.

Figure 8.9 shows the matrix of the scatter plots and allows us to evaluate the nature of the relationships between all pairs of variables. In particular, it highlights the existence of collinearity between some pairs of predictive attributes.

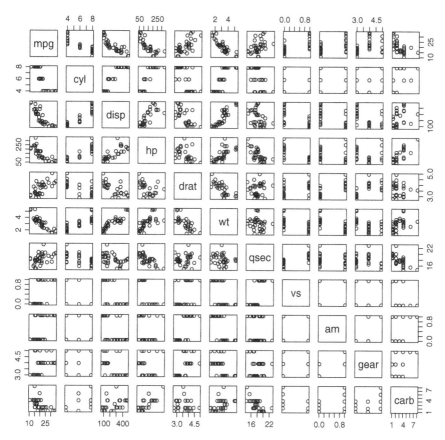

Figure 8.9   Matrix of scatter plots for the mtcars dataset

We will now briefly discuss four multiple regression models, A–D, described in Table 8.7, developed to explain the *mpg* response variable. Tables 8.8 and 8.9 provide the main summary statistics for the regression and the residuals, respectively, for the different models generated.

Table 8.7   Independent variables for the models developed for the *mtcars* dataset

model	dependent variable	independent variable
A	mpg	cyl, disp, hp, drat, wt, qsec, vs, am, gear, carb
B	mpg	(backward) wt, qsec, am
C	mpg	hp, wt, (hp/wt)
D	log(mpg)	log(hp), log(wt)

Table 8.8   Summary statistics for the models developed for the *mtcars* dataset

model	res. std error	mult. R-sq.	adj. R-sq	F-statistic	p-value
A	2.6500	0.8690	0.8066	13.93	< 0.0001
B	2.4590	0.8497	0.8336	52.75	< 0.0001
C	2.1530	0.8848	0.8724	71.66	< 0.0001
D	1.1112	0.8829	0.8748	109.30	< 0.0001

Table 8.9   Statistics of the residuals for the models developed for the *mtcars* dataset

model	minimum	lower quartile	median	upper quartile	maximum
A	−3.4506	−1.6044	−0.1196	1.2193	4.6271
B	−3.4811	−1.5555	−0.7257	1.4110	4.6610
C	−3.0632	−1.6491	−0.7362	1.4211	4.5513
D	0.8176	0.9235	1.0021	1.0820	1.2166

Table 8.10   Significance of the regression coefficients of model A for the *mtcars* dataset

predictor	value	std. error	t-value	$Pr > |t|$
(intercept)	12.30337	18.71788	0.657	0.5181
cyl	−0.11144	1.04502	−0.107	0.9161
disp	0.01334	0.01786	0.747	0.4635
hp	−0.02148	0.02177	−0.987	0.3350
drat	0.78711	1.63537	0.481	0.6353
wt	−3.71530	1.89441	−1.961	0.0633
qsec	0.82104	0.73084	1.123	0.2739
vs	0.31776	2.10451	0.151	0.8814
am	2.52023	2.05665	1.225	0.2340
gear	0.65541	1.49326	0.439	0.6652
carb	−0.19942	0.82875	−0.241	0.8122

Model A includes all the independent variables, and shows a lack of statistical significance, confirming what we have previously remarked about multicollinearity and attribute selection. Indeed, although Table 8.8 indicates a rather high value of the coefficient of determination and an F-value above the test threshold, the significance of the coefficients in model A is rather weak, as shown in Table 8.10. Only the coefficient of variable *wt*, which expresses the weight of the vehicle, can be considered significant. Table 8.11 on the analysis of variance for model A also shows a low significance compared to the majority of the predictors.

Table 8.11   Analysis of variance of model A for the *mtcars* dataset

predictor	df	sum of sq.	mean sq.	F-value	Pr > F
cyl	1	817.71	817.71	116.4245	< 0.0001
disp	1	37.59	37.59	5.3526	0.030911
hp	1	9.37	9.37	1.3342	0.261031
drat	1	16.47	16.47	2.3446	0.140644
wt	1	77.48	77.48	11.0309	0.003244
qsec	1	3.95	3.95	0.5623	0.461656
vs	1	0.13	0.13	0.0185	0.893173
am	1	14.47	14.47	2.0608	0.165858
gear	1	0.97	0.97	0.1384	0.713653
carb	1	0.41	0.41	0.0579	0.812179
residuals	21	147.49	7.02		
total	31	1126.04	985.57		

Table 8.12   Significance of the regression coefficients of model B for the *mtcars* dataset

predictor	value	std. error	t-value	Pr > |t|
(intercept)	9.6178	6.9596	1.382	0.1780
wt	−3.9165	0.7112	−5.507	< 0.0001
qsec	1.2259	0.2887	4.247	0.0002
am	2.9358	1.4109	2.081	0.0467

Table 8.13   Analysis of variance of model B for the *mtcars* dataset

predictor	df	sum of sq.	mean sq.	F-value	Pr > F
wt	1	847.73	847.73	140.2143	< 0.0001
qsec	1	82.86	82.86	13.7048	< 0.0001
am	1	26.18	26.18	4.3298	0.0467
residuals	28	169.29	6.05		
total	31	1126.06	962.82		

Model B has been obtained through a backward algorithm of automatic selection of the independent variables. As indicated in Table 8.7, the resulting model includes the three predictive variables *wt, qsec, am*, which express the weight, the time to travel 1/4 mile and the type of transmission. Table 8.8 shows an overall improvement in the summary indicators, with an increase in both the adjusted coefficient of determination and the total F-statistic. Also the significance of the regression coefficients and the analysis of variance, respectively shown in Tables 8.12 and 8.13, indicate in this case that the regression model is more robust and reliable.

Table 8.14   Significance of the regression coefficients of model C for the *mtcars* dataset

predictor	value	std. error	t-value	$Pr > |t|$
(intercept)	49.8084	3.60516	13.816	< 0.0001
hp	−0.12010	0.02470	−4.863	< 0.0001
wt	−8.21662	1.26971	−6.471	< 0.0001
hp/wt	0.02785	0.00742	3.753	< 0.0001

Table 8.15   Analysis of variance of model C for the *mtcars* dataset

predictor	df	sum of sq.	mean sq.	F-value	$Pr > F$
hp	1	678.37	678.37	146.380	< 0.0001
wt	1	252.63	252.63	54.512	< 0.0001
hp:wt	1	65.29	65.29	14.088	< 0.0001
residuals	28	129.76	4.63		
total	31	1126.05	1000.92		

As stated earlier, it is possible to obtain more accurate models through the inclusion of new variables by means of adequate transformations and the exclusion of some original variables. Model C is based on the predictors *hp* and *wt*, representing power and weight, and on the new variable expressed by the ratio *hp/wt*. As shown in Table 8.8, the adjusted coefficient of determination and the total F-statistic noticeably increase, while the standard deviation of the residuals decreases. The regression coefficients appear quite significant, as shown in Tables 8.14 and 8.15.

A further improvement is obtained by model D, which predicts the logarithm $\log(mpg)$ of the response variable using as predictors the new variables $\log(hp)$ and $\log(wt)$. As shown in Table 8.8, the adjusted coefficient of determination and the total F-statistic show a further increase, while the standard deviation of the residuals markedly decreases. Notice that the value of the standard error of the residuals is already corrected in the table, through the inverse transformation $\exp(\cdot)$, in order to bring back the scale of the errors to the scale of the response variable *mpg*. The original value of the deviation of the residuals is equal to 0.1054. Also the statistics for the quartiles of the residuals, which in turn had already been related to the *mpg* scale through the anti-transformation $\exp(\cdot)$, appear improved. The regression coefficients are significant, as shown in Tables 8.16 and 8.17.

The example described above confirms that the development of a regression model is a complex activity, particularly with respect to the selection of an optimal set of predictive variables, obtained through appropriate transformations of the original attributes. Automatic selection techniques may prove useful, as

Table 8.16   Significance of the regression coefficients of model D for the *mtcars* dataset

predictor	value	std. error	t-value	$Pr > |t|$
(intercept)	4.83469	0.22440	21.545	< 0.0001
log(hp)	−0.25532	0.05840	−4.372	< 0.0001
log(wt)	−0.56228	0.08742	−6.432	< 0.0001

Table 8.17   Analysis of variance of model D for the *mtcars* dataset

predictor	df	sum of sq.	mean sq.	F-value	$Pr > F$
log(hp)	1	1.96733	1.96733	177.178	< 0.0001
log(wt)	1	0.45939	0.45939	41.373	< 0.0001
residuals	29	0.32201	0.01110		
total	31	2.74873	2.43782		

in the case of model B, although in general experienced and insightful analysts may be able to obtain better results.

8.6   Notes and readings

There are many references on regression models. A basic bibliography includes some texts of a general nature, such as Kleinbaum et al. (1997), Ryan (1997), Montgomery et al. (2001) and Mendenhall and Sincich (2003). In particular, a useful reference for multiple regression is Allison (1998), while Scott Long (1997) discusses specific subjects, among them the treatment of categorical dependent variables, which will be considered at length in Chapter 10.

9

Time series

There are datasets in which the target attribute is time-dependent, since it is associated with a sequence of consecutive periods along the temporal dimension, and interest lies in the nature of such dependence. In such situations the values of the target variable are said to represent a *time series*.

The aim of models for time series analysis, described in this chapter, is to identify any regular pattern of observations relative to the past, with the purpose of making predictions for future periods. Time series analysis has many applications in business, financial, socio-economic, environmental and industrial domains. Depending on the specific application, predictions may refer to future sales of products and services, trends in economic and financial indicators, or sequences of measurements relative to ecosystems, for example.

We begin with a review of the basic concepts relating to time series, and introduce some examples. Then we describe the main statistical indicators for evaluating the accuracy of forecasting models. Next, we focus on decomposition methods, which are used to identify the basic components of time series. From there we move on to exponential smoothing models, the class of predictive methods considered among the most accurate based on empirical analyses, before considering the main characteristics of autoregressive models. Finally, we describe the methods based on combinations of predictive models, which are quite effective in practice, and we discuss the general criteria underlying the choice of a forecasting method.

9.1 Definition of time series

A *time series* is a sequence $\{y_t\}$ of values assumed by a quantity of interest that can be measured at specific time periods t. If the time indices t for which

Business Intelligence: Data Mining and Optimization for Decision Making   C. Vercellis
© 2009 John Wiley & Sons, Ltd

the measurements y_t are gathered belong to a discrete set $\mathcal{T}$, the time series $\{y_t\}$ is called a *discrete time series*; otherwise it is called a *continuous time series*. In discrete time series, to which our attention will be confined in this chapter, the periods consist of natural sequences along the time horizon and usually occur at uniform time intervals, such as minutes, hours, days, weeks, months, quarters or years.

Example 9.1 – Quarterly employment rate.   Figure 9.1 relates to the *employment* dataset described in Appendix B, and shows the percentage of employed individuals in the Italian population of youths of both sexes between 15 and 24 years of age, recorded every three months from the fourth quarter of 1992 to the fourth quarter of 2003.

Example 9.2 – Bimonthly electricity consumption.   Time series are commonly used to predict consumption in the utilities industry. The time series shown in Figure 9.2 shows the consumption of electricity in an Italian region expressed in tens of millions (10^8) of kilowatt-hours, recorded every two months from the first two-month period of 1998 to the sixth two-month period of 2003.

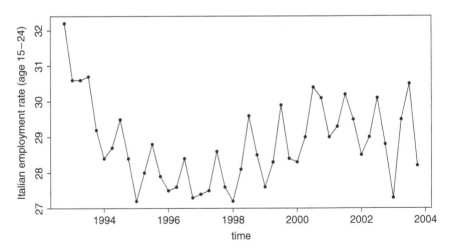

Figure 9.1   Time series of employment rate among the Italian youth population over 45 quarters

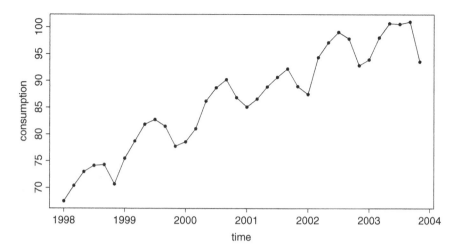

Figure 9.2   Time series of electricity consumption in an Italian region over 36 two-month periods

Example 9.3 – Monthly wine sales.   Figure 9.3 relates to the *wine* dataset described in Appendix B and shows sales of wine in Australia, expressed as numbers of bottles, recorded every month from January 1980 to August 1994.

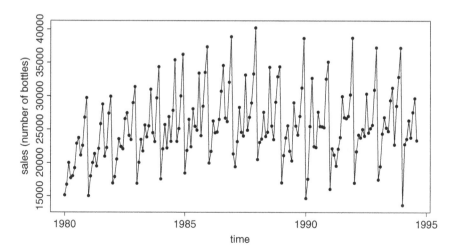

Figure 9.3   Time series of Australian wine sales over 176 one-month periods

To express the stochastic nature of the values taken by a time series, each observation y_t is assumed to represent a realization of a random variable Y_t. Consequently, a model of the time series $\{y_t\}$ would ideally involve assigning the joint probability distribution of the sequence of random variables $\{Y_t\}$, which represents a *stochastic process*. In most cases, only the expected value $E[Y_t]$ and the second-order moments $E[Y_{t+h}Y_t]$, $h \geq 1$, of the joint distribution are assigned. Throughout this chapter both $\{y_t\}$ and $\{Y_t\}$ will be used to denote a time series, the former referring to the sequence of observed values, and the latter to the probabilistic model of the time series.

Predictive models, also referred to as extrapolative models, analyze the observations of a time series $\{y_t\}$ to highlight regular patterns that have occurred in the past, in order to forecast the most likely behavior of the time series in the future. Let f_t be a prediction of the random value Y_t for period t of the time series $\{y_t\}$, supposing that t is the index of the current period and that the values $\{y_t, y_{t-1}, \ldots, y_{t-k+1}\}$ of the time series for k past periods are known. The general structure of a predictive model is

$$f_{t+1} = F(y_t, y_{t-1}, \ldots, y_{t-k+1}). \tag{9.1}$$

In order to develop a predictive model, it is first necessary to select the form of the function F which best represents the specific time series that one wishes to analyze. Having done so, one can then try to find values for the parameters that appear in the selected class of functions F. This procedure is similar to that described in Chapter 8 to determine the parameters of regression models.

In some situations one might wish to generate predictions for a given number of periods in the future and not just for the next period $t + 1$. Often it is necessary to predict the sales of a product for an entire year ahead. Forecasts made at time t for periods beyond $t + 1$ are based on the application of the model to the values that are known up to time t and to the predictions made by using the model itself for the subsequent periods, that is,

$$f_{t+h} = F(f_{t+h-1}, f_{t+h-2}, \ldots, f_{t+1}, y_t, y_{t-1}, \ldots, y_{t-k+1}). \tag{9.2}$$

Therefore, we can reasonably expect predictions to be less and less accurate as the prediction period $t + h$ gets further from t along the time horizon.

For the sake of simplicity, in what follows the time indices will correspond to the sequence of natural numbers $\{1, 2, 3, \ldots\}$, so that the first period of the time series is numbered 1 and so on.

9.1.1   Index numbers

Index numbers are dimensionless measures compared to a base value at a given moment which may prove useful for the analysis of a time series. Stock exchange listings in the financial industry and inflation indices in economics are examples of index numbers.

A *simple index number* I_t is defined as the ratio between the value of a single observation y_t of the time series at time t and its value y_0 at time t_0, multiplied by 100:

$$I_t = \frac{y_t}{y_0} \times 100. \tag{9.3}$$

The index number I_t therefore represents the percentage variation of the original time series compared to the value of a specific period taken as a base.

In some situations it is appropriate to define an index that describes the behavior of multiple time series simultaneously. For example, the FTSE 100 share index takes into account the prices of shares in 100 companies traded on the London Stock Exchange. Hence, when there are r time series $\{y_{1t}, y_{2t}, \ldots, y_{rt}\}$ a *composite index number* I_t^C will be used, which is defined as the ratio between the sum

$$y_t^C = y_{1t} + y_{2t} + \cdots + y_{rt}$$

of the values of the time series at time t and the corresponding sum

$$y_0^C = y_{10} + y_{20} + \cdots + y_{r0}$$

evaluated at time t_0, multiplied by 100:

$$I_t^C = \frac{y_t^C}{y_0^C} \times 100. \tag{9.4}$$

However, it may be the case that the time series are inhomogeneous and take values significantly different in magnitude. For instance, suppose that we wish to calculate an inflation rate indicator, taking into account several time series representing the prices of commodity goods. Certain goods, such as bread, are characterized by a low unit price and a high purchase frequency, while others, such as automobiles, typically have a high price and a very low purchase frequency. To offset such lack of homogeneity, it is appropriate to define a set of relative weights $\{w_1, w_2, \ldots, w_r\}$ for the various time series $\{y_{1t}, y_{2t}, \ldots, y_{rt}\}$ and to introduce a *composite weighted index* I_t^P, which is defined as the ratio between the weighted sum

$$y_t^P = w_1 y_{1t} + w_2 y_{2t} + \ldots + w_r y_{rt}$$

of the values of the time series at time t and the corresponding sum

$$y_0^P = w_1 y_{10} + w_2 y_{20} + \ldots + w_r y_{r0}$$

of the values at time t_0, multiplied by 100:

$$I_t^P = \frac{y_t^P}{y_0^P} \times 100. \tag{9.5}$$

9.2   Evaluating time series models

There are two main reasons for measuring the accuracy of predictions formulated using a time series model. First, during the development and identification stage of the model, accuracy measures are needed to compare alternative models with one another and to determine the value of the parameters that appear in the expression for the prediction function F. To identify the most accurate predictive model, each of the models considered is applied to past data, and the model with the minimum total error is selected. This procedure is similar to the minimization of the sum of squared errors SSE introduced in Chapter 8, the only difference being that for time series multiple criteria are considered.

Secondly, after a predictive model has been developed and used to generate predictions for future periods, it is necessary to periodically assess its accuracy, in order to detect any abnormality and inadequacy in the model that might arise at a later time. The evaluation of the accuracy of predictions at this stage makes it possible to determine if a model is still accurate or if a revision is required.

Given the observations y_t of a time series and the corresponding forecasts f_t for k past periods, the *prediction error* at time t, $t = 1, 2, \ldots, k$, is defined as

$$e_t = y_t - f_t,$$ (9.6)

and the *percentage prediction error* at time t as

$$e_t^P = \frac{y_t - f_t}{y_t} \times 100.$$ (9.7)

The percentage prediction error is independent of the scale on which the observations are measured and is therefore a more reliable measure, especially for comparing the errors generated in connection with different time series.

9.2.1   Distortion measures

Distortion indices are used to discriminate among predictive models based on signed mean errors. In particular, the *mean error* (ME) is defined as

$$\text{ME} = \frac{\sum_{t=1}^{k} e_t}{k} = \frac{\sum_{t=1}^{k} (y_t - f_t)}{k},$$ (9.8)

and the *mean percentage error* (MPE) as

$$\text{MPE} = \frac{\sum_{t=1}^{k} e_t^P}{k}.$$ (9.9)

Ideally one would choose among the alternatives the model that yields a mean error value closest to 0, and therefore a mean percentage error that in turn is closest to 0.

9.2.2   Dispersion measures

Sometimes distortion indices are insufficient to discriminate among candidate models. To see this, consider Figure 9.4, which shows the error density curves for three different predictive models, relating to predictions of the same time series. As we can see, models (a) and (c) may be preferred over model (b) as they indicate lower distortion. The mean error for (a) and (c) is zero, while that for model (b) equals 3. However, although their distortion is null, models (a) and (c) are not equivalent, since (a) generates errors that are smaller compared to (c), and is therefore preferable.

To discriminate among models with null distortion it is necessary to introduce *dispersion* indices, which are based on mean absolute errors. Hence, the *mean absolute deviation* (MAD) is defined as

$$\text{MAD} = \frac{\sum_{t=1}^{k} |e_t|}{k} = \frac{\sum_{t=1}^{k} |y_t - f_t|}{k}, \tag{9.10}$$

and the *mean absolute percentage error* (MAPE) as

$$\text{MAPE} = \frac{\sum_{t=1}^{k} |e_t^P|}{k}. \tag{9.11}$$

In addition to the previous indices, the most popular dispersion index should also be introduced, namely the *mean squared error* (MSE), given by

$$\text{MSE} = \frac{\sum_{t=1}^{k} e_t^2}{k} = \frac{\sum_{t=1}^{k} (y_t - f_t)^2}{k}, \tag{9.12}$$

which corresponds to the sum of squared errors SSE except for the factor k in the denominator. Unlike the mean absolute difference, which is a

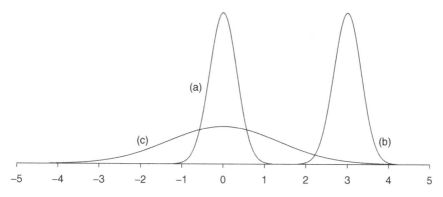

Figure 9.4   Distortion and dispersion for the errors density of three predictive models

non-differentiable function, the mean squared error is a quadratic function, therefore simplifying the structure of the minimization problem that must be solved to identify the parameters of the model. This is because optimization methods for quadratic problems are far more efficient than the corresponding solution methods for non-differentiable problems.

The mean squared error tends to amplify the effect of the largest errors, so that the *standard deviation of errors* (SDE) is introduced to obtain an index measured on the same scale as the original observations:

$$\text{SDE} = \sqrt{\frac{\sum_{t=1}^{k} e_t^2}{k}} = \sqrt{\frac{\sum_{t=1}^{k} (y_t - f_t)^2}{k}}. \qquad (9.13)$$

Out of all the models that show null distortion, or whose distortion is close to 0, the model with the least dispersion is usually preferred.

9.2.3   Tracking signal

Unlike the distortion and dispersion indicators, which can be used during both identification and control phases, the *tracking signal* TS_k is an index that is useful only in the monitoring stage of the forecasting process. It is defined as the ratio between the sum of signed errors and the sum of absolute errors:

$$TS_k = \frac{\sum_{t=1}^{k} e_t}{\sum_{t=1}^{k} |e_t|}. \qquad (9.14)$$

Consequently, it takes values in the interval $[-1, 1]$. If the tracking signal is close to -1 the model is distorted by excess, which means that it makes predictions that are mostly greater than the observed values, while if it is close to 1 it is distorted by defect. Values of TS_k close to 0 indicate a substantial lack of distortion.

Usually, during the monitoring of a forecasting model, a variant of the tracking signal at time k is considered, now in the interval $[0, 1]$, and given by

$$TS_k = \left| \frac{D_k}{G_k} \right|, \qquad (9.15)$$

where

$$D_k = \beta e_k + (1 - \beta) D_{k-1}, \qquad (9.16)$$

$$G_k = \beta |e_k| + (1 - \beta) G_{k-1}, \qquad (9.17)$$

where β, $0 < \beta < 1$, is a parameter that regulates the relative weight of the last realization of the error compared to the previous ones. Notice that D_k and

G_k represent approximations of the mean of the signed errors and the mean of the absolute errors, respectively. The recursive equalities (9.16) and (9.17) are a first example of the *exponential smoothing* update scheme, which will be extensively covered in Section 9.4.

Monitoring occurs by exception: the tracking signal, which ideally should be as close as possible to 0, is compared with a previously fixed threshold value $U \in [0, 1]$. If the condition $TS_k \leq U$ is violated, the abnormal occurrence is pointed out by some sort of alerting mechanism, and the model should be revised by using a new procedure for the identification of the parameters or, in the most critical situations, even by modifying the form of the function F itself.

The tracking signal may be computed periodically, each time a prediction is generated, and therefore is suited for the automation typical of monitoring procedures. This feature may prove useful when there are a large number of distinct forecasting models, required for instance to predict the demand for a company's entire range of products. In such circumstances, human inspection of the accuracy of each individual model would be cumbersome at best.

9.3   Analysis of the components of time series

Three main components are usually distinguished in time series: *trend*, *seasonality* and *random noise*.

Trend. A long-term *trend* component describes the average behavior of a time series over time, and it can be increasing, decreasing or stationary. In general, an effort is made to approximate the trend of a time series, denoted by M_t, using simple functions of linear, polynomial, exponential and logarithmic form.

Seasonality. The *seasonality* component, denoted by Q_t, is the result of wave-like short-term fluctuations of regular frequency that appear in the values of a time series, for example corresponding to days of the week, or to months or quarters of the year. These oscillations are usually persistent and are determined by the natural cycles by which demand develops, or by the seasonality of the products to which the time series refers.

Random noise. *Random noise* is a fluctuation component of a time series used to represent all irregular variations in the data that cannot be explained by the other components. In general, it is required that the time series $\{\varepsilon_t\}$, obtained from $\{y_t\}$ after the trend and the seasonality components have been identified and removed, represents a *white noise* process, indistinguishable from realizations of a sequence of independent random variables identically normally distributed with zero mean and constant variance.

Sometimes, a fourth component may be also considered, which reflects the wavelike oscillations that occur at irregular intervals throughout a time series, as a consequence of the *economic cycle*. The frequency of the economic cycle is usually a few years and therefore in short- and medium-term forecasts it is preferable to incorporate it into the trend component.

To identify and analyze the components of a time series one postulates that this latter can be expressed as a combination of its components

$$Y_t = g(M_t, Q_t, \varepsilon_t), \qquad (9.18)$$

where g represents an appropriate function to be selected. The most common forms of decomposition methods, reviewed in Section 9.3.2, postulate the existence of an additive or multiplicative function g expressing the relationship among the components of a time series. However, there are models that use more complex functions g, obtained for example as combinations of additive and multiplicative components, as in the case of exponential smoothing models, described in Section 9.4.

9.3.1 Moving average

The *moving average* $m_t(h)$ with parameter h at time t is defined as the arithmetic mean of h consecutive observations of the time series $\{y_t\}$, such that the time index t belongs to the indices of the h averaged observations. It is possible to compute different values of the moving average, depending on the position taken by the index t in the sequence of h contiguous observations used to calculate the average.

In particular, a *centered moving average* is the arithmetic mean of h observations such that t is the middle point of the set of periods corresponding to the observations, assuming that h is odd:

$$m_t(h) = \frac{y_{t+(h-1)/2} + y_{t+(h-1)/2-1} + \cdots + y_{t-(h-1)/2}}{h}. \qquad (9.19)$$

If h is even, a slightly more complex expression is used,

$$\begin{aligned} m_t(h) = &\frac{y_{t+h/2} + y_{t+h/2-1} + \cdots + y_{t-h/2+1}}{2h} \\ &+ \frac{y_{t+h/2-1} + y_{t+h/2-2} + \cdots + y_{t-h/2}}{2h}, \end{aligned} \qquad (9.20)$$

corresponding to a two-stage procedure. In the first stage, two moving averages virtually centered on the intermediate periods $t - 1/2$ and $t + 1/2$ are computed, and then the average centered on t of the two previous averages is evaluated.

The moving average can be used to identify a long-term trend component, by removing from the time series the random fluctuations and possibly the seasonality component. Suppose that a seasonality of length L can be identified in the time series, i.e. that there are L periods in each cycle; for example, four quarters or twelve months per year. In this case, the choice $h = L$ is made for the parameter, so that in the computation of the moving average the effects of seasonality are compensated, since for each term the average takes a whole cycle into account. The arithmetic mean operation also mitigates the effects of random fluctuations, and therefore the moving average reveals only the trend component.

Figure 9.5 shows the centered moving average with parameter $h = 6$ for the time series in Example 9.2. The moving average evidently mitigates the fluctuations of the time series due to seasonality and random components, thus highlighting the long-term trend, which looks slightly sublinear.

It may be useful to introduce a weighted variant of the moving average, whether centered or not, by associating a set of weights with the h observations of the time series that are taken into account to calculate the arithmetic mean, so as not to assign equal importance to the various periods.

The moving average can also be used to generate future predictions, by making period t correspond to the last of the h observations and by setting

$$f_{t+1} = \frac{y_t + y_{t-1} + \cdots + y_{t-h+1}}{h}, \qquad (9.21)$$

or, for a weighted moving average with weights w_i, $i = 1, 2, \ldots, h$,

$$f_{t+1} = \frac{w_1 y_t + w_2 y_{t-1} + \cdots + w_h y_{t-h+1}}{\sum_{i=1}^{h} w_i}. \qquad (9.22)$$

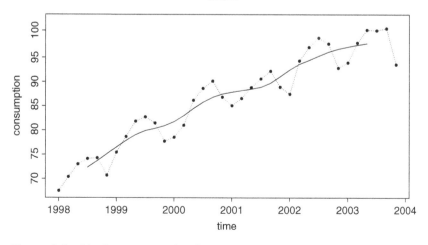

Figure 9.5   Moving average for the time series of electricity consumption

It can easily be verified that the following relation holds true:

$$f_{t+1} = f_t + \frac{y_t}{h} - \frac{y_{t-h}}{h}. \tag{9.23}$$

This means that the prediction for period $t + 1$ is equal to the prediction for period t with the addition of a corrective term, equal to $1/h$ times the difference between the most recent observation y_t and the observation y_{t-h}, dropped in the computation of the moving average for period t. Using the moving average as a predictive model, it is possible to obtain a level of accuracy that is generally lower than that achieved by using other simple methods, so that it is seldom adopted for practical purposes. However, the moving average provides a first example of the function F that expresses the dependence of the prediction f_{t+1} on past observations. In this case the model uses only the last h observations, even if a number $k > h$ of known past observations of the time series are available.

9.3.2 Decomposition of a time series

The *decomposition* of a time series involves identifying the three components of trend, seasonality and random fluctuation. The purpose of this is primarily to analyze and understand the structure of the time series, although it also allows the analyst to forecast future values.

To decompose a time series, it is first necessary to postulate the form of the function g that expresses the dependence of the time series on its components. A *multiplicative* model is usually assumed, such as

$$Y_t = M_t \times Q_t \times \varepsilon_t. \tag{9.24}$$

Alternatively, an *additive* model,

$$Y_t = M_t + Q_t + \varepsilon_t, \tag{9.25}$$

may be considered, although it is possible to formulate more complex relations. For instance, the Winters model, described in Section 9.4.3, postulates a mixed structure, additive in the trend component and multiplicative in the seasonality component.

We will now describe the steps that lead to the decomposition of a time series for a multiplicative model, with reference to Example 9.2.

Removal of the trend component

The first step is to remove the trend component from the time series to obtain a transformed time series that is *stationary*. This means that the transformed time series has no trend. The moving average centered on $h = L$ is then computed to remove the seasonality and the random fluctuation components, as shown in Figure 9.5. Hence, the approximate relation $M_t \approx m_t(L)$ holds true.

Subsequently, the product of the two components of seasonality and random fluctuation is identified by computing the time series $\{B_t\}$ given by

$$B_t = Q_t \varepsilon_t = \frac{Y_t}{M_t} \approx \frac{Y_t}{m_t(L)}. \tag{9.26}$$

As can be seen, by evaluating the ratio between the values of the time series and the moving average, which is known to approximate the trend, we have removed the trend component from $\{B_t\}$. Figure 9.6 shows the values of $Q_t \varepsilon_t$ obtained from (9.26) and confirms the lack of trend in the values of the new time series $\{B_t\}$ which oscillate around 1 showing a periodicity with parameter $L = 6$.

As an alternative to the method based on the computation of the moving average, it is possible to remove the trend component in a less intuitive although more effective way, by taking successive differences between adjacent values of the time series, as in

$$D_t = Y_t - Y_{t-1}, \quad t \geq 2. \tag{9.27}$$

Figure 9.7 shows the new time series $\{D_t\}$ obtained through differencing, together with the corresponding regression line having a null slope. If necessary, removal of the trend can be strengthened by taking differences of higher order, computing for example the differences between successive values of $\{D_t\}$.

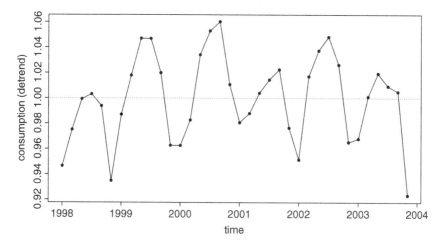

Figure 9.6   Removal of the multiplicative trend from the time series of electricity consumption

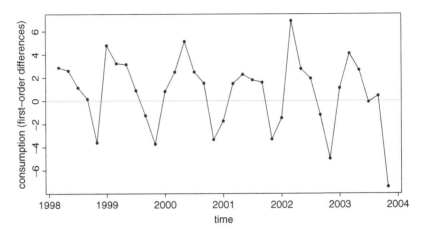

Figure 9.7   Removal of the trend from the time series of electricity consumption by differencing

Identification of the seasonality component

Before identifying the seasonality component, it is necessary to establish whether the time series actually shows a periodic regularity. Visual inspection of the graph of a time series is usually a good starting point for discovering its features and components. Furthermore, the value L of a hypothetical seasonality is often suggested by the specific nature of the data. For example, a series of monthly observations can be usually related to the value $L = 12$, just as daily values can be related to the values $L = 7$ or $L = 365$. There are, however, more rigorous investigation techniques for highlighting seasonal regularities and evaluating the length L of each cycle, based on the autocorrelation function (ACF) and the partial autocorrelation function (PACF), which will be described in the context of autoregressive methods in Section 9.5.

Once the actual existence of a seasonality component of length L has been verified, it is possible to calculate the seasonal indices Q_l, $l = 1, 2, \ldots, L$, obtained as the mean of $Q_t \varepsilon_t$ for the periods homologous to l, so that the effect of random fluctuations is removed. More specifically, let z_l be the set of indices of the periods homologous to l. If the seasonality corresponds, for instance, to the months of the year, the periods homologous to the month of January are represented by all January months included within the time horizon. For the time series of Example 9.2 all first two-month periods of the year, all second two-month periods and so on, are homologous to each other. We can therefore define

$$Q_l = \frac{\sum_{t \in z_l} Q_t \varepsilon_t}{|z_l|}. \tag{9.28}$$

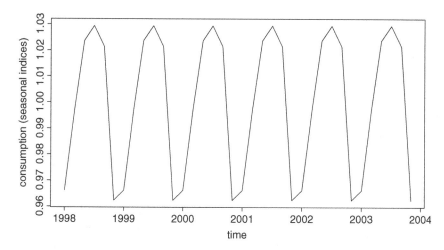

Figure 9.8   Seasonal indices for the time series of electricity consumption

Figure 9.8 shows the values of the seasonal indices, repeated for all six annual cycles included in the series of Example 9.2.

At this point, it is possible to *deseasonalize* the time series, by removing the seasonality component from each observation by computing the ratio between the original values and the corresponding seasonal index, that is,

$$\frac{y_t}{Q_{l(t)}} = M_t \varepsilon_t, \qquad (9.29)$$

where $l(t)$ indicates the type of period corresponding to t. Figure 9.9 shows the values of the deseasonalized time series together with the linear trend component.

Identification of the trend component

To isolate the trend component, we first need to identify a regression curve (linear, quadratic, exponential, logarithmic) that explains the values of the time series as a function of the time period, which plays the role of predictor variable. If a linear relationship is postulated for Example 9.2, the trend line, shown as a dotted line in Figure 9.9, can be calculated via a linear regression model with respect to the time dimension,

$$M_t = a + bt. \qquad (9.30)$$

The time series decomposition is now complete and the three multiplicative components have been isolated. Figure 9.10 shows the time series and the corresponding three components obtained through multiplicative decomposition.

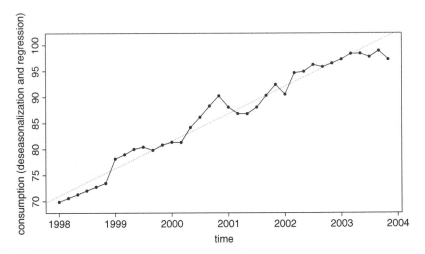

Figure 9.9   Deseasonalization and regression for the time series of electricity consumption

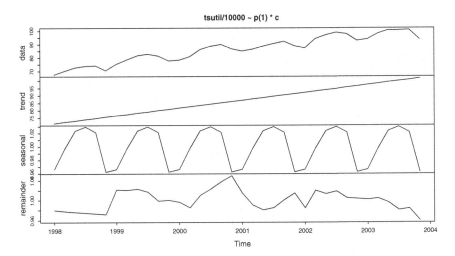

Figure 9.10   Multiplicative decomposition for the time series of electricity consumption

It is finally possible to derive predictions for the future periods based on the decomposition developed. To formulate the forecasts relative to the next L periods, it is first necessary to calculate the projection of the trend component in the future for an entire cycle, and then seasonalize the prediction through the seasonal indices, as in

$$f_{t+i} = M_{t+i} Q_{l(t+i)}, \quad i = 1, 2, \ldots, L. \tag{9.31}$$

9.4   Exponential smoothing models

Despite their simplicity, *exponential smoothing* models are among the most versatile and accurate predictive methods for time series analysis. They have proven particularly effective in the forecasting of economic phenomena. Although these models were originally formulated on an empirical and intuitive basis, recent advances in research have put them on a more sound theoretical footing. In this section we will review smoothing models of increasing complexity, aimed at identifying the different components of a time series, even in a self-correcting and adaptive form.

9.4.1   Simple exponential smoothing

The *simple exponential smoothing* model, also called the *Brown* model, relies on an estimator s_t of the average of the observations of the time series up until time t, called the *smoothed mean* and defined through the recursive expressions

$$s_t = \alpha y_t + (1 - \alpha) s_{t-1},$$
$$s_1 = y_1, \tag{9.32}$$

where $\alpha \in [0, 1]$ is a parameter that regulates the importance of the most recent value y_t with respect to the smoothed mean of the previous values. The prediction for period $t + 1$ is obtained by setting

$$f_{t+1} = s_t. \tag{9.33}$$

It can easily be verified that the following equalities apply for the exponential smoothing method:

$$s_1 = y_1,$$
$$s_2 = \alpha y_2 + (1 - \alpha)s_1,$$
$$s_3 = \alpha y_3 + (1 - \alpha)s_2, \tag{9.34}$$
$$\cdots,$$
$$s_t = \alpha y_t + (1 - \alpha)s_{t-1}.$$

By recursively applying equalities (9.34), one can derive the following relationship between the forecast for the next period and the observations in the past:

$$f_{t+1} = \alpha[y_t + (1 - \alpha)y_{t-1} + \cdots + (1 - \alpha)^{t-2}y_2] + (1 - \alpha)^{t-1}y_1. \tag{9.35}$$

Equation (9.35) indicates that for the exponential smoothing model the function F expresses the prediction as a linear convex combination of the observations of the time series, with weights that exponentially decrease as we go back in time. Equation (9.35) also allows us to derive an interpretation for the parameter α: if $\alpha \simeq 0$ the model has a greater inertia, in the sense that it assigns an almost constant weight to all past observations; if $\alpha \simeq 1$ the model is more responsive, as it assigns a larger weight to the most recent observations. The parameter α is chosen in such a way as to minimize the mean squared error.

Figures 9.11 and 9.12 show the values of the predictions computed for the exponential smoothing model applied to the time series of Example 9.2, corresponding to the values $\alpha = 0.5$ and $\alpha = 1.0$. The sums of squared errors are respectively SSE = 46.33 and SSE = 33.66, indicating that the second model is more accurate than the first. Notice that no prediction can be made for the first period, since the corresponding observation has been used to initialize the value of the smoothed mean.

9.4.2 Exponential smoothing with trend adjustment

Faced with a time series that includes a trend component, the simple exponential smoothing model inevitably fails to catch the trend, as confirmed by Figures 9.11 and 9.12. Consequently, when applied to a non-stationary time series with a trend component, the simple model always lags behind the actual observations and generates predictions distorted by excess or by defect. It is, however, possible to extend the simple model to incorporate a trend component, by means of the *Holt exponential smoothing* model.

To this end, besides the smoothed mean s_t, a linear *smoothed trend* m_t is also defined, which is an estimate of the trend additive component M_t. It is also possible to define more complex trend adjustments, for example by using quadratic or exponential functions. The recursive expression that defines the

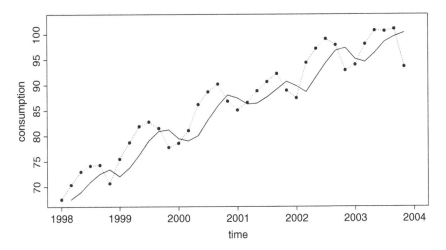

Figure 9.11   Simple exponential smoothing model with parameter $\alpha = 0.5$ for the time series of electricity consumption

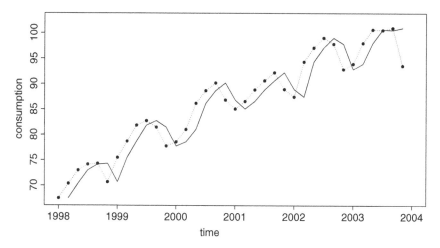

Figure 9.12   Simple exponential smoothing model with parameter $\alpha = 1.0$ for the time series of electricity consumption

smoothed mean must be adjusted to take into account the smoothed trend

$$s_t = \alpha y_t + (1 - \alpha)(s_{t-1} + m_{t-1}). \tag{9.36}$$

A corresponding recursive expression regulates the update of the smoothed trend

$$m_t = \beta(s_t - s_{t-1}) + (1 - \beta)m_{t-1}, \tag{9.37}$$

where $\beta \in [0, 1]$ is a parameter that modulates the importance of the most recent value of the trend, expressed as the difference between the smoothed means $(s_t - s_{t-1})$, with respect to the smoothed trend of previous periods, expressed by the iterated value m_{t-1}. If $\beta \simeq 0$ the model assigns an almost equal weight to trends exhibited in the past, while if $\beta \simeq 1$ the most recently exhibited trend is predominant. Observe that the difference between the smoothed means $(s_t - s_{t-1})$ is an estimate of the most recent trend that turns out to be much more robust than the difference $(y_t - y_{t-1})$ between the corresponding values of the time series, since this latter might be affected by seasonality and random fluctuations.

The prediction for period $t + 1$ is finally obtained by setting

$$f_{t+1} = s_t + m_t. \tag{9.38}$$

Here too the choice of the parameters α and β should be made in such a way as to minimize the sum of squared errors.

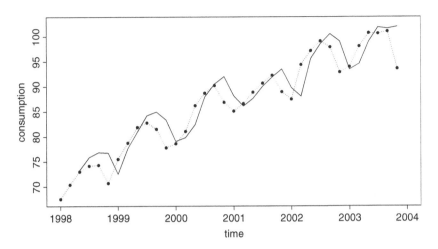

Figure 9.13   Holt exponential smoothing model with parameters $\alpha = 1.0$ and $\beta = 0.099$ for the time series of electricity consumption

Figure 9.13 shows the values of the predictions computed for the Holt exponential smoothing model applied to the time series of Example 9.2, with the parameters $\alpha = 1.0$ and $\beta = 0.099$. The sum of squared errors is equal to $SSE = 35.63$, and therefore the simple model with $\alpha = 1.0$ is preferable. Notice that there are no predictions for the first two periods since the corresponding observations have been used to initialize the value of the smoothed trend.

9.4.3   Exponential smoothing with trend and seasonality

If the time series includes also a seasonality component, then the exponential smoothing model needs to be further extended. In addition to the smoothed mean and the trend, the *Winters* model defines a *smoothed seasonal index* q_t, to approximate the multiplicative seasonality component Q_t.

Assuming that each seasonal cycle is composed of L periods, the updates of the smoothed mean and the trend are modified, and a recursive update is also introduced for the smoothed seasonal index q_t, according to

$$s_t = \alpha \frac{y_t}{q_{t-L}} + (1 - \alpha)(s_{t-1} + m_{t-1}), \tag{9.39}$$

$$m_t = \beta(s_t - s_{t-1}) + (1 - \beta)m_{t-1}, \tag{9.40}$$

$$q_t = \gamma \frac{y_t}{s_t} + (1 - \gamma)q_{t-L}. \tag{9.41}$$

Here $\gamma \in [0, 1]$ is a parameter that modulates the importance of the most recent value of the seasonality, expressed as the ratio between the most recent value y_t of the time series and the smoothed mean s_t, with respect to the smoothed seasonal index q_{t-L} of the previous period homologous to t. As for α and β, if $\gamma \simeq 0$ the model assigns an almost constant weight to the seasonality that occurred in the past, while if $\gamma \simeq 1$ the most recent seasonality is predominant. Notice that in recursion (9.39) the seasonal index q_{t-L} must be used, relative to the last past period homologous to t, because when computing s_t the smoothed index q_t has not been updated yet. Recursion (9.41) which defines q_t requires in fact that the new value of s_t be previously computed.

Finally, the prediction for period $t + 1$ is defined as

$$f_{t+1} = (s_t + m_t)q_{t-L+1}. \tag{9.42}$$

The choice of the parameters α, β and γ should be made in such a way as to minimize the sum of squared errors.

Figure 9.14 shows the values of the predictions computed for the Winters exponential smoothing model applied to the time series of Example 9.2, with parameters $\alpha = 0.8256$, $\beta = 0.0167$ and $\gamma = 1.0$. The sum of squared errors is equal to SSE = 12.03, which represents a substantial increase in accuracy with respect to previous models. The figure also shows the predictions for two future years, that is, for 12 periods.

9.4.4 Simple adaptive exponential smoothing

A further extension of the exponential smoothing model may be obtained by dynamic variation of the parameters of the model with respect to time t using

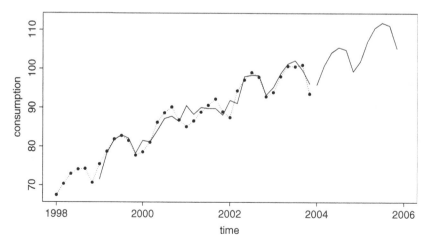

Figure 9.14 Winters exponential smoothing model with parameters $\alpha = 0.8256$, $\beta = 0.0167$ and $\gamma = 1.0$ for the time series of electricity consumption

adaptive update formulas. We will restrict our attention here to the *simple adaptive* smoothing model, and observe that it is also possible to define adaptive models that include trend and seasonality components.

The parameter α should be made dependent on the period t, and denoted as α_t, through the recursive equalities

$$s_t = \alpha_t y_t + (1 - \alpha_t) s_{t-1}, \tag{9.43}$$

$$\alpha_t = \left| \frac{d_t}{g_t} \right|, \tag{9.44}$$

$$d_t = \beta e_t + (1 - \beta) d_{t-1}, \tag{9.45}$$

$$g_t = \beta |e_t| + (1 - \beta) g_{t-1}, \tag{9.46}$$

where $\beta \in [0, 1]$. The update of s_t follows a recursive formula similar to (9.32), described for the simple smoothing model, while the update of α_t is based on the scheme already introduced for the tracking signal. From an intuitive viewpoint, this means that if the model is essentially undistorted the corresponding value of α_t is close to 0, whereas in the opposite case it increases and gets closer to 1.

The forecast for period $t + 1$ can be obtained as

$$f_{t+1} = s_t. \tag{9.47}$$

The choice of the parameters α_0 and β should be based on the minimization of dispersion measures.

9.4.5 Exponential smoothing with damped trend

The empirical evidence suggests that the trend component in a time series often reduces as the time passes. The sales of a new product may experience a period of growth after its introduction in the market, but in general the growing trend reduces until it reaches a peak in the product life cycle. This phenomenon should be taken into account in time series analysis, particularly in those situations where it is necessary to generate predictions over the medium to long term, that is, on a time horizon that stretches over one or two full cycles into the future. Exponential smoothing models with *damped trend* may offer a way to do this. The model presented here only includes the trend component, although it can also be extended to incorporate the seasonality component, if this is present in the time series.

The model consists of the automatic reduction of the trend component projected into the future through a parameter $\phi \in [0, 1]$ that adjusts the recursive equalities for updating the smoothed mean and smoothed trend

$$s_t = \alpha y_t + (1 - \alpha)(s_{t-1} + \phi m_{t-1}), \tag{9.48}$$

$$m_t = \beta(s_t - s_{t-1}) + (1 - \beta)\phi m_{t-1}. \tag{9.49}$$

The forecast for period $t + h$ is computed as

$$f_{t+h} = s_t + m_t \sum_{i=1}^{h} \phi^i. \tag{9.50}$$

In this case too the choice of the parameters ϕ, α and β should be based on the minimization of the sum of squared errors.

9.4.6   Initial values for exponential smoothing models

The application of exponential smoothing models requires that initial values be assigned for the recursive equalities to update the smoothed mean, trend and seasonal indices. If the time series analyzed extends over a large number of past periods, the observations of the first cycle should be used for the initial estimate of the parameters.

More precisely, in the simple exponential smoothing model it is enough to use the first observation for the initial estimate $s_1 = y_1$. In the Holt model, the first two observations provide the initial estimate of the trend m_1. In the Winters model, the seasonal indices for the entire first cycle may be estimated through the ratio between each of the values of the first cycle and the mean of the values in the cycle itself.

If the time series consists of a small number of cycles, it can be extended into the past by means of a technique called *backforecasting*. A predictive model is then applied to the time series considered over an inverse time horizon, that is from period t to period 1, and a prediction is generated for one or more cycles in the past, that is, for cycles preceding period 1. At the end of the procedure, a time series extending into the past is obtained, even though in a virtual fashion, with a profile that appears consistent with the observations actually recorded for the periods 1 through t.

9.4.7   Removal of trend and seasonality

Given a generic non-stationary time series $\{Y_t\}$, several techniques can be applied to obtain a transformed stationary time series $\{B_t\}$, that is, a series where the distribution of $\{B_t, B_{t+1} \ldots, B_{t+h}\}$ is the same as that of $\{B_{t+k}, B_{t+k+1} \ldots, B_{t+k+h}\}$ for each possible choice of t, k and h. In this hypothesis, the time series $\{B_t\}$ has a null trend component, represented by a horizontal line.

In Section 9.3, three methods for the removal of the trend component were described, which we briefly recap.

Moving average. On one hand, it is possible to divide the values of the time series $\{Y_t\}$ by the corresponding values of the moving average to derive a time series $\{B_t\}$ with no trend.

Trend. On the other hand, it is possible to identify the trend component M_t of the time series through additive or multiplicative decomposition, and therefore to remove it by subtraction or division, respectively.

Differencing. Finally, differences between adjacent values of the time series can be found, setting

$$B_t = \nabla Y_t = Y_t - Y_{t-1},$$
$$B_t(2) = \nabla^2 Y_t = Y_t - Y_{t-2},$$
$$\cdots,$$
$$B_t(h) = \nabla^h Y_t = Y_t - Y_{t-h}. \tag{9.51}$$

Differences of the first or at most of the second order are usually adequate to remove the trend component.

It is also possible to remove the seasonality component from a time series $\{Y_t\}$, that is to deseasonalize it. To this end, one has to first identify the additive or multiplicative seasonality component by decomposing the time series, and then subtract or divide it by the seasonal indices, respectively, to obtain a new time series with no seasonality.

As a consequence, given a time series $\{Y_t\}$ with both trend and seasonality components, there are at least three alternative methods to develop a forecasting model:

- applying the Winters model to the original time series;

- applying the Holt model to the time series after the data have been deseasonalized;

- applying the simple model to the time series obtained by deseasonalizing and removing the trend component.

Empirical analyses have shown that, in general, it is not possible to establish a priori which method is best with respect to the dispersion measures.

9.5 Autoregressive models

Autoregressive methods are based on the idea of identifying possible relationships between the observations of a time series at different periods by studying the autocorrelation between observations separated by a fixed time interval, called the *time lag*.

After a time lag h has been chosen, the new time series composed of lagged observations is defined via the *backshift* operator B:

$$B^h Y_t = Y_{t-h}, \quad t > h. \tag{9.52}$$

The correlation between the variables Y_t and $B^h Y_t$ is then analyzed. If Y_t has a seasonality component with period L, it is reasonable to expect the time series Y_t and $B^L Y_t = Y_{t-L}$ to be highly correlated.

In general, to develop autoregressive models the assumption is made that the time series is stationary. This means that its trajectory remains stable around a constant mean value. To obtain a stationary time series, the techniques described in Section 9.4.7 may be applied. In particular, to remove non-stationarity in the mean one may apply the differencing operator of order 1 between successive terms of the original time series:

$$\nabla Y_t = \nabla^1 Y_t = Y_t - Y_{t-1}, \quad t > 1. \tag{9.53}$$

It should be clear that the differencing operator ∇ and the backshift operator B are related to each other by the expression

$$\nabla = 1 - B.$$

Sometimes one may wish to consider differencing of higher order h, obtained from the original series by subtraction between terms separated by a time lag h:

$$\nabla^h Y_t = Y_t - Y_{t-h}, \quad t > h. \tag{9.54}$$

For example, differencing of order L may be useful to partially remove a seasonal component of cycle length L from a given time series.

Autoregressive models are more flexible and general than exponential smoothing models. However, empirical analyses conducted on a large number of benchmark time series, arising in different domains and characterized by different profiles, have shown that the greater effort required to develop and identify autoregressive models is not always rewarded by an improvement in prediction accuracy over simpler methods, particularly when dealing with dynamic time series, such as those arising in connection with economic phenomena.

The purpose of an *autoregressive* model AR(p) of order p is to establish a linear regression between the original time series and the time series obtained through the backshift operator up until order p, and takes the general form

$$Y_t = \gamma + \phi_1 Y_{t-1} + \phi_2 Y_{t-2} + \cdots + \phi_p Y_{t-p} + \varepsilon_t. \tag{9.55}$$

The parameters $\gamma, \phi_1, \phi_2, \ldots, \phi_p$ should be determined so as to minimize the sum of squared errors. The term ε_t is a random variable, referred to as *noise*, which represents random fluctuation and must follow a normal distribution with zero mean. The prediction for period t is then formulated as

$$f_{t+1} = \gamma + \phi_1 Y_t + \phi_2 Y_{t-1} + \cdots + \phi_p Y_{t-p+1}. \tag{9.56}$$

9.5.1   Moving average models

The aim of the *moving average* model MA(q) of order q is to establish a linear regression between the original time series and the time series composed by the prediction errors in the q previous periods; it takes the general form

$$Y_t = \gamma + \varepsilon_t - \vartheta_1 E_{t-1} - \vartheta_2 E_{t-2} - \cdots - \vartheta_q E_{t-q}, \qquad (9.57)$$

where the terms $E_{t-1}, E_{t-2}, \ldots, E_{t-q}$ represent the prediction errors in the q past periods. The parameters $\gamma, \vartheta_1, \vartheta_2, \ldots, \vartheta_q$ should be determined so as to minimize the sum of squared errors. The term ε_t represents the *noise*, which here too must follow a normal distribution with zero mean.

The prediction for period t is then formulated as

$$f_{t+1} = \gamma - \vartheta_1 E_t - \vartheta_2 E_{t-1} - \cdots - \vartheta_q E_{t-q+1}. \qquad (9.58)$$

9.5.2   Autoregressive moving average models

Sometimes a better approximation of the time series $\{Y_t\}$ can be achieved by using a hybrid forecasting model, containing both autoregressive and moving average terms. An *autoregressive moving average* model ARMA(p, q) of orders p and q takes the general form

$$\begin{aligned} Y_t = \gamma + \varepsilon_t + \phi_1 Y_{t-1} + \phi_2 Y_{t-2} + \cdots + \phi_p Y_{t-p} \\ - \vartheta_1 E_{t-1} - \vartheta_2 E_{t-2} - \cdots - \vartheta_q E_{t-q}. \end{aligned} \qquad (9.59)$$

The parameters $\gamma, \phi_1, \phi_2, \ldots, \phi_p, \vartheta_1, \vartheta_2, \ldots, \vartheta_q$ should be determined so as to minimize the sum of squared errors.

The prediction for period t is then formulated as

$$\begin{aligned} f_{t+1} = \gamma + \phi_1 Y_t + \phi_2 Y_{t-1} + \cdots + \phi_p Y_{t-p+1} \\ - \vartheta_1 E_t - \vartheta_2 E_{t-1} - \cdots - \vartheta_q E_{t-q+1}. \end{aligned} \qquad (9.60)$$

9.5.3   Autoregressive integrated moving average models

As noticed before, when the time series $\{Y_t\}$ is non-stationary it is possible to apply the ARMA(p, q) model to the time series $\{A_t\}$ obtained through differences of order d of the original time series, that is, by setting $A_t = \nabla^d Y_t$. The corresponding model is termed *autoregressive integrated moving average* and denoted by ARIMA(p, d, q):

$$\begin{aligned} A_t = \gamma + \varepsilon_t + \phi_1 A_{t-1} + \phi_2 A_{t-2} + \cdots + \phi_p A_{t-p} \\ - \vartheta_1 z_{t-1} - \vartheta_2 z_{t-2} - \cdots - \vartheta_q z_{t-q}, \end{aligned} \qquad (9.61)$$

where the terms $z_{t-1}, z_{t-2}, \ldots, z_{t-q}$ represent the prediction errors for the stationary time series $\{A_t\}$. Here again the search for the optimal parameters $\gamma, \phi_1, \phi_2, \ldots, \phi_p, \vartheta_1, \vartheta_2, \ldots, \vartheta_q$ is performed by minimizing the sum of squared errors.

The prediction for period t is then formulated as

$$
\begin{aligned}
f_{t+1} = {} & \gamma + \phi_1 A_t + \phi_2 A_{t-1} + \cdots + \phi_p A_{t-p+1} \\
& - \vartheta_1 z_t - \vartheta_2 z_{t-1} - \cdots - \vartheta_q z_{t-q+1}.
\end{aligned}
\tag{9.62}
$$

If the time series includes a component of seasonality, then it is often also appropriate to develop an ARIMA model for it. This model is usually denoted by ARIMA(P, D, Q) to distinguish its order from the corresponding order (p, d, q) of the non-seasonal component. Specifically, a model analogous to (9.61) is introduced, where the time series $\{A_t\}$ is obtained by successive differences of order DL, that is, using multiples of the length of each seasonality cycle.

9.5.4   Identification of autoregressive models

In order to develop an autoregressive model it is first necessary to select the type of model and its corresponding order (p, d, q), as well as the order (P, D, Q) for the seasonality component. Then, the parameters of the model chosen in the first stage should be determined by minimizing the sum of squared errors.

To choose the class of models to be used and to highlight the existence of a seasonality component, *autocorrelation* (ACF) and *partial autocorrelation* (PACF) diagrams should be used. The autocorrelation coefficient for lag h is defined as

$$
\text{ACF}_h = \text{corr}(Y_t, Y_{t-h}),
\tag{9.63}
$$

and indicates the degree of correlation between the values of the time series $\{Y_t\}$ and the values of the series $\{Y_{t-h}\}$, obtained by a backshift of order h.

The partial autocorrelation coefficient expresses the correlation between $\{Y_t\}$ and $\{Y_{t-h}\}$ that is not accounted for by shorter lags 1 through $h - 1$, and is defined as

$$
\text{PACF}_h = \text{corr}(Y_t, Y_{t-h} | Y_{t-1}, Y_{t-2}, \ldots, Y_{t-h+1}).
\tag{9.64}
$$

Different approximations have been proposed to calculate PACF_h in practice; see the references given at the end of chapter.

Procedure 9.1 lists the steps to be followed in order to identify the functional form and the parameters of an ARMA or an ARIMA model for approximating a given time series.

Procedure 9.1 – Identification of an appropriate autoregressive model

1. Obtain a stationary time series by removing the trend component. We will assume now that the time series $\{Y_t\}$ has no trend.

2. Examine the autocorrelation ACF and partial autocorrelation PACF diagrams.

3. If all coefficients are negligible, conclude that the series is composed of white noise and abandon the model.

4. If the ACF diagram slowly decreases while the PACF diagram shows a step decrease at lag p, then choose an AR(p) model.

5. If the ACF diagram shows a step decrease at lag q while PACF slowly decreases, then choose an MA(q) model.

6. If both diagrams slowly decrease, then choose an ARMA(p, q) model.

7. Repeat the same procedure only for those observations that correspond to the length L of the cycle, in order to determine the nature and order of any seasonality component.

By way of example, the method described will be applied to the time series of Example 9.2. Figure 9.15 shows the autocorrelation and partial autocorrelation diagrams for the original time series of electricity consumption. It should be borne in mind that the lags are expressed on the horizontal axis as fractions of the cycle and that the first vertical bar corresponds to a null lag. The autocorrelation function slowly decreases with the lag, due to the trend component which prevails in determining the autocorrelation effect. Actually, we first need to subtract adjacent terms of the original time series at least once, as suggested by Procedure 9.1, to obtain a stationary time series.

Figure 9.16 shows the autocorrelation and partial autocorrelation diagrams for the differencing operator of order 1 applied to the original time series, resulting in the series $\nabla^1 Y_t$. The existence of a significant positive partial autocorrelation (PACF diagram) for a lag of six periods confirms the seasonality of the time series. Notice also that there is a negative correlation for a lag of three periods, and this information is consistent with the bimonthly seasonality of the time series of energy consumption. Furthermore, the ACF diagram shows a slow decrease for lags separated by an entire cycle (i.e. (6, 18, 24) periods).

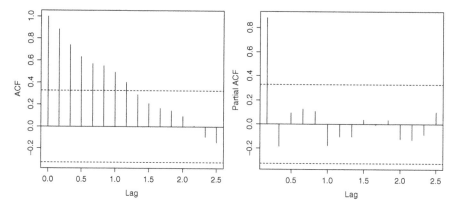

*Figure 9.15   Autocorrelation and partial autocorrelation diagrams for the time
series of electricity consumption*

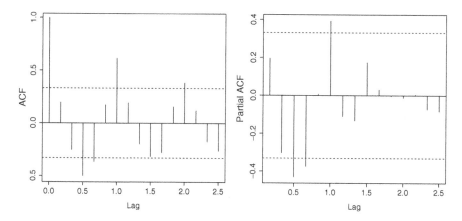

*Figure 9.16   Autocorrelation and partial autocorrelation diagrams for the
prime differences of the time series of electricity consumption*

Figure 9.17 shows the ACF and PACF diagrams for the series $\nabla^6 \nabla^1 Y_t$,
obtained by applying the first-order differences to remove the trend and then
the differences of lag 6 to analyze the seasonality. No more systematic effects
of variation in the autocorrelations can be found and it is therefore possible to
proceed with the determination of the order of the model.

Figure 9.18 shows the prediction for two cycles, that is, 12 two-month periods, obtained using an ARIMA$(0, 1, 1)(0, 1, 1)_6$ model. The values $d = 1$ and
$D = 1$ have necessarily been selected to remove the trend and the seasonality
of length 6, as shown by the ACF and PACF diagrams. The profiles of the two
diagrams lead us to prefer a MA(1) model, for both the level component and
the seasonal component, letting $p = P = 0$ and $q = Q = 1$.

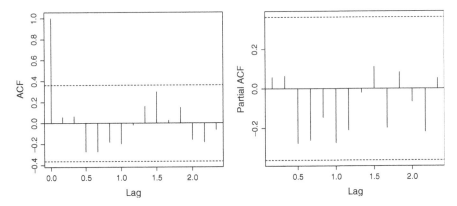

Figure 9.17   Autocorrelation and partial autocorrelation diagrams for the seasonal differences of prime differences $\nabla^6\nabla^1$ of the time series of electricity consumption

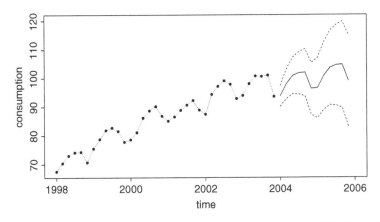

Figure 9.18   Prediction over two years (12 periods) with an ARIMA model for the time series of electricity consumption

9.6   Combination of predictive models

Empirical analysis of a large number of benchmark time series of economic origin has shown the combination of predictive models to be one of the most effective forecasting methods. A combined prediction is obtained as a weighted average of base predictors resulting from the use of different forecasting models. For example, one might use a hybrid combination of exponential smoothing models characterized by different values of the parameters together with autoregressive and moving-average models.

Given v prediction models $F_1, F_2, \ldots, F_v$ based on the same time series $\{Y_t\}$, we can derive a new model F_c as a *combination* of the v original predictors, through the use of v weights $w_1, w_2, \ldots, w_v$:

$$F_c = \frac{\sum_{i=1}^{v} w_i F_i}{\sum_{i=1}^{v} w_i}. \tag{9.65}$$

It can be shown that the predictor F_c is optimal, in the sense that it minimizes the mean squared error if and only if the weights are chosen as

$$\mathbf{w} = \frac{\mathbf{V}^{-1} \mathbf{e}}{\mathbf{e}' \mathbf{V}^{-1} \mathbf{e}}, \tag{9.66}$$

where $\mathbf{e}$ is a unit vector and $\mathbf{V}$ is the covariance matrix of the prediction errors corresponding to the v models. The estimate of the matrix $\mathbf{V}$ is problematic, so that it is usually assumed that the errors generated are independent, and the same weight is assigned to all models. Alternatively, it is possible to compute $\mathbf{w}$ by means of Bayesian statistical techniques.

The empirical evidence indicates that model F_c is better than each single component model F_i, $i = 1, 2, \ldots, v$, in terms of reduced measures of dispersion.

9.7   The forecasting process

In this section, we will review the main criteria that direct the choice of a time series model in connection with the features of the forecasting process. The term *forecasting process* refers here to the whole set of activities, more or less explicit, that lead to the generation of a forecast.

Notice that the terms *forecast* and *prediction* can be used interchangeably to a certain extent. However, we believe that the former is preferable when referring in a broad sense to the activities that are involved in the overall forecasting process, while the latter should be reserved for the attribution of specific numerical values to measurable quantities that will be observed in the future, possibly along with the corresponding confidence intervals.

9.7.1   Characteristics of the forecasting process

Forecasts are a major input to most decision-making processes in private and public organizations. For example, the decision to build a new manufacturing plant depends on forecasts of future demand, technological innovation scenarios, prices and costs, strategic plans of competitors and any legal regulations.

Forecasts of future events are required in order to make various kinds of decisions concerning strategy, research and development, administration and

control, marketing, production and logistics. We may therefore argue that forecasts play a key role at the heart of the entire decision-making process and
strongly affect its effectiveness.

Empirical evidence suggests that many decision-making processes in complex organizations are largely based on forecasts obtained from subjective
opinions and personal or team judgments, often formulated by experts in the
application domain to which the forecasts refer. Yet, subjective forecasts are
usually affected by a less precise and rigorous use of the available past data.
There are several experimental tests that have been conducted on groups of
people in order to show the arbitrary and inexact nature of subjective forecasts,
which often reveal an excess of confidence on the part of the individuals in
their own conclusions.

Planning horizon

The amplitude of the planning horizon is a factor that significantly affects the
forecasting process. Forecasts may be concerned with the immediate future
(e.g. sales of a product for the next two months), may address medium-term
planning issues (e.g. as part of the development of an annual financial plan) or
may stretch over a long time span (e.g. technological scenarios in an industry).
In each case it is clear that the goals of decision makers using such forecasts
are quite different.

Level of detail

The three hypothetical decision-making processes outlined above, besides
showing a difference of time horizon, also highlight variations in the level
of accuracy and detail that is required by forecasts. The object of forecasts
relating to short-term decisions may be the individual items produced by a
company and must achieve high accuracy, since the material procurement and
manufacturing plans depend on them. In the case of medium-term decisions,
the forecasts refer to production quantities aggregated by families of products.
Finally, strategic decisions relate to longer time spans, and the associated predictions are therefore even more aggregated and relative to the total production
quantity for a single enterprise or an entire industry, whereas the required level
of accuracy is necessarily lower.

A higher level of detail generally corresponds to a larger number of time
series for which the forecasts need to be generated. When the forecasts refer to
sales of the products marketed by a company, the number of time series may be
in the thousands and therefore imply a substantial effort for the development of
predictive models, which must necessarily be automated and kept under control
through monitoring procedures.

Aggregated time series show a more regular behavior over time, even though they are characterized by the existence of trend and seasonality components. Conversely, time series relative to a higher level of detail frequently modify their behavior and are therefore more difficult to predict.

9.7.2 Selection of a forecasting method

The choice of a forecasting method largely depends on the characteristics and objectives of the decision-making process for which the predictions will be used. The length of the forecasting horizon, as well as the availability and homogeneity of a large dataset of past observations, are some of the factors that influence the choice of a method for time series analysis.

In the initial phase of a product life cycle, sales data are not yet available, so that only market tests and opinion surveys can be used to forecast sales. On the other hand, in the maturity or decline phases, the dataset is richer and the use of quantitative models is usually easier and very useful.

A further aspect that needs to be considered is the trade-off between costs and benefits when evaluating the adoption of a particular class of methods. Several empirical analyses have been carried out to compare different forecasting methods, evaluating the accuracy of the various techniques with respect to thousands of benchmark time series obtained from real-world applications. In general, they indicate that the adoption of complicated forecasting methods is seldom justified since they are not very robust with respect to the dynamic nature of economic time series. Simpler methods, such as the various exponential smoothing techniques, are usually much more effective than complex filtering methods.

9.8   Notes and readings

An effective introduction to time series models is given in Chatfield (2003). Other interesting books on the subject are Harvey (1993), Hamilton (1994) and Brockwell and Davis (2002). A useful book devoted to methods based on the decomposition of a time series into its components, referred to as *state space* methods, is Durbin and Koopman (2001). For texts more oriented to the applications of time series in business the reader may wish to consult Makridakis *et al.* (1983), Makridakis and Wheelwright (1989) and Makridakis *et al.* (1998). The combination of forecasting methods is discussed in Winkler and Makridakis (1983). Autoregressive models are described in Box *et al.* (1994) and Jones (1980). Contributions of historical value are Holt (1957) and Winters (1960), in which the exponential smoothing methods were originally proposed.

10

Classification[1]

Classification models are supervised learning methods for predicting the value of a categorical target attribute, unlike regression models which deal with numerical attributes. Starting from a set of past observations whose target class is known, classification models are used to generate a set of rules that allow the target class of future examples to be predicted.

Classification holds a prominent position in learning theory due to its theoretical implications and the countless applications it affords. From a theoretical viewpoint, the development of algorithms capable of learning from past experience represents a fundamental step in emulating the inductive capabilities of the human brain.

On the other hand, the opportunities afforded by classification extend into several different application domains: selection of the target customers for a marketing campaign, fraud detection, image recognition, early diagnosis of diseases, text cataloguing and spam email recognition are just a few examples of real problems that can be framed within the classification paradigm.

In this chapter, we will describe the general characteristics of classification problems, discussing the main criteria used to evaluate and to compare different models. We will then review the major classification methods: classification trees, Bayesian methods, neural networks, logistic regression and support vector machines.

10.1 Classification problems

In a classification problem, we have a dataset D containing m observations described in terms of n *explanatory* attributes and a categorical *target* attribute.

[1]This chapter was co-written by Carlotta Orsenigo.

The explanatory attributes, also called *predictive variables*, may be partly categorical and partly numerical. The target attribute is also called a *class* or *label*, while the observations are also termed *examples* or *instances*. Unlike regression, for classification models the target variable takes a finite number of values. In particular, we have a *binary* classification problem if the instances belong to two classes only, and a *multiclass* or *multicategory* classification if there are more than two classes.

The purpose of a classification model is to identify recurring relationships among the explanatory variables which describe the examples belonging to the same class. Such relationships are then translated into *classification rules* which are used to predict the class of examples for which only the values of the explanatory attributes are known. The rules may take different forms depending on the type of model used.

Example 10.1 – Retention in the mobile phone industry.   Example 5.2 on the analysis of customer loyalty in the mobile phone industry is a binary classification problem in which the target attribute takes the value 1 if a customer has discontinued service and 0 otherwise. The features of each customer, described in Table 5.3, represent the predictive attributes. The purpose of a classification model is to derive general rules from the examples contained in the dataset and to apply these rules in order to assign the class to new instances for which the target value is unknown. In this way, the classification model may prove useful in identifying those customers who are likely to discontinue service, and therefore to drive a retention marketing campaign.

Example 10.2 – Segmentation of customers phoning a call center. Many services and manufacturing companies nowadays have a call center that their customers may call to request information or report problems. In order to size the staff and the activities of a call center and to verify the quality of the services offered, it is useful to classify customers based on the number of calls made to the call center. The target attribute may be obtained through a proper discretization of the numerical variable indicating the number of calls, setting for example: class $0 \equiv$ no calls, class $1 \equiv 1$ call, class $2 \equiv$ from 2 to 4 calls, class $4 \equiv$ more than 4 calls. Again predictive attributes are provided by the features of the customers. Hence, the segmentation of the customers with respect to the number of calls made to the call center is a multicategory classification problem.

From a mathematical viewpoint, in a classification problem m known examples are given, consisting of pairs $(\mathbf{x}_i, y_i)$, $i \in \mathcal{M}$, where $\mathbf{x}_i \in \mathbb{R}^n$ is the vector of the values taken by the n predictive attributes for the ith example and $y_i \in \mathcal{H} = \{v_1, v_2, \ldots, v_H\}$ denotes the corresponding target class. Each component x_{ij} of the vector $\mathbf{x}_i$ is regarded as a realization of the random variable X_j, $j \in \mathcal{N}$, which represents the attribute $\mathbf{a}_j$ in the dataset $\mathcal{D}$. In a binary classification problem one has $H = 2$, and the two classes may be denoted as $\mathcal{H} = \{0, 1\}$ or as $\mathcal{H} = \{-1, 1\}$, without loss of generality.

Let $\mathcal{F}$ be a class of functions $f(\mathbf{x}) : \mathbb{R}^n \mapsto \mathcal{H}$ called *hypotheses* that represent hypothetical relationships of dependence between y_i and $\mathbf{x}_i$. A *classification problem* consists of defining an appropriate hypothesis space $\mathcal{F}$ and an algorithm $A_{\mathcal{F}}$ that identifies a function $f^* \in \mathcal{F}$ that can optimally describe the relationship between the predictive attributes and the target class. The joint probability distribution $P_{\mathbf{x},y}(\mathbf{x}, y)$ of the examples in the dataset $\mathcal{D}$, defined over the space $\mathbb{R}^n \times \mathcal{H}$, is generally unknown and most classification models are nonparametric, in the sense that they do not make any prior assumption on the form of the distribution $P_{\mathbf{x},y}(\mathbf{x}, y)$.

The flow diagram shown in Figure 10.1 may clarify the probability assumptions concerning the three components of a classification problem: a *generator* of observations, a *supervisor* of the target class and a *classification algorithm*.

Generator. The task of the generator is to extract random vectors $\mathbf{x}$ of examples according to an unknown probability distribution $P_{\mathbf{x}}(\mathbf{x})$.

Supervisor. The supervisor returns for each vector $\mathbf{x}$ of examples the value of the target class according to a conditional distribution $P_{y|\mathbf{x}}(y|\mathbf{x})$ which is also unknown.

Algorithm. A classification algorithm $A_{\mathcal{F}}$, also called a *classifier*, chooses a function $f^* \in \mathcal{F}$ in the hypothesis space so as to minimize a suitably defined loss function.

Classification models, just like regression models, are suitable for both interpretation and prediction, as observed in Section 5.1 with general reference to

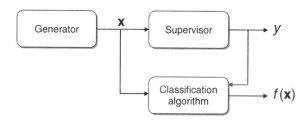

Figure 10.1   Probabilistic structure of the learning process for classification

supervised learning methods. Simpler models usually yield intuitive classification rules that can be easily interpreted, while more advanced models derive less intelligible rules although they usually achieve better prediction accuracy.

A portion of the examples in the dataset $\mathcal{D}$ is used for *training* a classification model, that is, for deriving the functional relationship between the target variable and the explicative variables expressed by means of the hypothesis $f^* \in \mathcal{F}$. What remains of the available data is used later to evaluate the accuracy of the generated model and to select the best model out of those developed using alternative classification methods.

The development of a classification model consists therefore of three main phases.

Training phase. During the *training* phase, the classification algorithm is applied to the examples belonging to a subset $\mathcal{T}$ of the dataset $\mathcal{D}$, called the *training set*, in order to derive classification rules that allow the corresponding target class y to be attached to each observation $\mathbf{x}$.

Test phase. In the *test* phase, the rules generated during the training phase are used to classify the observations of $\mathcal{D}$ not included in the training set, for which the target class value is already known. To assess the accuracy of the classification model, the actual target class of each instance in the *test set* $\mathcal{V} = \mathcal{D} - \mathcal{T}$ is then compared with the class predicted by the classifier. To avoid an overestimate of the model accuracy, the training set and the test set must be disjoint.

Prediction phase. The *prediction* phase represents the actual use of the classification model to assign the target class to new observations that will be recorded in the future. A prediction is obtained by applying the rules generated during the training phase to the explanatory variables that describe the new instance.

Figure 10.2 shows the logical flow of the learning process for a classification algorithm. In addition to the phases described above, Figure 10.2 also shows a portion of the training data, called the *tuning set*, which is used by some classification algorithms to identify the optimal value of some parameters appearing in the functions $f \in \mathcal{F}$.

10.1.1   Taxonomy of classification models

Before describing the different types of classifiers in the next sections, it may be useful to provide a taxonomy of classification models in order to place each individual algorithm in a broader framework. We can distinguish four main categories of classification models.

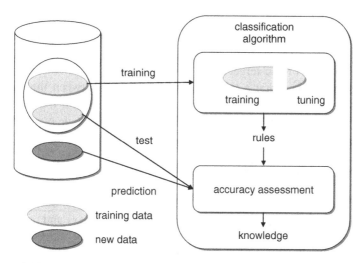

Figure 10.2   Phases of the learning process for a classification algorithm

Heuristic models. Heuristic methods make use of classification procedures based on simple and intuitive algorithms. This category includes *nearest neighbor* methods, based on the concept of distance between observations, and *classification trees*, which use *divide-and-conquer* schemes to derive groups of observations that are as homogeneous as possible with respect to the target class.

Separation models. Separation models divide the attribute space $\mathbb{R}^n$ into H disjoint regions $\{S_1, S_2, \ldots, S_H\}$, separating the observations based on the target class. The observations x_i in region S_h are assigned to class $y_i = v_h$. Each region S_h may comprise a composite set obtained from set-theoretic operations of union and intersection applied to regions of an elementary form, such as halfspaces or hyperspheres. However, the resulting regions should be kept structurally simple, to avoid compromising the generalization capability of the model. In general, it is hard to determine a collection of simple regions that exactly subdivide the observations of the dataset based on the value of the target class. Therefore, a loss function is defined to take the misclassified points into account, and an optimization problem is solved in order to derive a subdivision into regions that minimizes the total loss. The various classification models that belong to this category differ from one another in terms of the type of separation regions, the loss function, the algorithm used to solve the optimization problem. The most popular separation techniques include *discriminant analysis*, *perceptron methods*, *neural networks* and *support vector machines*. Some variants of classification trees can also be placed in this category.

Regression models. Regression models, described in Chapter 8 for the prediction of continuous target variables, make an explicit assumption concerning the functional form of the conditional probabilities $P_{y|\mathbf{x}}(y|\mathbf{x})$, which correspond to the assignment of the target class by the supervisor. For linear regression models, it is assumed that a linear relationship exists between the dependent variable and the predictors, and this leads to the derivation of the value of the regression coefficients. We will see in Section 10.5 that *logistic regression* is an extension of linear regression suited to handling binary classification problems.

Probabilistic models. In probabilistic models, a hypothesis is formulated regarding the functional form of the conditional probabilities $P_{\mathbf{x}|y}(\mathbf{x}|y)$ of the observations given the target class, known as *class-conditional probabilities*. Subsequently, based on an estimate of the *prior* probabilities $P_y(y)$ and using Bayes' theorem, the *posterior* probabilities $P_{y|\mathbf{x}}(y|\mathbf{x})$ of the target class assigned by the supervisor can be calculated. The functional form of $P_{\mathbf{x}|y}(\mathbf{x}|y)$ may be parametric or nonparametric. Naive Bayes classifiers and Bayesian networks are well-known families of probabilistic methods.

Besides the taxonomy described above, it should be mentioned that most classifiers also generate a *score function* $g(\mathbf{x}) : \mathbb{R}^n \mapsto \mathbb{R}$ which associates each observation $\mathbf{x}$ with a real number. Possibly after standardization, the score function may be interpreted as an estimate of the probability that the class predicted by the classifier for the observation $\mathbf{x}$ is correct. This property is evident in probabilistic models. In separation models, the score of an observation may be set equal to its distance from the boundary of the region corresponding to its target class value. Finally, in classification trees the score function is associated with the density of observations having the same target class which have been included in the leaf node to which the observation itself has been assigned, as we will see in detail in Section 10.3.

Starting from the score function g, it is possible to derive a classification rule to predict the target class of the observation $\mathbf{x}$. For example, for the separation models applied to binary classification problems one may set $f(\mathbf{x}) = \text{sgn}(g(\mathbf{x}))$, where $\text{sgn}(\cdot)$ is a function indicating the sign of its argument, and therefore takes the values of the two classes $\{-1, 1\}$.

10.2   Evaluation of classification models

Within a classification analysis it is usually advisable to develop alternative models and then select the method affording the best prediction accuracy. To obtain alternative models one may vary the method, by using for instance classification trees, neural networks, Bayesian techniques or support vector machines, and also modify the values of the parameters involved. Classification methods can be evaluated based on several criteria, as follows.

Accuracy. Evaluating the accuracy of a classification model is crucial for two main reasons. First, the accuracy of a model is an indicator of its ability to predict the target class for future observations. Based on their accuracy values, it is also possible to compare different models in order to select the classifier associated with the best performance. Let $\mathcal{T}$ be the training set and $\mathcal{V}$ the test set, and t and v be the number of observations in each subset, respectively. The relations $\mathcal{D} = \mathcal{T} \cup \mathcal{V}$ and $m = t + v$ obviously hold. The most natural indicator of the accuracy of a classification model is the proportion of observations of the test set $\mathcal{V}$ correctly classified by the model. If y_i denotes the class of the generic observation $\mathbf{x}_i \in \mathcal{V}$ and $f(\mathbf{x}_i)$ the class predicted through the function $f \in \mathcal{F}$ identified by the learning algorithm $A = A_{\mathcal{F}}$, the following loss function can be defined:

$$L(y_i, f(\mathbf{x}_i)) = \begin{cases} 0, & \text{if } y_i = f(\mathbf{x}_i), \\ 1, & \text{if } y_i \neq f(\mathbf{x}_i). \end{cases} \tag{10.1}$$

The accuracy of model A can be evaluated as

$$\text{acc}_A(\mathcal{V}) = \text{acc}_{A_{\mathcal{F}}}(\mathcal{V}) = 1 - \frac{1}{v} \sum_{i=1}^{v} L(y_i, f(\mathbf{x}_i)). \tag{10.2}$$

In some cases, it is preferable to use an alternative performance indicator given by the proportion of errors made by the classification algorithm:

$$\text{err}_A(\mathcal{V}) = \text{err}_{A_{\mathcal{F}}}(\mathcal{V}) = 1 - \text{acc}_{A_{\mathcal{F}}}(\mathcal{V}) = \frac{1}{v} \sum_{i=1}^{v} L(y_i, f(\mathbf{x}_i)). \tag{10.3}$$

Speed. Some methods require shorter computation times than others and can handle larger problems. However, classification methods characterized by longer computation times may be applied to a small-size training set obtained from a large number of observations by means of random sampling schemes. It is not uncommon to obtain more accurate classification rules in this way.

Robustness. A classification method is *robust* if the classification rules generated, as well as the corresponding accuracy, do not vary significantly as the choice of the training set and the test set varies, and if it is able to handle missing data and outliers.

Scalability. The scalability of a classifier refers to its ability to learn from large datasets, and it is inevitably related to its computation speed. Therefore, the remarks made in connection with sampling techniques for data reduction, which often result in rules having better generalization capability, also apply in this case.

Interpretability. If the aim of a classification analysis is to interpret as well as predict, then the rules generated should be simple and easily understood by knowledge workers and experts in the application domain.

10.2.1   Holdout method

The *holdout* estimation method involves subdividing the m observations available into two disjoint subsets T and V, for training and testing purposes respectively, and then evaluating the accuracy of the model through the accuracy $acc_A(V)$ on the test set. In general, T is obtained through a simple sampling procedure, which randomly extracts t observations from D, leaving the remaining examples for the test set V. The portion of data used for training may vary based on the size of the dataset D. Usually the value t falls somewhere between one half and two thirds of the total number m of observations.

The accuracy of a classification algorithm evaluated via the holdout method depends on the test set V selected, and therefore it may over- or underestimate the actual accuracy of the classifier as V varies. To obtain a more robust estimate, and consequently to evaluate the accuracy of the classification model with greater reliability, different alternative strategies can be followed as described in the following sections.

10.2.2   Repeated random sampling

The *repeated random sampling* method involves replicating the holdout method a number r of times. For each repetition a random independent sample T_k is extracted, which includes t observations, and the corresponding accuracy $acc_A(V_k)$ is evaluated, where $V_k = D - T_k$. At the end of the procedure, the accuracy of the classifier A_F is estimated using the sample mean

$$acc_A = acc_{A_F} = \frac{1}{r} \sum_{k=1}^{r} acc_{A_F}(V_k). \tag{10.4}$$

The total number r of repetitions may be evaluated a priori using techniques from statistical inference for determining the appropriate sample size.

The method of repeated random sampling is preferred over the holdout method since it yields a more reliable estimate. However, there is no control over the number of times that each observation may appear in the training or in the test set. In this way, a dominant observation containing outliers for one or more attributes may cause undesired effects on both the classification rules generated and the resulting accuracy estimate.

10.2.3   Cross-validation

The method of *cross-validation* offers an alternative to repeated random sampling techniques and guarantees that each observation of the dataset $\mathcal{D}$ appears the same number of times in the training sets and exactly once in the test sets.

The cross-validation scheme is based on a partition of the dataset $\mathcal{D}$ into r disjoint subsets $\mathcal{L}_1, \mathcal{L}_2, \ldots, \mathcal{L}_r$, and requires r iterations. At the kth iteration of the procedure, subset $\mathcal{L}_k$ is selected as the test set and the union of all the other subsets in the partition as the training set, that is,

$$\mathcal{V}_k = \mathcal{L}_k, \qquad \mathcal{T}_k = \bigcup_{l \neq k} \mathcal{L}_l. \tag{10.5}$$

Once the partition has been generated, by randomly extracting the observations for each set $\mathcal{L}_k$, the classification algorithm $A_{\mathcal{F}}$ is applied r times, using each of the r training sets in turn and evaluating the accuracy each time on the corresponding test set. At the end of the procedure, the overall accuracy is computed as the arithmetic mean of the r individual accuracies, as in expression (10.4).

If $r = 2$, each observation is used precisely once for training and once for evaluating the accuracy on the test set. However, higher values of r are usually preferred in order to obtain a more robust estimate of the accuracy. A popular choice in practice is *tenfold cross-validation*, whereby the dataset $\mathcal{D}$ is partitioned into 10 subsets. Figure 10.3 illustrates the tenfold cross-validation scheme.

A different evaluation method, called *leave-one-out*, is obtained by choosing $r = m$. In this case each of the m test sets includes only one observation, and each example is used in turn to measure the accuracy. By virtue of a greater

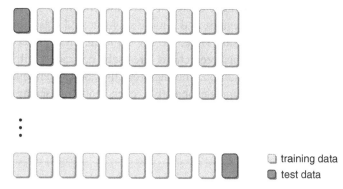

Figure 10.3   Evaluation of accuracy by means of tenfold cross-validation

computational effort, due to m executions of the classifier, the leave-one-out method affords the advantage of carrying out the training operation with a larger number of observations, equal to $m - 1$ for each iteration.

It is usually required that each subset $\mathcal{L}_k$ in the partition contains the same proportion of observations belonging to each target class. If for example the binary target class indicates by the value $\{1\}$ that a customer has churned, it is appropriate that each subset $\mathcal{L}_k$ contain the same proportion of observations having the target class value $\{1\}$ and that this proportion be the same as for the full dataset $\mathcal{D}$. In order to preserve in the partition the proportions originally contained in $\mathcal{D}$, stratified random sampling is used to generate the subsets $\mathcal{L}_k$.

10.2.4 Confusion matrices

The accuracy measurement methods described above are not always adequate for discriminating among models, and in some instances they may even yield paradoxical results, as shown in the following two examples.

Example 10.3 – Early medical diagnosis. Consider a binary classification problem in the field of medical diagnosis where observations referred to patients may belong to a binary target class: the value $\{1\}$ means that a patient has a given medical condition, whereas the value $\{-1\}$ indicates that a patient does not have the condition. The explanatory variables express the results of clinical tests, and the objective of the model is to classify the health status of patients based on test results, in order to obtain an early diagnosis of the condition to which the examples refer. Assume that the patients who have the condition represent 2% of the observations available in the whole dataset $\mathcal{D}$. The accuracy of the trivial rule claiming that 'the patient does not have the condition' is equal to 0.98, irrespective of test results, and a more sophisticated classifier could hardly achieve a better result. Here, the 2% set made up of observations corresponding to erroneous predictions precisely coincides with those patients who have the condition, and therefore the above-mentioned rule fails to meet the real diagnostic objective, even though it seems highly accurate.

Example 10.4 – Customer retention. In classification applications aiming at customer retention, the value $\{1\}$ for the target class means that a customer has canceled the service or discontinued the purchase of goods from the enterprise, while the value {0} indicates that a customer is still active. The explanatory variables refer to transactions recording purchases or usage of the service, and the objective of the model is to predict the future class of customers ahead of time, to allow marketing for retention purposes to be carried out. Assume again that the customers abandoning the service constitute 2% of the available examples in the dataset. The trivial rule claiming that 'all customers are loyal' achieves an accuracy of 0.98, irrespective of the predictive variables, and again it seems hard to beat. Yet, the marketing objective is to identify the highest proportion of customers at risk of churning, which can be found in the 2% of the customer base.

The above examples show that it is not enough to consider the number of accurate predictions, but also the *type* of error committed should be accounted for. For this purpose, it is useful to resort to decision tables, usually called *confusion matrices*, which for the sake of simplicity we will only describe in connection with binary classification, though they can be easily extended to multicategory classification.

Let us assume that we wish to analyze a binary classification problem where the values taken by the target class are $\{-1, 1\}$. We can then consider a 2×2 matrix whose rows correspond to the observed values and whose columns are associated with the values predicted using a classification model, as shown in Table 10.1.[2]

The elements of the confusion matrix have the following meanings: p is the number of correct predictions for the negative examples, called *true negatives*; u is the number of incorrect predictions for the positive examples, called *false negatives*; q is the number of incorrect predictions for the negative examples, called *false positives*; and v is the number of correct predictions for the positive

Table 10.1   Confusion matrix for a binary target attribute encoded with the class values $\{-1, 1\}$

		predictions		
		−1 (negative)	+1 (positive)	total
	−1 (negative)	p	q	$p + q$
examples	+1 (positive)	u	v	$u + v$
	total	$p + u$	$q + v$	m

---

[2]Definitions of the confusion matrix can also be found where rows and columns are transposed with respect to our definition.

examples, called *true positives*. Using these elements, further indicators useful for validating a classification algorithm can be defined.

Accuracy. The accuracy of a classifier may be expressed as

$$\text{acc} = \frac{p + v}{p + q + u + v} = \frac{p + v}{m}. \tag{10.6}$$

True negatives rate. The true negatives rate is defined as

$$\text{tn} = \frac{p}{p + q}. \tag{10.7}$$

False negatives rate. The false negatives rate is defined as

$$\text{fn} = \frac{u}{u + v}. \tag{10.8}$$

False positives rate. The false positives rate is defined as

$$\text{fp} = \frac{q}{p + q}. \tag{10.9}$$

True positives rate. The true positives rate, also known as *recall*, is defined as

$$\text{tp} = \frac{v}{u + v}. \tag{10.10}$$

Precision. The precision is the proportion of correctly classified positive examples, and is given by

$$\text{prc} = \frac{v}{q + v}. \tag{10.11}$$

Geometric mean. The geometric mean is defined as

$$\text{gm} = \sqrt{\text{tp} \times \text{prc}}, \tag{10.12}$$

and sometimes also as

$$\text{gm} = \sqrt{\text{tp} \times \text{tn}}. \tag{10.13}$$

In both cases, the geometric mean is equal to 0 if all the predictions are incorrect.

F-measure. The F-measure is defined as

$$F = \frac{(\beta^2 - 1)\, \text{tp} \times \text{prc}}{\beta^2\, \text{prc} + \text{tp}}, \tag{10.14}$$

where $\beta \in [0, \infty)$ regulates the relative importance of the precision with respect to the true positives rate. The F-measure is also equal to 0 if all the predictions are incorrect.

With some classifiers it is possible to assign a matrix of costs for incorrectly classified examples, which we will refer to as *misclassification costs*. Obviously, the cost associated with the $p + v$ correctly classified observations is zero. By adequately regulating the relative weight of the two costs of misclassification corresponding to the u false negatives and to the q false positives, it is possible to direct a classifier toward the actual objectives of the decision makers. In Example 10.3 the cost of false negatives should be much higher than the cost of false positives, so as to identify the maximum number of patients who have the medical condition and apply preventive therapies to them. By the same token, in Example 10.4 the cost of false positives must be much higher than the cost of false negatives, so as to identify the maximum number of customers at risk of churning to whom a preventive retention action can be addressed.

10.2.5 ROC curve charts

Receiver operating characteristic (ROC) curve charts allow the user to visually evaluate the accuracy of a classifier and to compare different classification models. They visually express the information content of a sequence of confusion matrices and allow the ideal trade-off between the number of correctly classified positive observations and the number of incorrectly classified negative observations to be assessed. In this respect, they are an alternative to the assignment of misclassification costs.

An ROC chart is a two-dimensional plot with the proportion of false positives fp on the horizontal axis and the proportion of true positives tp on the vertical axis. The point $(0,1)$ represents the ideal classifier, which makes no prediction error since its proportion of false positives is null (fp $= 0$) and its proportion of true positives is maximum (tp $= 1$). The point $(0,0)$ corresponds to a classifier that predicts the class $\{-1\}$ for all the observations, while the point $(1,1)$ corresponds to a classifier predicting the class $\{1\}$ for all the observations.

Most classifiers allow a few parameters in the optimal hypothesis $f^* \in \mathcal{F}$ to be adjusted so as to increase the number of true positives tp, albeit at the expense of a corresponding increase in the number of false positives fp. To obtain the trajectory of the ROC curve for a specific classifier, it is therefore necessary to use pairs of values (fp, tp) which have been empirically obtained for different values of the parameters in $f^* \in \mathcal{F}$. A classifier with no parameters to be tuned yields only one point in the diagram. Figure 10.4 shows an example of ROC curves chart. The area beneath the ROC curve gives a concise measurement comparing the accuracy of various classifiers: the classifier associated with the ROC curve with the greatest area is then preferable.

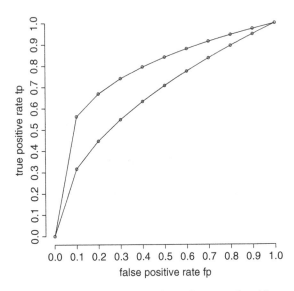

Figure 10.4   ROC curve chart for two classifiers

10.2.6   Cumulative gain and lift charts

The *lift* measure corresponds to the intuitive idea of evaluating the accuracy of a classifier based on the density of positive observations inside the set that has been identified based on model predictions.

To further clarify this point, let S be a subset of observations of cardinality s selected according to the classifier. For example, the set S may include those patients deemed potentially at risk of a given medical condition based on the scoring derived by a classification model. Analogously, S may contain those customers deemed at higher risk of churning for a company analyzing the loyalty of its customers. The lift is defined as the ratio between the proportion b/s of positive observations existing in S and the proportion a/m of positive observations in the whole dataset $\mathcal{D}$, that is,

$$\text{lift} = \frac{b/s}{a/m}, \tag{10.15}$$

where b and a are the number of positive observations in the sets S and $\mathcal{D}$, respectively.

Cumulative gain and lift diagrams allow the user to visually evaluate the effectiveness of a classifier. To describe the procedure for constructing the charts and to understand their meaning, consider an example of binary classification arising in a relational marketing analysis, in which we are interested in identifying a subset of customers to be recipients of a cross-selling campaign aimed at promoting a new service.

The company has a total of $m = 1\,000\,000$ customers and, based on similar past campaigns, estimates that the proportion of customers who might respond to the promotion by subscribing to the new service is equal to 2%. A campaign addressed to a random sample of s customers, where $s \in [0, 1\,000\,000]$, would therefore yield a number of positive responses equal to approximately $0.02s$, as an effect of the law of large numbers. The procedure for randomly selecting the recipients is clearly a benchmark to assess the effectiveness of any classifier, which must necessarily be more accurate than random extraction to be of any practical use.

The cumulative gain chart in Figure 10.5 shows on the horizontal axis the number of customers receiving the promotion and on the vertical the percentage of expected positive responses. The straight line in the diagram corresponds to the procedure for the selection of recipients based on random sampling.

If we use a classifier associated with a score function, the customers can be ranked by decreasing score, so that the first customers in the list have a greater probability of subscribing to the offer according to the prediction provided by the classifier. For each value s, consider the set S consisting of the first s customers of the list ordered by decreasing scores, correspondingly placing the percentage of positive responses along the vertical axis. Usually, the curve is obtained by considering the deciles on the horizontal axis, i.e. dividing into ten equal parts the interval between 0 and the size m of the entire population, represented in our example by the interval $[0, 1\,000\,000]$.

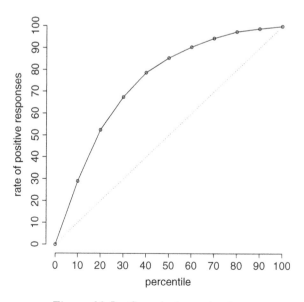

Figure 10.5   Cumulative gain chart

Table 10.2   Number of positive responses at customer base deciles

number of customers contacted	number of positive responses
0	0
100 000	5 800
200 000	10 500
300 000	13 500
400 000	15 900
500 000	17 100
600 000	18 100
700 000	18 900
800 000	19 500
900 000	19 700
1 000 000	20 000

Suppose that we have followed the procedure described above in order to evaluate the predictive effectiveness of a given classifier, obtaining Table 10.2 which shows the number of positive responses existing in subsets S whose cardinalities equal the deciles. The curve in Figure 10.5 represents the cumulative gain for the classifier.

To derive the corresponding lift curve we calculate for each value available on the horizontal axis, expressing the deciles in our example, the ratio between the proportion of positive responses obtained according to the scoring generated by the classifier and the proportion of positive responses obtained from random sampling. For each value on the horizontal axis, the lift curve is therefore obtained as the ratio between the two values on the cumulative gain curves. Figure 10.6 shows the lift curve for our example.

Two criteria based on cumulative gain and lift curves allow different classifiers to be compared. On the one hand, the area between the cumulative gain curve and the line corresponding to random sampling is evaluated: a greater area corresponds to a classification method that is more effective overall. However, it may be the case that a method producing a greater area is associated with a lower lift value than a different method at a specific value s on the horizontal axis that indicates the actual number of recipients of the marketing campaign, determined according to the available budget. The second criterion is therefore based on the maximum lift value at a specific value s on the horizontal axis.

10.3   Classification trees

Classification trees are perhaps the best-known and most widely used learning methods in data mining applications. The reasons for their popularity lie in their conceptual simplicity, ease of usage, computational speed, robustness with

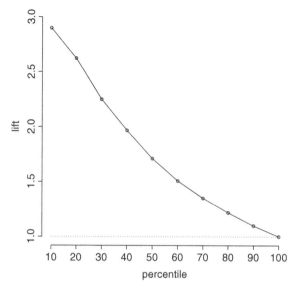

Figure 10.6 Lift chart

respect to missing data and outliers and, most of all, the interpretability of the rules they generate. To separate the observations belonging to different classes, methods based on trees obtain simple and explanatory rules for the relationship existing between the target variable and predictive variables.

The development of a classification tree corresponds to the training phase of the model and is regulated by a recursive procedure of heuristic nature, based on a divide-and-conquer partitioning scheme referred to as *top-down induction of decision trees*, described by Procedure 10.1. Some of the mechanisms governing the development of a tree may be implemented by following different approaches, so that classification trees actually represent a broad class of methods that we will first describe in general terms and then illustrate in detail in some specific cases.

Procedure 10.1 – Top-down induction of decision trees

1. In the initialization phase, each observation is placed in the root node of the tree. The root is included in the list L of active nodes.

2. If the list L is empty the procedure is stopped, otherwise a node J belonging to the list L is selected, is removed from the list and is used as the node for analysis.

3. The optimal rule to split the observations contained in J is then determined, based on an appropriate preset criterion. The splitting rule generated in this way is then applied, and descendant nodes are constructed by subdividing the observations contained in J. For each descendant node the conditions for stopping the subdivision are verified. If these are met, node J becomes a leaf, to which the target class is assigned according to the majority of the observations contained in J. Otherwise, the descendant nodes are added to the list L. Finally, step 2 is repeated.

The observations of the training set initially contained in the *root node* of the tree are divided into disjoint subsets that are tentatively placed in two or more descendant nodes (*branching*). At each of the nodes determined in this way, a check is applied to verify if the conditions for stopping the development of the node are satisfied. If at least one of these conditions is met, no further subdivision is performed and the node becomes a *leaf* of the tree. Otherwise, the actual subdivision of the observations contained in the node is carried out. At the end of the procedure, when no tree node can be further subdivided, each leaf node is labeled with the value of the class to which the majority of the observations in the node belong, according to a criterion called *majority voting*.

The subdivision of the examples in each node is carried out by means of a *splitting rule*, also termed a *separating rule*, to be selected based upon a specific evaluation function. By varying the metrics used to identify the splitting rule, different versions of classification trees can be obtained. Most of the proposed evaluation criteria share the objective of maximizing the uniformity of the target class for the observations that are placed in each node generated through the separation. The different splitting rules are usually based on the value of some explanatory variables that describe the observations, according to the specific methods that will be discussed in the next section.

At the end of the procedure, the set of splitting rules that can be found along the path connecting the tree root to a leaf node constitutes a *classification rule*.[3]

During the prediction phase, in order to assign the target class to a new observation, a path is followed from the root node to a leaf node by obeying the sequence of rules applied to the values of the attributes of the new observation. The predicted target class then coincides with the class by which the leaf node so reached has been labeled during the development phase, that is, with the class of the majority of the observations in the training set falling in that leaf node.

---

[3] A characteristic property of trees guarantees that there is one and only one path connecting the root node to each other node, in particular to each leaf node.

We observed in Section 10.1 that most classifiers associate with each observation a score function, which is then converted into a prediction of the target class. This is also true for classification trees, which associate each observation contained in a leaf node with the highest proportion of the target class for the observations contained in the leaf itself, which also determines its labeling by majority voting. For example, if in a leaf node there are 100 customers of a company and 85 of these have positively responded in the past to a marketing campaign, then the value 85% can be interpreted as the probability that a customer falling in that leaf node, based on the values of its attributes and the generated classification rules, positively responds to a similar campaign to be conducted in the future.

Starting from a training dataset it is possible to construct an exponential number of distinct classification trees. It can be shown that the problem of determining the optimal tree is NP-hard, that is, computationally difficult. As a consequence, the methods for developing classification trees are heuristic in nature.

As observed, the scheme for generating classification trees described in Procedure 10.1 is a general framework and requires that some steps be specified before deriving an implementable classification algorithm. In the following sections we will examine the following components of the top-down induction of decision trees procedure.

Splitting rules. For each node of the tree it is necessary to specify the criteria used to identify the optimal rule for splitting the observations and for creating the descendant nodes. As shown in the next section, there are several alternative criteria, which differ in the number of descendants, the number of attributes and the evaluation metrics.

Stopping criteria. At each node of the tree different *stopping* criteria are applied to establish whether the development should be continued recursively or the node should be considered as a leaf. In this case too, various criteria have been proposed, which result in quite different topologies of the generated trees, all other elements being equal.

Pruning criteria. Finally, it is appropriate to apply a few *pruning* criteria, first to avoid excessive growth of the tree during the development phase (*pre-pruning*), and then to reduce the number of nodes after the tree has been generated (*post-pruning*).

Figure 10.7 shows a classification tree obtained for the dataset described in Example 5.2, concerning the analysis of loyalty in the mobile phone industry, which will be analyzed in detail in Section 10.3.3.

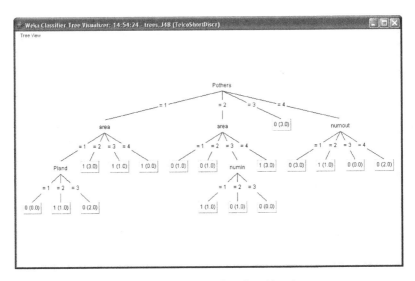

Figure 10.7   Example of a classification tree

10.3.1   Splitting rules

Classification trees can be divided into *binary* and *general* trees based on the maximum number of descendants that each node is allowed to generate.

Binary trees. A tree is said to be *binary* if each node has at most two branches. Binary trees represent in a natural way the subdivision of the observations contained at a node based on the value of a binary explanatory attribute. For example, the customers who have authorized mailing of promotional communications may be placed in the right descendant node and customers who have refused such communications in the left node. When dealing with categorical attributes with more than two classes, binary trees should necessarily form two groups of categories in order to perform a split. For example, the customers residing in areas {1, 2} may be placed in the right branch and those residing in areas {3, 4} in the left branch. Numerical attributes can be separated based on a threshold value. For example, customers whose age is less than 45 may be placed in the right node and older customers in the left one. Finally, binary trees may also be used to develop multicategory classification.

Multi-split classification trees. A tree is said to be *multi-split* if each node has an arbitrary number of branches. This allows multi-valued categorical explanatory attributes to be handled more easily. On the other hand, with numerical attributes, it is again necessary to group together adjacent values. The latter operation is basically equivalent to discretization, obtained in a dynamic way by the algorithm itself during the tree development phase.

Based on the empirical evidence, no significant differences seem to emerge in the predictive accuracy of classification trees in connection with the maximum number of descendant nodes.

Another relevant distinction within classification tree methods concerns the way the explanatory attributes contribute to the definition of the splitting rule at each node. In particular, we may distinguish between *univariate* and *multivariate* trees.

Univariate trees. For *univariate* trees the splitting rule is based on the value assumed by a single explanatory attribute X_j. If the selected attribute is categorical, the observations at a given node are divided by means of conditions of the form $X_j \in B_k$, where the collection $\{B_k\}$ is composed of disjoint and exhaustive subsets of the set of values assumed by the attribute X_j. For example, for a binary attribute taking the values $\{0, 1\}$, the two subsets B_0 and B_1 correspond to the values $\{0\}$ and $\{1\}$, as shown in Figure 10.8. Figure 10.9 shows the partition for a non-binary categorical attribute. If X_j is a numerical attribute, the univariate partition consists of a rule in the form $X_j \le b$ or $X_j > b$, as shown in Figure 10.10 for the age of a customer, depending on a threshold value b determined by the algorithm itself. Univariate trees are also referred to as *axis-parallel* trees, since the splitting rules induce a partition of the space of

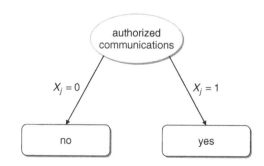

Figure 10.8   Univariate split for a binary attribute

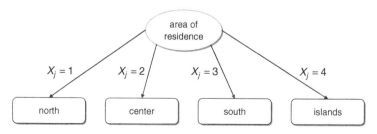

Figure 10.9   Univariate split for a nominal attribute

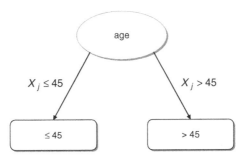

Figure 10.10   Univariate split for a numerical attribute

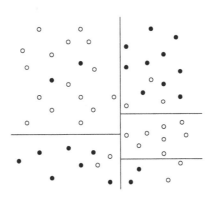

Figure 10.11   Classification   by   means   of   univariate   splitting   rules (axis-parallel). Each line corresponds to a splitting rule generated at a node in the development of the tree

the observations into hyper-rectangles, determined by the intersection of half-spaces whose support hyperplanes are parallel to the components of the vector of instances, as shown in Figure 10.11.

Multivariate trees. For multivariate trees, the partition of the observations at a given node is based on the value assumed by a function $\varphi(x_1, x_2, \ldots, x_n)$ of the attributes and leads to a rule of the form $\varphi(\mathbf{x}) \leq b$ or $\varphi(\mathbf{x}) > b$. Different methods have been proposed whereby the function φ represents a linear combination of the explanatory variables. In this case, the expression that leads to the separation of the observations takes the form

$$\sum_{j=1}^{n} w_j x_j \leq b, \tag{10.16}$$

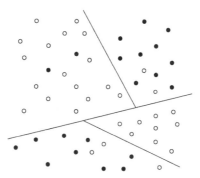

Figure 10.12   Classification by means of multivariate splitting rules (oblique). Each straight line corresponds to a splitting rule generated at a node in the development of the tree

where the threshold value b and the coefficients $w_1, w_2, \ldots, w_n$ of the linear combination have to be determined, for example by solving an optimization problem for each node, just like for classification trees generated by means of discrete variants of support vector machines. Multivariate trees are also referred to as *oblique decision trees*, as they generate polygonal partitions of the space of the observations by means of separating hyperplanes, as shown in Figure 10.12.

Oblique trees are usually characterized by greater predictive accuracy than univariate trees, against a lower interpretability of the classification rules generated. In many cases, a limited number of separating hyperplanes may be enough to classify with high accuracy the instances, whereas to achieve the same result an axis-parallel tree would require the partition of the space of the observations in several hyper-rectangles. Figure 10.13 shows an example of a two-dimensional dataset for which the two target classes may be easily separated by a single oblique line, while they would require a high number of univariate rules.

Notice also that the number of regions generated is a function of the number of leaves in the tree, and therefore of its depth. Accuracy being equal, oblique trees usually generate a lower number of classification rules with respect to univariate trees.

10.3.2   Univariate splitting criteria

Although they are usually characterized by a lower accuracy, the algorithms that develop classification trees based on univariate rules are more popular than their multivariate counterpart, partly because of the simplicity and interpretability of the rules generated and partly because they were proposed first.

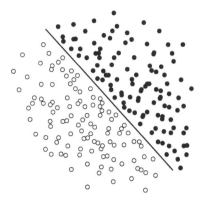

Figure 10.13 Classification by means of a single multivariate (oblique) splitting rule for a dataset that cannot be accurately split by univariate (axis-parallel) rules

The main component that differentiates the variants of univariate classification trees proposed so far is the splitting rule, used to identify the best explanatory attribute out of those available and to select the most effective partitioning criterion among the ones it induces. Usually, both choices are performed by calculating an *evaluation function*, for each attribute and for each possible partition, which provides a heterogeneity measure in the values of the target class between the examples belonging to the parent node and those belonging to the descendants. The maximization of the evaluation function therefore identifies the partition that generates descendant nodes that are more homogeneous within themselves than the parent node is.

Let p_h be the proportion of examples of target class v_h, $h \in \mathcal{H}$, at a given node q and let Q be the total number of instances at q. We have

$$\sum_{h=1}^{H} p_h = 1. \tag{10.17}$$

The heterogeneity index $I(q)$ of a node is usually a function of the relative frequencies p_h, $h \in \mathcal{H}$, of the target class values for the examples at the node, and it has to satisfy three requirements: it must take its maximum value when the examples at the node are distributed homogeneously among all the classes; it must take its minimum value when all the instances at the node belong to the same class; and it must be a symmetric function with respect to the relative frequencies p_h, $h \in \mathcal{H}$.

Among the heterogeneity indices of a node q that satisfy these properties, also referred to as *impurity* or *inhomogeneity* measures, the most popular are the *misclassification index*, the *entropy index* and the *Gini index*.

Misclassification index. The misclassification index is defined as

$$\text{Miscl}(q) = 1 - \max_h p_h, \qquad (10.18)$$

and measures the proportion of misclassified examples when all the instances at node q are assigned to the class to which the majority of them belong, according to the majority voting principle.

Entropy index. The entropy is defined as

$$\text{Entropy}(q) = -\sum_{h=1}^{H} p_h \log_2 p_h; \qquad (10.19)$$

note that, by convention, $0 \log_2 0 = 0$.

Gini index. The Gini index is defined as

$$\text{Gini}(q) = 1 - \sum_{h=1}^{H} p_h^2. \qquad (10.20)$$

In a binary classification problem the impurity measures defined above reach their maximum value when $p_1 = p_2 = 1 - p_1 = 0.5$, and are 0 when $p_1 = 0$ or $p_1 = 1$, as shown in Figure 10.14.

The univariate splitting criteria based on the *information gain* compare one of the impurity indices evaluated for the parent node with the same index

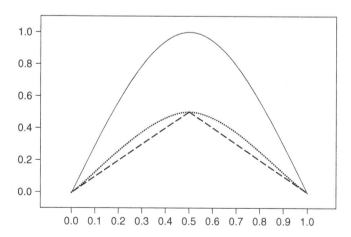

Figure 10.14   Graph of the misclassification index (dashed line), the Gini index (dotted line) and the entropy (full line) for a binary target attribute as the frequency of the examples in one class varies

computed for the set of descendant nodes, and then choose the attribute and the corresponding partition that maximize the difference. Since the value of the impurity index for the parent node is independent of the partition, the choice may be equivalently done by minimizing the overall impurity index for the descendant nodes.

Let $I(\cdot)$ be one of the impurity indices previously defined, and suppose that a splitting rule separates the examples contained at node q into K descendant nodes $\{q_1, q_2, \ldots, q_K\}$, each containing Q_k instances. If we analyze the partition originated by a categorical attribute X_j taking H_j distinct values, the set of examples at q can be separated into H_j disjoint subsets, as shown in Figure 10.9. In this way we have $K = H_j$, and the descendant node q_j contains the examples for which the explanatory variable X_j takes the value v_j. If $H_j > 2$, this partition is possible only if the tree is of general type. If a binary tree is developed, it is necessary to divide the H_j values into two non-empty sets and then compute the heterogeneity indices for all 2^{H_j-1} possible partitions. Finally, if the attribute X_j is numerical, the examples can be subdivided by intervals of values, as shown in Figure 10.10. To avoid the analysis of all possible threshold values for the separation, the algorithm operates a binary search among the values actually taken by the attribute X_j in the dataset $\mathcal{D}$.

The impurity of the descendant nodes, and therefore the impurity of the splitting rule, is defined as

$$I(q_1, q_2, \ldots, q_K) = \sum_{k=1}^{K} \frac{Q_k}{Q} I(q_k). \tag{10.21}$$

Hence, the impurity of the partition is expressed by a weighted sum of the impurities of each descendant node, where each weight equals the percentage of examples from the parent node that are placed in the corresponding descendant.

The algorithms for the development of univariate trees select for each node the rule and the corresponding attribute that determine the minimum value of expression (10.21). This choice is equivalent to maximizing the information gain $\Delta(\cdot)$, defined as

$$\Delta(q, q_1, q_2, \ldots, q_K) = I(q) - I(q_1, q_2, \ldots, q_K)$$
$$= I(q) - \sum_{k=1}^{K} \frac{Q_k}{Q} I(q_k). \tag{10.22}$$

10.3.3 Example of development of a classification tree

In this section we will describe the development of two classification trees obtained by using respectively the entropy and the Gini index for the small dataset from Example 5.2 comprising 23 observations, drawn as part of a retention analysis in the mobile phone industry.

Discretization of the dataset

Although it would be possible to apply the two classifiers to datasets containing numerical attributes, in order to make the description of the two algorithms easier it is appropriate to perform a discretization of the numerical predictive variables, as a preliminary step to the development of the classification trees.

The numerical attributes in the dataset have been discretized by subdividing the examples into equally wide classes based on the ranges of values. Table 10.3 shows the classes obtained, and Table 10.4 shows the new dataset consisting solely of categorical variables, obtained at the end of the discretization step.

Entropy index

By referring to Table 10.4 and using the definition of the entropy index in (10.19), it is possible to compute the amount of information required to establish if a generic example belongs to class $\{0\}$ or to class $\{1\}$ corresponding to the root node for the entire dataset. Indeed, the entropy of the root q is equal to

$$I_E(q) = \text{Entropy}(q) = -\frac{13}{23} \log_2 \frac{13}{23} - \frac{10}{23} \log_2 \frac{10}{23} = 0.988.$$

By splitting on the attribute *area*, the root node would be subdivided into four child nodes $\{q_1, q_2, q_3, q_4\}$ that correspond to the values $\{1, 2, 3, 4\}$, as shown in Figure 10.10. Hence, associated with the four descendant nodes are the following proportions of membership in the two target classes $\{0, 1\}$:

$$p_0(q_1) = 5/6, \quad p_0(q_2) = 2/5, \quad p_0(q_3) = 5/7, \quad p_0(q_4) = 1/5, \quad (10.23)$$

$$p_1(q_1) = 1/6, \quad p_1(q_2) = 3/5, \quad p_1(q_3) = 2/7, \quad p_1(q_4) = 4/5. \quad (10.24)$$

Table 10.3   Subdivision into classes for the discretization of the numerical attributes in Example 5.2

attribute	class 1	class 2	class 3	class 4
numin	[0,20)	[20,40)	[40, ∞)	
timein	[0,10 000)	[10 000,20 000)	[20 000,30 000)	[30 000, ∞)
numout	[0,30)	[30,60)	[60,90)	[90, ∞)
Pothers	[0,0.1)	[0.1,0.2)	[0.2,0.3)	[0.3, ∞)
Pmob	[0,0.2)	[0.2,0.4)	[0.4,0.6)	[0.6, ∞)
Pland	[0,0.25)	[0.25,0.5)	[0.5, ∞)	
numsms	[0,1)	[1,10)	[10,20)	[20, ∞)
numserv	[0,1)	[1,2)	[2,3)	[3, ∞)
numcall	[0,1)	[1,3)	[3, ∞)	

Table 10.4 Discretized input data for Example 5.2

area	numin	timein	numout	Pothers	Pmob	Pland	numsms	numserv	numcall	diropt	churner
2	1	1	2	1	4	1	3	2	2	0	1
1	1	3	3	2	4	1	4	2	3	0	0
3	2	1	2	2	4	1	3	2	1	0	0
1	2	3	2	3	4	1	1	2	1	0	0
2	3	4	4	4	1	1	3	2	1	0	0
3	3	4	1	4	2	1	4	3	1	0	0
3	3	3	4	4	3	1	4	3	1	1	0
1	1	1	1	1	3	2	1	1	1	0	1
2	2	2	2	2	3	2	2	3	1	1	1
4	2	1	3	2	3	2	2	2	1	0	1
3	1	1	2	2	2	2	2	4	1	1	1
4	3	4	4	3	1	2	2	4	1	0	1
2	1	1	3	2	3	2	2	2	1	1	0
4	2	1	2	3	2	2	2	4	1	0	0
3	3	4	4	4	2	2	4	2	1	1	0
1	1	2	1	1	1	3	1	1	1	0	0
4	1	1	2	1	1	3	1	1	1	0	1
3	3	1	1	1	1	3	1	2	1	0	0
2	3	4	3	2	2	3	4	1	1	0	1
1	2	3	3	2	2		1	1	1	0	1
4	2	2	2	4	1				1	0	0
3	3	2	1	4			1	1	1	0	0

The entropy index, which measures the heterogeneity of the four descendant nodes, is given by

$$I_E(q_1, q_2, q_3, q_4) = \frac{6}{23}I_E(q_1) + \frac{5}{23}I_E(q_2) + \frac{7}{23}I_E(q_3) + \frac{5}{23}I_E(q_4)$$

$$= \frac{6}{23}0.650 + \frac{5}{23}0.971 + \frac{7}{23}0.863 + \frac{5}{23}0.722 = 0.8.$$

Therefore, the gain achieved by separating the examples through the attribute *area* is

$$\Delta_E(area) = \Delta_E(q, q_1, q_2, q_3, q_4) = I_E(q) - I_E(q_1, q_2, q_3, q_4)$$

$$= 0.988 - 0.8 = 0.188.$$

Similarly, it is possible to compute the gain associated with the separation performed on the remaining attributes of the dataset:

$$\Delta_E(numin) = 0.057, \qquad \Delta_E(Pland) = 0.125,$$
$$\Delta_E(timein) = 0.181, \qquad \Delta_E(numsms) = 0.080,$$
$$\Delta_E(numout) = 0.065, \qquad \Delta_E(numserv) = 0.057,$$
$$\Delta_E(Pothers) = 0.256, \qquad \Delta_E(numcall) = 0.089,$$
$$\Delta_E(Pmob) = 0.043, \qquad \Delta_E(diropt) = 0.005.$$

The maximum information gain is achieved when the examples are separated by means of the attribute *Pothers*, which measures the proportion of calls placed to other mobile phone companies. By recursively performing the subsequent iterations in an analogous way, we obtain the classification tree depicted in Figure 10.7.

Gini index

The Gini index evaluated for the root q, corresponding to the entire dataset described in Table 10.4, is equal to

$$I_G(q) = \text{Gini}(q) = 1 - \left(\frac{13}{23}\right)^2 - \left(\frac{10}{23}\right)^2 = 0.491.$$

As outlined above, the attribute *area* separates the examples at the root node q into four descendant nodes $\{q_1, q_2, q_3, q_4\}$ corresponding to the values $\{1, 2, 3, 4\}$, and associated with the proportions of membership in the two target classes $\{0, 1\}$ shown in (10.23) and (10.24).

The Gini index, which measures the heterogeneity of the four descendant nodes, is equal to

$$I_G(q_1, q_2, q_3, q_4) = \frac{6}{23}I_G(q_1) + \frac{5}{23}I_G(q_2) + \frac{7}{23}I_G(q_3) + \frac{5}{23}I_G(q_4)$$

$$= \frac{6}{23}0.278 + \frac{5}{23}0.480 + \frac{7}{23}0.408 + \frac{5}{23}0.528 = 0.320.$$

Therefore, the gain achieved by separating the examples through the attribute *area* is

$$\Delta_G(area) = \Delta_G(q, q_1, q_2, q_3, q_4) = I_G(q) - I_G(q_1, q_2, q_3, q_4)$$

$$= 0.491 - 0.370 = 0.121.$$

Similarly, it is possible to compute the gain associated with the separation performed on the remaining attributes of the dataset:

$$\Delta_G(numin) = 0.037, \quad \Delta_G(Pland) = 0.078,$$
$$\Delta_G(timein) = 0.091, \quad \Delta_G(numsms) = 0.052,$$
$$\Delta_G(numout) = 0.043, \quad \Delta_G(numserv) = 0.038,$$
$$\Delta_G(Pothers) = 0.146, \quad \Delta_G(numcall) = 0.044,$$
$$\Delta_G(Pmob) = 0.028, \quad \Delta_G(diropt) = 0.003.$$

As for the entropy index, the examples at the root of the tree are best separated by means of the attribute *Pothers*, for which the increase in homogeneity is maximum.

10.3.4 Stopping criteria and pruning rules

Stopping criteria are a set of rules used at each node during the development of a tree in order to determine whether it is appropriate to create more branches and generate descendant nodes or whether the current node should become a leaf. There are two main reasons to limit the growth of a classification tree. First, a tree with too many ramifications usually achieves better accuracy with the training set but causes larger errors when used to make predictions on the test set or on future data. From an intuitive point of view, one may think that a tree with many branches excessively reflects the peculiarity of the examples in the training set and is therefore less capable of generalization. This phenomenon, usually referred to as *overfitting*, can be well explained from a theoretical perspective within the framework of statistical learning theory: a tree with too many ramifications actually represents a broader space of hypotheses, and therefore reduces the empirical error for the training set, but at the same time increases the generalization error.

Furthermore, a more ramified tree implies a proliferation of leaves and hence generates deep classification rules, obtained by combining a large number of splitting rules along the path leading from the root node to a leaf. This reduces the overall interpretability of the resulting classification model.

In principle, node splitting might be stopped when there is only one observation in the node. The stopping criteria followed in practice prevent this situation from happening, by introducing some restrictions on the minimum number of observations that a node must contain to be partitioned. Additionally, it can be assigned a minimum threshold on the uniformity, sometimes called *purity*, of the target class proportion within a node to be split.

More precisely, a node becomes a leaf of the tree when at least one of the following conditions occurs.

Node size. The node contains a number of observations that is below a preset minimum threshold value.

Purity. The proportion of observations at the node and belonging to the same class is above a preset maximum threshold value that corresponds to the accuracy that one wishes to achieve.

Improvement. The possible subdivision of the node would generate a gain $\Delta(\cdot)$ that is below a preset minimum threshold value.

The stopping criteria described above are also referred to as *pre-pruning* rules because their aim is to prune a priori the tree by limiting its growth. However, there are other *post-pruning*, or simply *pruning*, techniques which are applied upon completion of tree development to reduce the number of ramifications without worsening, and hopefully even improving, the predictive accuracy of the resulting model. At each iteration during the post-pruning phase, the possible advantage deriving from the removal of a given branch is evaluated, by comparing the predictive accuracy of the original tree with that of the reduced tree in classifying the observations of the tuning set. At the end of the evaluation, the reduced tree associated with the minimum prediction error is selected.

10.4   Bayesian methods

Bayesian methods belong to the family of probabilistic classification models. They explicitly calculate the *posterior* probability $P(y|\mathbf{x})$ that a given observation belongs to a specific target class by means of Bayes' theorem, once the *prior* probability $P(y)$ and the class conditional probabilities $P(\mathbf{x}|y)$ are known.

Unlike other methods described in this chapter, which are not based on probabilistic assumptions, Bayesian classifiers require the user to estimate the probability $P(\mathbf{x}|y)$ that a given observation may occur, provided it belongs to a specific class. The learning phase of a Bayesian classifier may therefore be identified with a preliminary analysis of the observations in the training set, to derive an estimate of the probability values required to perform the classification task.

Let us consider a generic observation $\mathbf{x}$ of the training set, whose target variable y may take H distinct values denoted as $\mathcal{H} = \{v_1, v_2, \ldots, v_H\}$.

Bayes' theorem is used to calculate the posterior probability $P(y|\mathbf{x})$, that is, the probability of observing the target class y given the example $\mathbf{x}$:

$$P(y|\mathbf{x}) = \frac{P(\mathbf{x}|y)P(y)}{\sum_{l=1}^{H} P(\mathbf{x}|y)P(y)} = \frac{P(\mathbf{x}|y)P(y)}{P(\mathbf{x})}. \qquad (10.25)$$

In order to classify a new instance $\mathbf{x}$, the Bayes classifier applies a principle known as the *maximum a posteriori hypothesis* (MAP), which involves calculating the posterior probability $P(y|\mathbf{x})$ using (10.25) and assigning the example $\mathbf{x}$ to the class that yields the maximum value $P(y|\mathbf{x})$, that is,

$$y_{\text{MAP}} = \arg\max_{y \in \mathcal{H}} P(y|\mathbf{x}) = \arg\max_{y \in \mathcal{H}} \frac{P(\mathbf{x}|y)P(y)}{P(\mathbf{x})}. \qquad (10.26)$$

Since the denominator $P(\mathbf{x})$ is independent of y, in order to maximize the posterior probability it is enough to maximize the numerator of (10.26). Therefore, the observation $\mathbf{x}$ is assigned to the class v_h if and only if

$$P(\mathbf{x}|y = v_h)P(y = v_h) \geq P(\mathbf{x}|y = v_l)P(y = v_l), \quad l = 1, 2, \ldots, H.$$

The prior probability $P(y)$ can be estimated using the frequencies m_h with which each value of the target class v_h appears in the dataset $\mathcal{D}$, that is,

$$P(y = v_h) = \frac{m_h}{m}. \qquad (10.27)$$

Given a sufficiently large sample the estimates of the prior probabilities obtained through (10.27) will be quite accurate.

Unfortunately an analogous sample estimate of the class conditional probabilities $P(\mathbf{x}|y)$ cannot be obtained in practice due to the computational complexity and the huge number of sample observations that it would require. To see this, consider the following hypothetical situation: if the dataset includes 50 binary categorical attributes, there would be $2^{50} \approx 10^{15}$ possible combinations of attribute values that constitute the cells of the joint density of $\mathbf{x}$ for which the relative frequencies should be estimated. To obtain a reliable estimate it would be necessary to have at least 10 observations for each combination. Hence, the dataset $\mathcal{D}$ should contain at least 10^{16} observations. If the attributes were nominal categorical, or numerical, this number would grow exponentially fast.

To overcome the computational difficulty described, it is possible to introduce two simplifying hypotheses that lead to *naive Bayesian* classifiers and to *Bayesian networks* respectively, described next.

10.4.1   Naive Bayesian classifiers

Naive Bayesian classifiers are based on the assumption that the explanatory variables are conditionally independent given the target class. This hypothesis allows us to express the probability $P(\mathbf{x}|y)$ as

$$P(\mathbf{x}|y) = P(x_1|y) \times P(x_2|y) \times \cdots \times P(x_n|y) = \prod_{j=1}^{n} P(x_j|y). \qquad (10.28)$$

The probabilities $P(x_j|y)$, $j \in \mathcal{N}$, can be estimated using the examples from the training set, depending on the nature of the attribute considered.

Categorical or discrete numerical attributes. For a categorical or discrete numerical attribute $\mathbf{a}_j$ which may take the values $\{r_{j1}, r_{j2}, \ldots, r_{jK_j}\}$, the probability $P(x_j|y) = P(x_j = r_{jk}|y = v_h)$ is evaluated as the ratio between the number s_{jhk} of instances of class v_h for which the attribute $\mathbf{a}_j$ takes the value r_{jk}, and the total number m_h of instances of class v_h in the dataset $\mathcal{D}$, that is,

$$P(x_j|y) = P(x_j = r_{jk}|y = v_h) = \frac{s_{jhk}}{m_h}. \qquad (10.29)$$

Numerical attributes. For a numerical attribute $\mathbf{a}_j$, the probability $P(x_j|y)$ is estimated assuming that the examples follow a given distribution. For example, one may consider a Gaussian density function, for which

$$P(x_j|y = v_h) = \frac{1}{\sqrt{2\pi}\sigma_{jh}} e^{-\frac{(x_j - \mu_{jh})^2}{2\sigma_{jh}^2}}, \qquad (10.30)$$

where μ_{jh} and σ_{jh} respectively denote the mean and standard deviation of the variable X_j for the examples of class v_h, and may be estimated on the basis of the examples contained in $\mathcal{D}$.

In spite of the simplifying assumption of conditional independence of the attributes, which makes it easy to compute the conditional probabilities, the empirical evidence shows that Bayesian classifiers are often able to achieve accuracy levels which are not lower than those provided by classification trees or even by more complex classification methods.

10.4.2 Example of naive Bayes classifier

In order to describe the usage of the naive Bayes classifier, consider again the dataset in Table 10.4, obtained by discretization of the numerical attributes in the small dataset comprising the 23 observations of Example 5.2, concerning a retention analysis in the mobile phone industry.

The relative frequencies of the sample attribute values given the target class are as follows:

area

$$P\,(area = 1|\,churner = 0) = \frac{5}{13}, \qquad P\,(area = 1|\,churner = 1) = \frac{1}{10},$$

$$P\,(area = 2|\,churner = 0) = \frac{2}{13}, \qquad P\,(area = 2|\,churner = 1) = \frac{3}{10},$$

$$P\,(area = 3|\,churner = 0) = \frac{5}{13}, \qquad P\,(area = 3|\,churner = 1) = \frac{2}{10},$$

$$P\,(area = 4|\,churner = 0) = \frac{1}{13}, \qquad P\,(area = 4|\,churner = 1) = \frac{4}{10}.$$

numin

$$P\,(numin = 1|\,churner = 0) = \frac{4}{13}, \qquad P\,(numin = 1|\,churner = 1) = \frac{5}{10},$$

$$P\,(numin = 2|\,churner = 0) = \frac{3}{13}, \qquad P\,(numin = 2|\,churner = 1) = \frac{3}{10},$$

$$P\,(numin = 3|\,churner = 0) = \frac{6}{13}, \qquad P\,(numin = 3|\,churner = 1) = \frac{2}{10}.$$

timein

$$P\,(timein = 1|\,churner = 0) = \frac{4}{13}, \qquad P\,(timein = 1|\,churner = 1) = \frac{6}{10},$$

$$P\,(timein = 2|\,churner = 0) = \frac{2}{13}, \qquad P\,(timein = 2|\,churner = 1) = \frac{2}{10},$$

$$P\,(timein = 3|\,churner = 0) = \frac{4}{13}, \qquad P\,(timein = 3|\,churner = 1) = 0,$$

$$P\,(timein = 4|\,churner = 0) = \frac{3}{13}, \qquad P\,(timein = 4|\,churner = 1) = \frac{2}{10}.$$

numout

$$P\,(numout = 1|\,churner = 0) = \frac{4}{13}, \qquad P\,(numout = 1|\,churner = 1) = \frac{2}{10},$$

$$P\,(numout = 2|\,churner = 0) = \frac{3}{13}, \qquad P\,(numout = 2|\,churner = 1) = \frac{5}{10},$$

$$P\,(numout = 3|\,churner = 0) = \frac{3}{13}, \qquad P\,(numout = 3|\,churner = 1) = \frac{2}{10},$$

$$P\,(numout = 4|\,churner = 0) = \frac{3}{13}, \qquad P\,(numout = 4|\,churner = 1) = \frac{1}{10}.$$

Pothers

$$P\,(Pothers = 1|\,churner = 0) = \frac{2}{13}, \qquad P\,(Pothers = 1|\,churner = 1) = \frac{5}{10},$$

$$P\,(Pothers = 2|\,churner = 0) = \frac{3}{13}, \qquad P\,(Pothers = 2|\,churner = 1) = \frac{4}{10},$$

$$P\,(Pothers = 3|\;churner = 0) = \frac{3}{13}, \qquad P\,(Pothers = 3|\;churner = 1) = 0,$$

$$P\,(Pothers = 4|\;churner = 0) = \frac{5}{13}, \qquad P\,(Pothers = 4|\;churner = 1) = \frac{1}{10}.$$

Pmob

$$P\,(Pmob = 1|\;churner = 0) = \frac{3}{13}, \qquad P\,(Pmob = 1|\;churner = 1) = \frac{3}{10},$$

$$P\,(Pmob = 2|\;churner = 0) = \frac{5}{13}, \qquad P\,(Pmob = 2|\;churner = 1) = \frac{3}{10},$$

$$P\,(Pmob = 3|\;churner = 0) = \frac{2}{13}, \qquad P\,(Pmob = 3|\;churner = 1) = \frac{3}{10},$$

$$P\,(Pmob = 4|\;churner = 0) = \frac{3}{13}, \qquad P\,(Pmob = 4|\;churner = 1) = \frac{1}{10}.$$

Pland

$$P\,(Pland = 1|\;churner = 0) = \frac{6}{13}, \qquad P\,(Pland = 1|\;churner = 1) = \frac{1}{10},$$

$$P\,(Pland = 2|\;churner = 0) = \frac{4}{13}, \qquad P\,(Pland = 2|\;churner = 1) = \frac{6}{10},$$

$$P\,(Pland = 3|\;churner = 0) = \frac{3}{13}, \qquad P\,(Pland = 3|\;churner = 1) = \frac{3}{10}.$$

numsms

$$P\,(numsms = 1|\;churner = 0) = \frac{3}{13}, \qquad P\,(numsms = 1|\;churner = 1) = \frac{5}{10},$$

$$P\,(numsms = 2|\;churner = 0) = \frac{4}{13}, \qquad P\,(numsms = 2|\;churner = 1) = \frac{3}{10},$$

$$P\,(numsms = 3|\;churner = 0) = \frac{2}{13}, \qquad P\,(numsms = 3|\;churner = 1) = \frac{1}{10},$$

$$P\,(numsms = 4|\;churner = 0) = \frac{4}{13}, \qquad P\,(numsms = 4|\;churner = 1) = \frac{1}{10}.$$

numserv

$$P\,(numserv = 1|\;churner = 0) = \frac{2}{13}, \qquad P\,(numserv = 1|\;churner = 1) = \frac{4}{10},$$

$$P\,(numserv = 2|\;churner = 0) = \frac{7}{13}, \qquad P\,(numserv = 2|\;churner = 1) = \frac{4}{10},$$

$$P\,(numserv = 3|\;churner = 0) = \frac{2}{13}, \qquad P\,(numserv = 3|\;churner = 1) = \frac{1}{10},$$

$$P\,(numserv = 4|\;churner = 0) = \frac{2}{13}, \qquad P\,(numserv = 4|\;churner = 1) = \frac{1}{10}.$$

numcall

$$P\,(numcall = 1|\,churner = 0) = \frac{12}{13}, \quad P\,(numcall = 1|\,churner = 1) = \frac{9}{10},$$

$$P\,(numcall = 2|\,churner = 0) = 0, \quad P\,(numcall = 2|\,churner = 1) = \frac{1}{10},$$

$$P\,(numcall = 3|\,churner = 0) = \frac{1}{13}, \quad P\,(numcall = 3|\,churner = 1) = 0.$$

diropt

$$P\,(diropt = 0|\,churner = 0) = \frac{10}{13}, \quad P\,(diropt = 0|\,churner = 1) = \frac{7}{10},$$

$$P\,(diropt = 1|\,churner = 0) = \frac{3}{13}, \quad P\,(diropt = 1|\,churner = 1) = \frac{3}{10}.$$

Once the conditional probabilities of each attribute given the target class have been estimated, suppose that we wish to predict the target class of a new observation, represented by the vector $\mathbf{x} = (1, 1, 1, 2, 1, 4, 2, 1, 2, 1, 0)$. With this aim in mind, we compute the posterior probabilities $P(\mathbf{x}|0)$ and $P(\mathbf{x}|1)$:

$$P\,(\mathbf{x}|0) = \frac{5}{13} \cdot \frac{4}{13} \cdot \frac{4}{13} \cdot \frac{3}{13} \cdot \frac{2}{13} \cdot \frac{3}{13} \cdot \frac{4}{13} \cdot \frac{3}{13} \cdot \frac{7}{13} \cdot \frac{12}{13} \cdot \frac{10}{13} = 0.81 \cdot 10^{-5},$$

$$P\,(\mathbf{x}|1) = \frac{1}{10} \cdot \frac{5}{10} \cdot \frac{6}{10} \cdot \frac{5}{10} \cdot \frac{5}{10} \cdot \frac{1}{10} \cdot \frac{6}{10} \cdot \frac{5}{10} \cdot \frac{4}{10} \cdot \frac{9}{10} \cdot \frac{7}{10} = 5.67 \cdot 10^{-5}.$$

Since the relative frequencies of the two classes are given by

$$P\,(churner = 0) = \frac{13}{23} = 0.56, \quad P\,(churner = 1) = \frac{10}{23} = 0.44,$$

we have

$$P(churner = 0|\mathbf{x}) = P(\mathbf{x}|0)P\,(churner = 0)$$

$$= 0.81 \cdot 10^{-5} \cdot 0.56 = 0.46 \cdot 10^{-5},$$

$$P(churner = 1|\mathbf{x}) = P(\mathbf{x}|1)P\,(churner = 1)$$

$$= 5.67 \cdot 10^{-5} \cdot 0.44 = 2.495 \cdot 10^{-5}.$$

The new example $\mathbf{x}$ is then labeled with the class value $\{1\}$, since this is associated with the maximum a posteriori probability.

10.4.3   Bayesian networks

Bayesian networks, also called *belief networks*, allow the hypothesis of conditional independence of the attributes to be relaxed, by introducing some reticular hierarchical links through which it is possible to assign selected stochastic dependencies that experts of the application domain deem relevant.

A Bayesian network comprises two main components. The first is an *acyclic oriented graph* in which the nodes correspond to the predictive variables and the arcs indicate relationships of stochastic dependence. In particular, it is assumed that the variable X_j associated with node $\mathbf{a}_j$ in the network is dependent on the variables associated with the predecessor nodes of $\mathbf{a}_j$, and conditionally independent of the variables associated with the nodes that are not directly reachable from $\mathbf{a}_j$.

The second component consists of a table of conditional probabilities assigned for each variable. In particular, the table associated with the variable X_j indicates the conditional distribution of $P(X_j|C_j)$, where C_j represents the set of explanatory variables associated with the predecessor nodes of node $\mathbf{a}_j$ in the network and is estimated based on the relative frequencies in the dataset. The complexity required to compute all possible combinations of predictor values for estimating the conditional probabilities, already pointed out above, is limited in Bayesian networks, since the calculation simply reduces to those conditioning relationships determined by the precedence links included in the network. Consequently, the number of precedence links should be restricted to avoid the overwhelming computational effort mentioned above.

10.5 Logistic regression

Logistic regression is a technique for converting binary classification problems into linear regression ones, described in Chapter 8, by means of a proper transformation.

Suppose that the response variable y takes the values $\{0,1\}$, as in a binary classification problem. The logistic regression model postulates that the posterior probability $P(y|\mathbf{x})$ of the response variable conditioned on the vector $\mathbf{x}$ follows a *logistic function*, given by

$$P(y = 0|\mathbf{x}) = \frac{1}{1 + e^{\mathbf{w}'\mathbf{x}}}, \tag{10.31}$$

$$P(y = 1|\mathbf{x}) = \frac{e^{\mathbf{w}'\mathbf{x}}}{1 + e^{\mathbf{w}'\mathbf{x}}}. \tag{10.32}$$

Here we suppose that the matrix $\mathbf{X}$ and the vector $\mathbf{w}$ have been extended to include the intercept as described in Section 8.3. The *standard logistic function* $S(t)$, also known as the *sigmoid* function, can be found in many applications of statistics in the economic and biological fields and is defined as

$$S(t) = \frac{1}{1 + e^{-t}}. \tag{10.33}$$

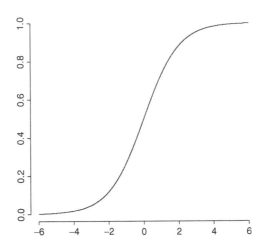

Figure 10.15   Graph of the standard logistic function (sigmoid)

The function $S(t)$ has the graphical shape shown in Figure 10.15.

By inverting expressions (10.31) and (10.32), we observe that the logarithm of the ratio between the conditional probabilities of the two classes depends linearly on the predictive variables, that is

$$\log \frac{P(y=1|\mathbf{x})}{P(y=0|\mathbf{x})} = \mathbf{w}'\mathbf{x}. \qquad (10.34)$$

Consequently, by setting

$$z = \log \frac{P(y=1|\mathbf{x})}{P(y=0|\mathbf{x})}, \qquad (10.35)$$

the binary classification problem is traced back to the identification of a linear regression model between the dependent variable z and the original explanatory attributes. Once the linear regression coefficients have been calculated and the significance of the model verified, using the validation tests described in Section 8.4, one may use the model to predict the target class of a new observation $\mathbf{x}$. The coefficients $\mathbf{w}$ are computed using an iterative method, usually aimed at maximizing the likelihood, by minimizing the sum of logarithms of predicted probabilities.

In general, logistic regression models present the same difficulties described in connection with regression models, from which they derive. To avoid multi-collinearity phenomena that jeopardize the significance of the regression coeffi-cients it is necessary to proceed with attribute selection. Moreover, the accuracy of logistic regression models is in most cases lower than that obtained using other classifiers and usually requires a greater effort for the development of the model. Finally, it appears computationally cumbersome to treat large datasets, both in terms of number of observations and number of attributes.

10.6   Neural networks

Neural networks are intended to simulate the behavior of biological systems composed of neurons. Since the 1950s, when the simplest models were proposed, neural networks have been used for predictive purposes, not only for classification but also for regression of continuous target attributes.

A neural network is an oriented graph consisting of nodes, which in the biological analogy represent neurons, connected by arcs, which correspond to dendrites and synapses. Each arc is associated with a *weight*, while at each node an *activation function* is defined which is applied to the values received as input by the node along the incoming arcs, adjusted by the weights of the arcs. The training stage is performed by analyzing in sequence the observations contained in the training set one after the other and by modifying at each iteration the weights associated with the arcs.

10.6.1   The Rosenblatt perceptron

The *perceptron*, shown in Figure 10.16, is the simplest form of neural network and corresponds to a single neuron that receives as input the values $(x_1, x_2, \ldots, x_n)$ along the incoming connections, and returns an output value $f(\mathbf{x})$. The input values coincide with the values of the explanatory attributes, while the output value determines the prediction of the response variable y. Each of the n input connections is associated with a weight w_j. An activation function g and a constant ϑ, called the *distortion*, are also assigned.

Suppose that the values of the weights and the distortion have already been determined during the training phase. The prediction for a new observation $\mathbf{x}$ is then derived by performing the following steps.

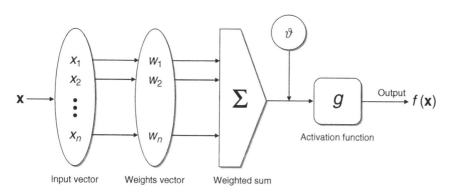

Figure 10.16   Operation of a single unit in a neural network

First, the weighted linear combination of the values of the explanatory variables for the new observation is calculated and the distortion is subtracted from it:

$$w_1 x_1 + w_2 x_2 + \cdots + w_n x_n - \vartheta = \mathbf{w}' \mathbf{x} - \vartheta. \qquad (10.36)$$

The prediction $f(\mathbf{x})$ is then obtained by applying the activation function g to the linear combination of the predictors:

$$f(\mathbf{x}) = g(w_1 x_1 + w_2 x_2 + \cdots + w_n x_n - \vartheta) = g(\mathbf{w}' \mathbf{x} - \vartheta). \qquad (10.37)$$

The purpose of the function g is to map the linear combination into the set $\mathcal{H} = \{v_1, v_2, \ldots, v_H\}$ of the values assumed by the target variable, usually by means of a sigmoid profile. For binary classification problems we have $\mathcal{H} = \{-1, 1\}$, so that one may select $g(\cdot) = \text{sgn}(\cdot)$, making the prediction coincide with the sign of the weighted sum in (10.36):

$$f(\mathbf{x}) = \text{sgn}(w_1 x_1 + w_2 x_2 + \cdots + w_n x_n - \vartheta) = g(\mathbf{w}' \mathbf{x} - \vartheta). \qquad (10.38)$$

An iterative algorithm is then used to determine the values of the weights w_j and the distortion ϑ, examining the examples in sequence, one after the other. For each example $\mathbf{x}_i$ the prediction $f(\mathbf{x}_i)$ is calculated, and the value of the parameters is then updated using recursive formulas that take into account the error $y_i - f(\mathbf{x}_i)$.

For binary classification problems it is possible to give a geometrical interpretation of the prediction obtained using a Rosenblatt perceptron. Indeed, if we place the m observations of the training dataset in the space $\mathbb{R}^n$, the weighted linear combination in (10.36) calculated for $\mathbf{x}_i$ expresses the slack between the observation and the hyperplane:

$$z = w_1 x_1 + w_2 x_2 + \cdots + w_n x_n - \vartheta = \mathbf{w}' \mathbf{x} - \vartheta. \qquad (10.39)$$

The purpose of the activation function $g(\cdot) = \text{sgn}(\cdot)$ is therefore to establish if the point associated with the example $\mathbf{x}_i$ is placed in the lower or upper halfspace with respect to the separating hyperplane. Hence, the Rosenblatt perceptron corresponds to a linear separation of the observations based on the target class. The aim of the iterative procedure is therefore to determine the coefficients of the separating hyperplane. Notice, however, that these coefficients can be derived more efficiently by solving a single optimization problem, as we will show in Section 10.7 devoted to support vector machines.

10.6.2   Multi-level feed-forward networks

A *multi-level feed-forward* neural network, shown in Figure 10.17, is a more complex structure than the perceptron, since it includes the following components.

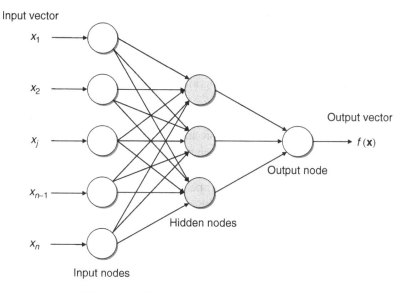

Figure 10.17   Example of neural network

Input nodes. The purpose of the input nodes is to receive as input the values of the explanatory attributes for each observation. Usually, the number of input nodes equals the number of explanatory variables.

Hidden nodes. Hidden nodes apply given transformations to the input values inside the network. Each node is connected to incoming arcs that go from other hidden nodes or from input nodes, and it is connected with outgoing arcs to output nodes or to other hidden nodes.

Output nodes. Output nodes receive connections from hidden nodes or from input nodes and return an output value that corresponds to the prediction of the response variable. In classification problems, there is usually only one output node.

Each node of the network basically operates as a perceptron, in the sense that given weights are associated with the input arcs, while each node is associated with a distortion coefficient and an activation function. In general, the activation function may assume forms that are more complex than the sign function sgn(·), such as a linear function, a sigmoid or a hyperbolic tangent.

The method that determines the weights of all the arcs and the distortions at the nodes, called the *backpropagation* algorithm, follows a logic that is not dissimilar from that described for the single perceptron. The weights are initialized in an arbitrary way, for instance by setting their value equal to randomly generated numbers. The examples of the training set are therefore examined in sequence, using at each iteration the current values of the weights, in order to calculate the prediction and the corresponding misclassification error. This latter is used to recursively correct the values of the weights, used at a later time to analyze the subsequent example within the procedure. The weights are updated using a descent algorithm, which is a variant of the gradient method.

One of the strengths of neural networks is that they are a learning mechanism applicable to both classification and regression problems. Furthermore, they perform attribute selection automatically, since irrelevant or redundant variables can be excluded from the analysis by looking at the coefficients assuming negligible values. However, neural networks require very long times for model training, provide results with modest interpretability that are dependent upon the order in which the examples are analyzed, and also present a lower robustness with respect to data affected by noise.

10.7   Support vector machines

Support vector machines are a family of separation methods for classification and regression developed in the context of statistical learning theory. They have been shown to achieve better performance in terms of accuracy with respect to other classifiers in several application domains, and to be efficiently scalable for large problems. A further important feature is concerned with the interpretation of the classification rules generated. Support vector machines identify a set of examples, called *support vectors*, which appear to be the most representative observations for each target class. In a way, they play a more critical role than the other examples, since they define the position of the separating surface generated by the classifier in the attribute space.

10.7.1   Structural risk minimization

As already observed, a classification algorithm $A_{\mathcal{F}}$ defines an appropriate hypothesis space $\mathcal{F}$ and a function $f^* \in \mathcal{F}$ which optimally describes the relationship between the class value y and the vector of explanatory variables $\mathbf{x}$. In order to describe the criteria for selecting the function f^*, let $V(y, f(\mathbf{x}))$ denote a loss function which measures the discrepancy between the values returned by the predictive function $f(\mathbf{x})$ and the actual values of the class y.

To select an optimal hypothesis $f^* \in \mathcal{F}$, decision theory suggests minimizing the *expected risk* functional, defined as

$$R(f) = \frac{1}{2} \int V(y, f(\mathbf{x})) dP(\mathbf{x}, y), \qquad (10.40)$$

where $P(\mathbf{x}, y) = P_{\mathbf{x},y}(\mathbf{x}, y)$ denotes the joint probability distribution over $\mathbb{R}^n \times \mathcal{H}$ of the examples $(\mathbf{x}, y)$ from which the instances in the dataset $\mathcal{D}$ are assumed to be independently drawn.

Since the distribution $P(\mathbf{x}, y)$ is generally unknown, in place of the expected risk one is naturally led to minimize the *empirical risk* over the training set $\mathcal{T}$, defined as

$$R_{emp}(f) = \frac{1}{t} \sum_{i=1}^{t} V(y_i, f(\mathbf{x}_i)). \tag{10.41}$$

However, this induction principle for selecting the function f^*, called *empirical risk minimization* (ERM), suffers from two critical drawbacks, *overfitting* and *ill-posedness*, as described in what follows.

If the space $\mathcal{F}$ is chosen too broad, the empirical error can be significantly reduced, eventually to zero, but the optimal hypothesis f^* has a low generalization capability in predicting the output value of the instances in the test set $\mathcal{V}$ or of future unseen examples. In these cases, the expected risk may still be large even if the empirical risk approaches 0, since the minimization of $R_{emp}(f)$ does not necessarily imply the minimization of $R(f)$.

This phenomenon, known as *overfitting*, has been investigated in detail in the framework of statistical learning theory, leading to probability bounds on the difference between expected risk and empirical risk. In general, these bounds are inversely dependent on the size t of the training set $\mathcal{T}$ and directly dependent on the *capacity*, known as the *Vapnik–Chervonenkis dimension* (VC), which measures the complexity of the hypothesis space $\mathcal{F}$.

The VC dimension of the hypothesis space $\mathcal{F}$ is defined as the maximum number of training points that can be correctly classified by a function from $\mathcal{F}$, wherever the points are placed in the space and whatever their binary class values $\{-1, 1\}$ are. If the VC dimension is h, then there exists at least one set of $h + 1$ points that cannot be correctly classified by the hypotheses belonging to the space $\mathcal{F}$. For instance, if the training examples are represented by points in two-dimensional space, the class of functions represented by oriented straight lines has dimension VC $= 3$, since there exist sets of four points which cannot be *shattered* by means of a single separating line, as depicted in Figure 10.18. More generally, it has been shown that the class of separating hyperplanes in the space $\mathbb{R}^n$ has dimension VC $= n + 1$.

According to the most common bound on the expected risk, for a binary classification problem the inequality

$$R(f) \leq \sqrt{\frac{\gamma \log(2t/\gamma) + 1 - \log(\eta/4)}{t}} + R_{emp}(f)$$

$$= \Psi(t, \gamma, \eta) + R_{emp}(f), \tag{10.42}$$

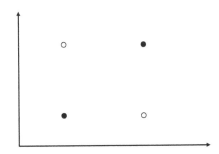

Figure 10.18   Four points in a plane which cannot be shattered by a straight line. The binary class value of each example is provided by its color

holds with probability $1 - \eta$, where $0 \leq \eta \leq 1$ is a predefined confidence level, γ is the VC dimension of the chosen hypothesis space $\mathcal{F}$, and $\Psi(t, \gamma, \eta)$ is a function called the *VC confidence* which represents the generalization error.

As a consequence, in order to control the classification error against broader hypothesis space, a larger training set is required, according to a proportion which is regulated by theoretical bounds on the error, cast in the form (10.42). Hence, in order to identify a hypothesis $f^* \in \mathcal{F}$ capable of achieving a high level of accuracy on the training set and a good generalization capability on the test set and on future sets of examples, one may resort to the *structural risk minimization* (SRM) principle, which formally establishes the concept of choosing the function $f^* \in \mathcal{F}$ which minimizes the right hand-side in (10.42).

Furthermore, the minimization of the empirical risk is an ill-conditioned problem, in the sense that the parameters of the optimal hypothesis f^* may vary significantly for small changes in the input data.

A way to circumvent the second weakness has been devised in the context of regularization theory and also applied in statistical learning theory to prevent overfitting. It relies on the solution of a well-posed problem which corresponds to the structural risk minimization principle, and chooses as the optimal hypothesis $f^* \in \mathcal{F}$ the one that minimizes a modified risk functional

$$\hat{R}(f) = \frac{1}{t} \sum_{i=1}^{t} V(y_i, f(\mathbf{x}_i)) + \lambda \| f \|_K^2, \tag{10.43}$$

where K is a given symmetric positive definite function called *kernel*, $\| f \|_K^2$ denotes the norm of f in the reproducing kernel Hilbert space[4] induced by K and λ is a parameter that controls the trade-off between the empirical error

---

[4]A *Hilbert space* is a linear space in which an inner product is defined, so that a norm is assigned which is also *complete* – that is, such that every Cauchy sequence is convergent to a point in the space. The simplest example of a finite-dimensional Hilbert space is the Euclidean space $\mathbb{R}^n$. However, there are also Hilbert spaces of infinite dimension.

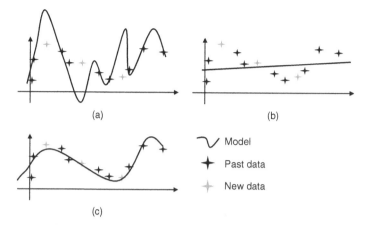

Figure 10.19   Finding the best trade-off between empirical error and generalization capability

and the generalization capability. The parameter λ can be interpreted as a penalty against a hypothesis f with high complexity VC and low generalization capability, and also as a smoothing regularizer for transforming the variational problem into a well-posed one. Figure 10.19 provides an intuitive explanation of how it is possible to find an ideal trade-off between the accuracy on the training set and the generalization capability to future unseen examples. The dark gray points represent the training examples, while the light gray ones correspond to new instances. The hypothesis shown Figure 10.19(a) belongs to a too broad space $\mathcal{F}$, which may be composed, for example, of high order polynomials; it is associated with a negligible classification error on the training set but it is not able to generalize well to future instances. On the other hand, the hypothesis in Figure 10.19(b) originates from a too narrow space, whereas in Figure 10.19(c) an ideal trade-off is achieved between the ability to provide an accurate classification of the training examples and the future generalization capability.

Figure 10.20 shows the conflicting effects produced on the two right-hand-side terms in (10.42) by an increase in the complexity of the hypothesis space $\mathcal{F}$, measured by the VC dimension along the horizontal axis: an increase in the complexity corresponds to a decrease in the empirical error but, at the same time, to an increase in the VC confidence which represents the generalization error.

The classical theory of support vector machines developed in the theoretical framework outlined above is based on the so called *soft-margin* loss function, given by

$$V(y_i, f(\mathbf{x}_i)) = |1 - y_i g(\mathbf{x}_i)|_+, \qquad (10.44)$$

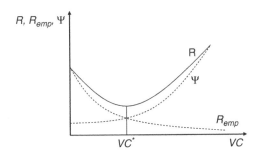

Figure 10.20   Searching for the optimal VC dimension

where g is a score function such that $f(\mathbf{x}) = \mathrm{sgn}(g(\mathbf{x}))$ and $|t|_+ = t$ if t is positive and zero otherwise. In particular, the loss function (10.44) computes the sum of the slacks of the misclassified examples from the separating surface.

However, one may resort to alternative loss functions. For example, a different family of classification models, termed *discrete support vector machines*, is motivated by the following loss function that counts the number of misclassified examples:

$$V(y_i, f(\mathbf{x}_i)) = c_i \theta(-y_i g(\mathbf{x}_i)), \qquad (10.45)$$

where $\theta(t) = 1$ if t is positive and zero otherwise, while c_i is a penalty for the misclassification of the example $\mathbf{x}_i$.

10.7.2   Maximal margin hyperplane for linear separation

Once the theoretical framework for the supervised learning process has been established, we are able to provide a geometrical interpretation of the structural risk minimization principle, and to formulate an optimization problem which aims to determine a linear separating surface for solving binary classification problems.

For linear separating functions represented by the hyperplanes in the space $\mathbb{R}^n$, the minimization of the right-hand side in (10.42) can be traced back to the maximization of the *margin of separation*, which will be described with reference to Figures 10.21 and 10.22.

Two sets of points belonging to binary target classes, like the ones depicted in Figure 10.21, are said to be *linearly separable* if there exists a hyperplane capable of separating them in the space $\mathbb{R}^n$, reducing to a line in the two-dimensional case. As shown in Figure 10.21, there are an infinite number of linear separating functions that appear equivalent to each other in terms of empirical error on the training set, which is equal to zero for all the lines drawn in the figure. However, lines (a) and (c), which lie near the training examples,

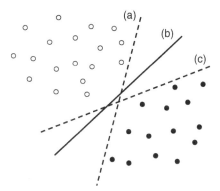

Figure 10.21   The lines represent possible separating hyperplanes in the two-dimensional space for the points in the diagram

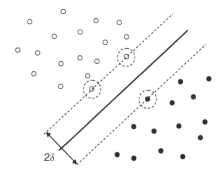

Figure 10.22   Maximal margin separating hyperplanes and canonical hyperplanes for a linearly separable dataset

have lower generalization capability with respect to a line, such as (b), that is equally distant from the nearest points of the two sets.

The optimal separating line, endowed with the maximum generalization capability and the minimum empirical error, is drawn in Figure 10.22. The margin of separation is defined as the distance between the pair of parallel *canonical supporting hyperplanes*, as shown in the same figure. Hence, the margin is equal to twice the minimum distance between the training points and the separating hyperplane. Those training points which are at the minimum distance from the separating hyperplane, and thus lie on the canonical hyperplanes, are called *support vectors* and play a more prominent role with respect to the other examples, since it is precisely these points that determine the classification rule.

Letting $\mathbf{w}$ denote the vector of the hyperplane coefficients and b the intercept, the separating hyperplane is given by

$$\mathbf{w}'\mathbf{x} = b, \tag{10.46}$$

while the two canonical supporting hyperplanes are

$$\mathbf{w}'\mathbf{x} - b - 1 = 0, \qquad \mathbf{w}'\mathbf{x} - b + 1 = 0. \tag{10.47}$$

The margin of separation δ is defined as

$$\delta = \frac{2}{\|w\|}, \quad \text{where} \quad \|w\| = \sqrt{\sum_{j \in \mathcal{N}} w_j^2}. \tag{10.48}$$

In order to determine the coefficients $\mathbf{w}$ and b of the optimal separating hyperplane, a quadratic optimization problem with linear constraints can be solved:

$$\min_{\mathbf{w},b} \quad \frac{1}{2}\|\mathbf{w}\|^2 \tag{10.49}$$

$$\text{s.to} \quad y_i(\mathbf{w}'\mathbf{x}_i - b) \geq 1, \quad i \in \mathcal{M}. \tag{10.50}$$

The aim of the objective function is to maximize the margin of separation through the minimization of its reciprocal, while the constraints (10.50) force each point $\mathbf{x}_i$ to lie in the halfspace corresponding to the class value y_i.

It is most likely that the m points of a dataset cannot be linearly separable, as for the set of examples depicted in Figure 10.23. In these cases, it is necessary to relax the constraints (10.50) by replacing them with weaker conditions which allow for the presence of possible misclassification errors, and to modify the objective function of the optimization problem. This can be done by introducing a new set of *slack variables* d_i, $i \in \mathcal{M}$, which measure the positive differences between the values of the misclassified examples on the vertical axis and the ordinate values along the canonical hyperplane that defines the region associated with the class value y_i, as geometrically illustrated in Figure 10.23.

In the non-separable case, we can therefore formulate the following optimization problem:

$$\min_{\mathbf{w},b,d} \quad \frac{1}{2}\|\mathbf{w}\|^2 + \lambda \sum_{i=1}^{m} d_i \tag{10.51}$$

$$\text{s.to} \quad y_i(\mathbf{w}'\mathbf{x}_i - b) \geq 1 - d_i, \quad i \in \mathcal{M}, \tag{10.52}$$

$$\qquad\qquad d_i \geq 0, \qquad\qquad\qquad i \in \mathcal{M}. \tag{10.53}$$

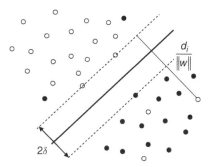

Figure 10.23 Maximal margin separating hyperplanes and canonical hyperplanes for a nonlinearly separable dataset

The objective function is composed of the weighted sum of two terms representing respectively the reciprocal of the margin of separation and the empirical error. The parameter λ is introduced in order to regulate the trade-off between the generalization capability, represented by the reciprocal of the margin, and the accuracy on the training set, evaluated as the sum of the slack variables.

The quadratic problem (10.51) can be solved via Lagrangian duality. Among other advantages, this allows us to identify the support vectors, which are associated with positive Lagrange multipliers in the optimal solution of the dual problem.

Denote by $\alpha_i \geq 0$ the Lagrangian multipliers of the constraints (10.52) and by $\mu_i \geq 0$ the multipliers of the constraints (10.53). The Lagrangian function of problem (10.51) is given by

$$
L(\mathbf{w}, b, \mathbf{d}, \boldsymbol{\alpha}\,\boldsymbol{\mu}) = \frac{1}{2}\|\mathbf{w}\|^2 + \lambda \sum_{i=1}^{m} d_i
$$

$$
- \sum_{i=1}^{m} \alpha_i [y_i(\mathbf{w}'\mathbf{x}_i - b) - 1 + d_i] - \sum_{i=1}^{m} \mu_i d_i. \tag{10.54}
$$

In order to find the optimal solution, the derivatives with respect to the variables $\mathbf{w}, \mathbf{d}, b$ of the primal problem (10.51) must be set to 0,

$$
\frac{\partial L(\mathbf{w}, b, \mathbf{d}, \boldsymbol{\alpha}\,\boldsymbol{\mu})}{\partial \mathbf{w}} = \mathbf{w} - \sum_{i=1}^{m} \alpha_i y_i \mathbf{x}_i = \mathbf{0}, \tag{10.55}
$$

$$
\frac{\partial L(\mathbf{w}, b, \mathbf{d}, \boldsymbol{\alpha}\,\boldsymbol{\mu})}{\partial \mathbf{d}} = \lambda - \alpha_i - \mu_i = 0, \tag{10.56}
$$

$$
\frac{\partial L(\mathbf{w}, b, \mathbf{d}, \boldsymbol{\alpha}\,\boldsymbol{\mu})}{\partial b} = \sum_{i=1}^{m} \alpha_i y_i = 0, \tag{10.57}
$$

leading to the conditions

$$\mathbf{w} = \sum_{i=1}^{m} \alpha_i y_i \mathbf{x}_i, \tag{10.58}$$

$$\lambda = \alpha_i + \mu_i, \tag{10.59}$$

$$\sum_{i=1}^{m} \alpha_i y_i = 0. \tag{10.60}$$

By substituting these equality constraints into the Lagrangian function (10.54), we obtain the objective function of the dual problem

$$L(\mathbf{w}, b, \mathbf{d}, \boldsymbol{\alpha}\,\boldsymbol{\mu}) = \sum_{i=1}^{m} \alpha_i - \frac{1}{2} \sum_{i=1}^{m} \sum_{h=1}^{m} y_i y_h \alpha_i \alpha_h \mathbf{x}_i' \mathbf{x}_h, \tag{10.61}$$

with the additional constraints $\alpha_i \leq \lambda$, $i \in \mathcal{M}$. The Karush–Kuhn–Tucker complementarity conditions applied to the pair of primal–dual problems lead to the equalities

$$\alpha_i[y_i(\mathbf{w}'\mathbf{x}_i - b) - 1 + d_i] = 0, \qquad i \in \mathcal{M}, \tag{10.62}$$

$$\mu_i(\alpha_i - \lambda) = 0, \qquad i \in \mathcal{M}. \tag{10.63}$$

In particular, conditions (10.63) allow us to identify the support vectors, which are the most representative points in determining the classification rules learned from the training set. Indeed, the examples $\mathbf{x}_i$ whose Lagrangian multipliers satisfy the condition $0 < \alpha < \lambda$ are at a distance $1/\|\mathbf{w}\|$ from the separating hyperplane, and thus are located on one of the two canonical hyperplanes.

Finally, observe that for support vector machines the decision function which classifies a new observation $\mathbf{x}$ is given by

$$f(\mathbf{x}) = \text{sgn}(\mathbf{w}'\mathbf{x} - b). \tag{10.64}$$

By substituting (10.58) into (10.64), we can reformulate the decision function as

$$f(\mathbf{x}) = \text{sgn}\left(\sum_{i=1}^{m} \alpha_i y_i \mathbf{x}'\mathbf{x}_i - b \right). \tag{10.65}$$

10.7.3   Nonlinear separation

Linear separating functions are not able to perform accurate classifications when the set of examples is intrinsically characterized by a nonlinear pattern, like that shown in Figure 10.24(a).

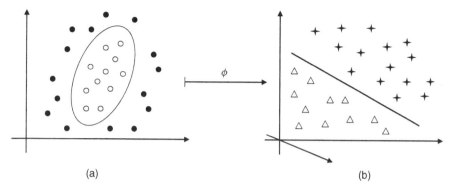

Figure 10.24   Nonlinear separation achieved by means of transformations in the feature space

In such cases, one may resort to mappings of the attributes which allow one to obtain linearly separable datasets in the transformed space, which is called the *feature space*. Figure 10.24 shows a possible transformation enjoying this property. In Figure 10.24(a) the examples of opposite classes can be separated by an elliptical function represented by a second-order polynomial in the explanatory variables. If the original two-dimensional space is transformed into a five-dimensional space which contains the second-order monomials $x_1 x_2, x_1^2, x_2^2$, in addition to the original attributes x_1 and x_2, it is possible to linearly separate the transformed points by means of a hyperplane defined in the space $\mathbb{R}^5$. Notice that the linear separation achieved in this way is made possible by an increase in the number of attributes in the feature space, which may be quite relevant for some applications.

Besides the intuitive justification based on Figure 10.24, the mapping from the original attribute space to a feature space of high – sometimes even infinite – dimensionality can also be justified from a theoretical point of view by Cover's theorem. We already know that m points in a general position in the n-dimensional space can be linearly separated if $m \leq n + 1$, since the VC dimension of the class of hyperplanes is $n + 1$, so that the number of possible linear separations is 2^m. If, however, $m > n + 1$, Cover's theorem states that the number of linear separations is given by

$$2 \sum_{i=0}^{n} \binom{m-1}{i}. \tag{10.66}$$

As n increases, the number of terms in the sum increases in turn, so that there are more linear separations.

This line of reasoning may, however, entail some practical difficulties. Indeed, the transformation must be carried out very efficiently even if the feature space is of high or infinite dimension. Fortunately, there exists a wide class of mappings which can be evaluated in a very efficient way, consisting of *kernel functions*, for which the mapping of the original observations into the feature space is not explicitly computed. The use of kernels is based on the dual formulation of problem (10.51), and implicitly involves a linear separation of the examples in high-dimensional functional spaces – or even infinite-dimensional for some families of kernels.

Specifically, consider a map $\Phi : \mathbb{R}^n \mapsto \mathcal{B}$ that transforms the examples given in the space $\mathbb{R}^n$ of the original attributes into a Hilbert space $\mathcal{B}$, called the *feature space*.

Observe now that the arguments set out in Section 10.7.2 to derive the maximal margin hyperplane can easily be adapted to achieve an optimal linear separation in the feature space. To do this, we simply have to substitute $\Phi(\mathbf{x})$ wherever we wrote $\mathbf{x}$ in expressions (10.46) through (10.65). In particular, the Lagrangian dual problem takes the form

$$L(\mathbf{w}, b, \mathbf{d}, \boldsymbol{\alpha} \, \boldsymbol{\mu}) = \sum_{i=1}^{m} \alpha_i - \frac{1}{2} \sum_{i=1}^{m} \sum_{h=1}^{m} y_i y_h \alpha_i \alpha_h \Phi(\mathbf{x}_i)' \Phi(\mathbf{x}_h), \qquad (10.67)$$

and we obtain the following decision function in the feature space:

$$f(\mathbf{x}) = \text{sgn} \left(\sum_{i=1}^{m} \alpha_i y_i \Phi(\mathbf{x})' \Phi(\mathbf{x}_i) - b \right). \qquad (10.68)$$

Now, the key point is that the data appear in the Lagrangian dual (10.67) and in the corresponding decision function (10.68) only in the form of inner products $\Phi(\mathbf{x}_i)' \Phi(\mathbf{x}_h)$ and $\Phi(\mathbf{x})' \Phi(\mathbf{x}_i)$. Hence, the potentially expensive computation of the images $\Phi(\mathbf{x}_i)$, $i \in \mathcal{M}$, of the observations under the mapping Φ can be dramatically simplified if there exists a positive definite *kernel function* k such that

$$\Phi(\mathbf{x}_i)' \Phi(\mathbf{x}_h) = k(\mathbf{x}_i, \mathbf{x}_h). \qquad (10.69)$$

If such a kernel function to represent the mapping Φ exists, then the dual problem takes the form

$$L(\mathbf{w}, b, \mathbf{d}, \boldsymbol{\alpha} \, \boldsymbol{\mu}) = \sum_{i=1}^{m} \alpha_i - \frac{1}{2} \sum_{i=1}^{m} \sum_{h=1}^{m} y_i y_h \alpha_i \alpha_h k(\mathbf{x}_i, \mathbf{x}_h), \qquad (10.70)$$

with the additional constraints $\alpha_i \leq \lambda$, $i \in \mathcal{M}$. Since k is positive definite, the matrix $\mathbf{Q} = [Q_{ij}] = [y_i y_k k(\mathbf{x}_i, \mathbf{x}_h)]$ is positive definite in turn, so that problem (10.70) is convex and can be solved efficiently to derive the dual multipliers

α_i, $i \in \mathcal{M}$. Therefore, this allows us to identify the support vectors, and to express the decision function in the form

$$f(\mathbf{x}) = \operatorname{sgn}\left(\sum_{i=1}^{m} \alpha_i y_i k(\mathbf{x}, \mathbf{x}_i) - b\right).\qquad(10.71)$$

The linear separation in the feature space hence corresponds to a nonlinear separation in the space of the original attributes, which by means of kernel functions can be computed with basically the same effort as a linear separation in the original space.

There remains the question of the existence of meaningful kernel functions that enable accurate nonlinear separations to be computed. Sufficient conditions for the existence of pairs (k, Φ) consisting of a kernel function and a map Φ represented by k, are provided by Mercer's theorem, whose formulation goes beyond the scope of our discussion and for which the reader is directed to the references suggested at the end of this chapter. Mercer's theorem also implies that the linear combination of kernel functions is in turn a kernel function, so that it provides a rule for deriving more complex kernels starting from simple ones.

Many kernels have been proposed in the literature, among which the most popular parametric families are *polynomial* kernels of degree d,

$$k(\mathbf{x}_i, \mathbf{x}_h) = (\mathbf{x}_i' \mathbf{x}_h + 1)^d;\qquad(10.72)$$

radial basis function kernels, also called *Gaussian* kernels, of width $\sigma > 0$,

$$k(\mathbf{x}_i, \mathbf{x}_h) = \exp\left(\frac{-||\mathbf{x}_i - \mathbf{x}_h||^2}{2\sigma^2}\right);\qquad(10.73)$$

and *neural networks* kernels with a hyperbolic tangent activation function, $\kappa > 0$,

$$k(\mathbf{x}_i, \mathbf{x}_h) = \tanh(\kappa \mathbf{x}_i' \mathbf{x}_h - \delta).\qquad(10.74)$$

Notice that nonlinear separations can be also attained in the original space of attributes where the examples are defined, without resorting to mappings carried out by means of kernel functions.

For example, some authors have proposed polyhedral methods, described in Figure 10.25, which have been shown to perform very accurate classifications even in the presence of a small number of training examples. Alternatively, nonlinear separations may be achieved by means of hybrid classification methods, which recursively partition the data with linear classifiers embedded within classification trees, as depicted in Figure 10.26. These methods appear to be quite effective for solving multicategory classification problems.

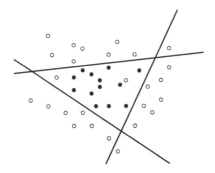

Figure 10.25   Classification by means of polyhedral separating regions

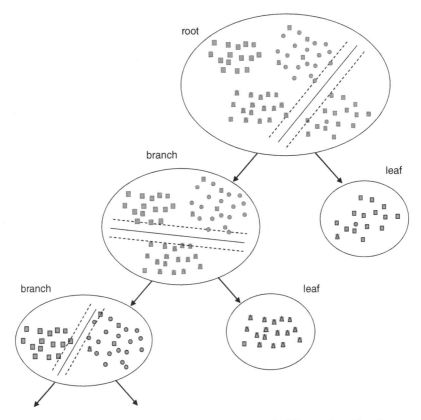

Figure 10.26   Nonlinear separation by means of oblique classification trees

In conclusion, the theoretical developments in statistical learning theory and the classification algorithms based on support vector machines have played a prominent role in defining a rigorous methodological framework for supervised learning methods. The extent of the application domain and the considerable classification accuracy make support vector machines and their variants the most effective methods currently available for supervised learning analysis.

10.8 Notes and readings

The first works on classification go back to early classical statistics, under the heading of *discriminant analysis* (Fisher, 1936).

Classification models play a prominent role in artificial intelligence, where they are designated as *pattern recognition* and *machine learning*. Several classical problems in artificial intelligence can be traced back to the classification of objects: from voice and sound recognition, to character and text recognition, to static and dynamic image recognition. There are several texts devoted to pattern recognition, among which Duda *et al.* (2001), Fukunaga (1990), Baldi and Brunak (2001), Cherkassky and Mulier (1998), Jensen and Cohen (2000) and Kulkarni *et al.* (1998) are of particular note.

For axis-parallel classification trees the reader can refer, for example, to Breiman *et al.* (1984), Quinlan (1993), Breslow and Aha (1997), Murthy (1998) and Safavian and Landgrebe (1998). For oblique trees see Murthy *et al.* (1994) and Orsenigo and Vercellis (2003), Orsenigo and Vercellis (2006). For logistic regression refer to Pampel (2000) and Scott Long (1997). For Bayesian classifiers see Ramoni and Sebastiani (2001) and Domingos and Pazzani (1997). Thorough descriptions of neural networks can be found in Fausett (1994), Bishop (1995), Anderson (1995), Ripley (1996), Schurmann (1996), Haykin (1998) and Raudys (2001); see also Rosenblatt (1962) for its historical value.

The literature on support vector machines is very broad; see Vapnik (1996, 1998), Cristianini and Shawe-Taylor (2000, 2004), Schölkopf and Smola (2001) and Burges (1998). The relationship between statistical learning theory and regularization theory is clearly analyzed in Evgeniou *et al.* (2002). Discrete support vector machines have been proposed and developed in Orsenigo and Vercellis (2004, 2007, 2009). Mangasarian (1997), Bradley *et al.* (1999) and Mirkin (1996) focus on the role of optimization models in classification.

For a discussion of the criteria for evaluating and comparing different classifiers see Kohavi (1995).

11

Association rules

In this chapter we will describe a class of unsupervised learning models that can be used when the dataset of interest does not include a target attribute. These are methods that derive *association rules* the aim of which is to identify regular patterns and recurrences within a large set of transactions. They are fairly simple and intuitive and are frequently used to investigate sales transactions in market basket analysis and navigation paths within websites.

In Section 11.1 we will review the structure and evaluation criteria of association rules. Then, in Section 11.2, we will introduce simple association rules that address a single dimension of analysis and asymmetric binary attributes. Section 11.3 will be devoted to describing the *Apriori* algorithm, which is the most popular method of generating association rules, both in its original form and in a number of variants. Finally, in Section 11.4, we will briefly discuss the main issues that arise in connection with the extraction of association rules of a general nature, characterized by a more complex structure.

11.1 Motivation and structure of association rules

In several areas of application, such as those described in the following examples, the systematic collection of data gives rise to massive lists of transactions that lend themselves to analysis through association rules in order to identify possible recurrences in the data.

Market basket analysis. When a customer makes a purchase at a point of sale and receives a receipt for her payment, this transaction is recorded by the information system of the retailer that manages the point of sale. For each transaction recorded, a list of purchased items is stored along with the price, time and place of the transaction. These transactions are gathered into a massive

Business Intelligence: Data Mining and Optimization for Decision Making   C. Vercellis
© 2009 John Wiley & Sons, Ltd

dataset, at least for large retailers, which can be exploited to perform a data mining analysis aimed at identifying recurrent rules that relate the purchase of a product, or group of products, to the purchase of another product, or group of products. For example, it might be possible to obtain association rules of the form 'a customer buying breakfast cereals will also buy milk with a probability of 0.68'. As shown in greater detail in Chapter 13, the association rules for market basket analysis can be quite useful for marketing managers in planning promotional initiatives or defining the assortment and location of products on the shelves.

Web mining. Within web mining analyses it is particularly useful to understand the pattern of navigation paths and the frequency with which combinations of web pages are visited by a given individual during a single session or consecutive sessions. In this case too the list of pages visited during a session is recorded as a transaction, possibly matched with a sequence number and the time of visit. It is therefore interesting to identify regular patterns possibly hidden in the data that allow the association of one or more pages being viewed with visits to other pages. The rules may assume a form such as 'if an individual visits the site *timesonline.co.uk* then within a week she will also visit the site *economist.com* with a probability of 0.87'. Association rules of this kind may influence the structure of the links between pages, in order to ease the navigation and to recommend specific navigation paths, or to place advertisement banners and other promotional messages.

Purchases with a credit card. Association rules are also used to analyze the purchases made by credit card holders in order to direct future promotions. In this case each transaction consists of the purchases and the payments made by a credit card holder. Notice that products and services that can potentially be accessed by the credit card owner are virtually infinite in this situation.

Fraud detection. In fraud identification, transactions consist of incident reports and applications for compensation for the damage suffered. The existence of special combinations may reveal potentially fraudulent behaviors and therefore justify an in-depth investigation on the part of the insurance company.

Given two propositions Y and Z, which may be true or false, we can state in general terms that a *rule* is an implication of the type $Y \Rightarrow Z$ with the following meaning: if Y is true then Z is also true. A rule is called *probabilistic* if the validity of Z is associated with a probability p: if Y is true then Z is also true with probability p.

Rules represent a classical paradigm for knowledge representation, popular because of their simple and intuitive structure, which makes them easily understandable and similar to logical schemes typical of human reasoning.

Rules aimed at extracting knowledge for a business intelligence analysis should be non-trivial and interpretable, so that they may potentially be useful for knowledge workers and easy to translate into concrete action plans. On the one hand, it may be the case that a marketing analysis generates a rule that only reflects the effects of promotional advertising campaigns carried out in the past, or that merely states the obvious. On the other hand, rules may sometimes reverse the causal relationship of an implication. Consider, for instance, a rule stating that 'buyers of an insurance policy will also buy a car with a probability of 0.98'. It is clear that this rule confuses the cause with the effect and is therefore useless for marketing managers: indeed, having purchased a car, the likelihood that customers will subsequently buy an insurance policy is higher.

To formally represent association rules it is convenient to introduce some notation. Let

$$\mathcal{O} = \{o_1, o_2, \ldots, o_n\} \qquad (11.1)$$

be a set of n *objects*. A generic subset $L \subseteq \mathcal{O}$ is called an *itemset*. An itemset that contains k objects is called a *k-itemset*. A *transaction* represents a generic itemset that has been recorded in a database in conjunction with an activity or cycle of activities. The dataset $\mathcal{D}$ is therefore composed of a list of m transactions T_i, each associated with a unique identifier, denoted by t_i, as shown in the example of Table 11.1.

With reference to market basket analysis, the objects represent the items available from a retailer, and each transaction corresponds to the items listed in a sales receipt, as shown in Table 11.1. Similarly, in web mining the objects represent the pages in the website of interest and each transaction corresponds to the list of pages visited by a navigator during one session.

Table 11.1 Example of a dataset consisting of transactions defined over the set of objects $\mathcal{O} = \{a, b, c, d, e\} = \{$bread, milk, cereals, coffee, tea$\}$

identifier t_i	transaction T_i
001	$\{a, c\}$
002	$\{a, b, d\}$
003	$\{b, d\}$
004	$\{b, d\}$
005	$\{a, b, c\}$
006	$\{b, c\}$
007	$\{a, c\}$
008	$\{a, b, e\}$
009	$\{a, b, c, e\}$
010	$\{a, e\}$

Notice also that a dataset composed of transactions can be represented by a two-dimensional matrix $\mathbf{X}$, where the n objects of the set $\mathcal{O}$ correspond to the columns of the matrix, the m transactions T_i to the rows, and the generic element of $\mathbf{X}$ is defined as

$$x_{ij} = \begin{cases} 1 & \text{if object } o_j \text{ belongs to transaction } T_i, \\ 0 & \text{otherwise.} \end{cases} \tag{11.2}$$

Table 11.2 shows the matrix $\mathbf{X}$ for the example shown in Table 11.1.

The representation described above could be generalized, assuming that each object o_j appearing in a transaction T_i is associated with a number f_{ij} representing the frequency with which the object appears in the same transaction. In this way it is possible to fully describe multiple sales of a given item during a single transaction, or repeated visits to the same web page during a single navigation session.

Let $L \subseteq \mathcal{O}$ be a given set of objects. A transaction T is said to *contain* the set L if the relationship $L \subseteq T$ holds. In the example of Table 11.1, the 2-itemset $L = \{a, c\}$ is contained in the transaction with identifier $t_i = 005$, but it is not contained in the transaction with identifier $t_i = 006$.

The *empirical frequency* $f(L)$ of an itemset L is defined as the number of transactions T_i existing in the dataset $\mathcal{D}$ that contain the set L, that is,

$$f(L) = \text{card}\{T_i : L \subseteq T_i, i = 1, 2, \ldots, m\}. \tag{11.3}$$

When dealing with a large sample (i.e. as m increases), the ratio $f(L)/m$ between the empirical frequency and the total number of transactions approximates the probability $\Pr(L)$ of occurrence of the itemset L, interpreted as the probability that L is contained in a new transaction T recorded in the database.

For the dataset in Table 11.1, the set of objects $L = \{a, c\}$ has a frequency $f(L) = 4$ and its probability of occurrence is therefore estimated as $\Pr(L) = 4/10 = 0.4$.

Table 11.2   Matrix $\mathbf{X}$ for the example of Table 11.1

identifier t_i	a	b	c	d	e
001	1	0	1	0	0
002	1	1	0	1	0
003	0	1	0	1	0
004	0	1	0	1	0
005	1	1	1	0	0
006	0	1	1	0	0
007	1	0	1	0	0
008	1	1	0	0	1
009	1	1	1	0	1
010	1	0	0	0	1

11.2   Single-dimension association rules

Given two itemsets $L \subset \mathcal{O}$ and $H \subset \mathcal{O}$ such that $L \cap H = \emptyset$ and a transaction T, an *association rule* is a probabilistic implication denoted by $L \Rightarrow H$ with the following meaning: if L is contained in T, then H is also contained in T with a given probability p, termed the *confidence* of the rule in $\mathcal{D}$ and defined as

$$p = \mathrm{conf}\{L \Rightarrow H\} = \frac{f(L \cup H)}{f(L)}. \tag{11.4}$$

The set L is called the *antecedent* or *body* of the rule, and H is the *consequent* or *head*. The confidence of the rule indicates the proportion of transactions containing the set H among those that include the set L, and therefore expresses the inferential reliability of the rule. As the number m of transactions increases, the confidence approximates the conditional probability that H belongs to a transaction T given that L does belong to T, that is,

$$\Pr\{H \subseteq T | L \subseteq T\} = \frac{\Pr\{\{H \subseteq T\} \cap \{L \subseteq T\}\}}{\Pr\{L \subseteq T\}}. \tag{11.5}$$

Consequently, a higher confidence corresponds to a greater probability that the itemset H exists in a transaction that also contains the itemset L.

The rule $L \Rightarrow H$ is said to have a *support* s in $\mathcal{D}$ if the proportion of transactions containing both L and H is equal to s, that is, if

$$s = \mathrm{supp}\{L \Rightarrow H\} = \frac{f(L \cup H)}{m}. \tag{11.6}$$

The support of a rule expresses the proportion of transactions that contain both the body and the head of the rule, and therefore measures the frequency with which an antecedent–consequent pair appear together in the transactions of a dataset. A low support suggests that a rule may have occurred occasionally. Low-support rules are usually of little interest to decision makers, and therefore the support is used to discard rules of scant significance. As the size of the dataset increases, the support approximates the probability that both H and L are contained in some future transaction.

In the example of Table 11.1, given the itemsets $L = \{a, c\}$ and $H = \{b\}$, for the rule $L \Rightarrow H$ we have

$$p = \mathrm{conf}\{L \Rightarrow H\} = \frac{f(L \cup H)}{f(L)} = \frac{2}{4} = \frac{1}{2} = 0.5, \tag{11.7}$$

$$s = \mathrm{supp}\{L \Rightarrow H\} = \frac{f(L \cup H)}{m} = \frac{2}{10} = 0.2. \tag{11.8}$$

The notions of support and confidence lead to a formal treatment of the analysis of association rules. Once a dataset $\mathcal{D}$ of m transactions has been assigned and minimum threshold values s_{min} for the support and p_{min} for the confidence have been fixed, all *strong* association rules should be determined, characterized by a support $s \geq s_{min}$ and by a confidence $p \geq p_{min}$.

However, when dealing with a large dataset, extracting all association rules through a complete enumeration procedure is prohibitive in terms of computation time. The number N_T of possible association rules increases exponentially as the number n of objects increases, according to the formula

$$N_T = 3^n - 2^{n+1} + 1. \tag{11.9}$$

On the other hand, most of these rules are not strong, in the sense that they do not fulfill the requirements of exceeding the predetermined minimum thresholds for support and confidence. An enumerative algorithm that generates all the rules, calculates the support and confidence for each of them, and finally discards unfeasible rules, carries out a lot of useless operations. It is therefore appropriate to devise a method capable of deriving strong rules only, implicitly filtering out the rules that do not fulfill the minimum threshold requirements.

The problem of generating strong association rules may be divided into two successive phases: first, the generation of *frequent itemsets*; and second, the generation of *strong rules*.

Generation of frequent itemsets. As we can infer from definition (11.6), the support of a rule only depends on the union of the itemsets L and H, and not on the actual distribution of the objects between the body and the head of the rule. For example, all six rules that can be generated using the objects $\{a, b, c\}$ from Table 11.1 have support $s = 2/10 = 0.20$:

$$\{a, b\} \Rightarrow \{c\}, \qquad \{a, c\} \Rightarrow \{b\},$$
$$\{b, c\} \Rightarrow \{a\}, \qquad \{a\} \Rightarrow \{b, c\},$$
$$\{b\} \Rightarrow \{a, c\}, \qquad \{c\} \Rightarrow \{a, b\}.$$

If the threshold value for the support is $s_{min} = 0.25$ (i.e. greater than the threshold value $s = 0.20$), the six rules can be implicitly eliminated solely based on the analysis of the union set $\{a, b, c\}$, without needing to separately consider each of them to also compute the confidence.

Based on this observation, the aim of generating *frequent itemsets* is to extract all sets of objects whose relative frequency is greater than the assigned minimum support s_{min}. This phase is more burdensome from a computational viewpoint than the subsequent phase of rule generation. Therefore, several algorithms have been proposed for obtaining the frequent itemsets in an efficient manner. In the following section we will describe the most popular one, known as the *Apriori* algorithm.

Generation of strong rules. After all frequent itemsets have been generated, we can proceed to the next phase of identifying *strong rules*. It is necessary to separate the objects contained in each frequent itemset according to all possible combinations of head and body of the rule and verify if the confidence of the rule in turn exceeds the minimum threshold p_{min}. If, for example, the itemset $\{a, b, c\}$ is frequent, it is possible to obtain from it the six rules previously listed, although only the rules that show a confidence higher than the threshold p_{min} can be considered strong.

Lift of a rule

Strong rules are not always meaningful or of interest to decision makers. To see this, consider a retailer wishing to analyze a set of transactions to identify associations between sales of color printers and sales of digital cameras. Let us assume that there are a total of 1000 transactions, of which 600 include cameras, 750 include printers and 400 include both. If the threshold values $s_{min} = 0.3$ and $p_{min} = 0.6$ are fixed in advance for the support and the confidence, the rule $\{camera\} \Rightarrow \{printer\}$ is selected among the strong rules, since it has support $s = 400/1000 = 0.4$ and confidence $p = 400/600 = 0.66$, exceeding the minimum thresholds. The rule seems to suggest that the purchase of a camera also induces the purchase of a printer. However, a more careful analysis reveals that the probability of purchasing a printer is equal to $750/1000 = 0.75$ and is greater than 0.66, which is the probability of purchasing a printer conditioned on the purchase of a camera. In fact, sales of printers and cameras show a negative correlation, since the purchase of a camera reduces the probability of buying a printer.

The example described above highlights a shortcoming in the evaluation of the effectiveness of a rule based only on its support and confidence. A third measure of the significance of an association rule is the *lift*, defined as

$$l = \text{lift}\{L \Rightarrow H\} = \frac{\text{conf}\{L \Rightarrow H\}}{f(H)} = \frac{f(L \cup H)}{f(L)f(H)}. \tag{11.10}$$

Lift values greater than 1 indicate that the rule being considered is more effective than the relative frequency of the head in predicting the probability that the head is contained in some transaction of the dataset. In this case body and head of the rule are positively associated. Conversely, if the lift is less than 1 the rule is less effective than the estimate obtained through the relative frequency of the head. In this case body and head are negatively associated. If the lift is less than 1, the rule that negates the head, expressed as $\{L \Rightarrow (\mathcal{O} - H)\}$, is more effective than the original rule, since it has confidence $1 - \text{conf}\{L \Rightarrow H\}$ and therefore a lift greater than 1.

11.3   Apriori algorithm

A dataset $\mathcal{D}$ of m transactions defined over a set $\mathcal{O}$ of n objects may contain up to $2^n - 1$ frequent itemsets, excluding the empty set from the collection of all subsets of $\mathcal{O}$. Consequently, since in real-world applications the cardinality n is at least of the order of a few dozen objects, the number of itemsets increases exponentially and makes it impractical to use an exhaustive generation method based on a complete enumeration of the itemsets to extract the frequent sets.

The *Apriori* algorithm is a more efficient method of extracting strong rules contained in a set of transactions. During the first phase the algorithm generates the frequent itemsets in a systematic way, without exploring the space of all candidates, while in the second phase it extracts the strong rules.

The theoretical assumption on which the Apriori algorithm is based consists of a property called the *Apriori principle*.

Theorem 11.1 (Apriori principle). *If an itemset is frequent, then all its subsets are also frequent.*

To illustrate the Apriori principle, assume that the itemset $\{a, b, c\}$ is frequent. It is clear that each transaction containing $\{a, b, c\}$ should also contain each of its six proper subsets, including the 2-itemsets $\{a, b\}$, $\{a, c\}$, $\{b, c\}$ and the 1-itemsets $\{a\}$, $\{b\}$, $\{c\}$. Therefore, these six itemsets are also frequent in turn. Actually, there is a consequence of the Apriori principle that is fundamental in reducing the search space of the algorithm. Indeed, if we suppose that the itemset $\{a, b, c\}$ is not frequent, then each of the itemsets containing it must in turn not be frequent. In this way, once a non-frequent itemset has been identified in the course of the algorithm, it is possible to implicitly eliminate all itemsets with a greater cardinality that contain it, significantly increasing the overall efficiency.

11.3.1   Generation of frequent itemsets

The Apriori algorithm generates frequent itemsets iteratively, starting with the frequent 1-itemsets and then determining the frequent k-itemsets based upon the frequent $(k - 1)$-itemsets generated during the previous step. The total number of iterations is equal to $k_{max} + 1$, where k_{max} denotes the maximum cardinality of a frequent itemset in the dataset $\mathcal{D}$.

Procedure 11.1 describes the phase of frequent itemset generation for the Apriori algorithm.

Procedure 11.1 – Apriori algorithm: generation of frequent itemsets

1. The transactions in the dataset are scanned to compute the relative frequency of each object. Objects with a frequency lower than the support threshold s_{min} are discarded. At the end of this step, the collection of all frequent 1-itemsets has been generated, and the iteration counter is set to $k = 2$.

2. The candidate k-itemsets are iteratively generated starting from the $(k - 1)$-itemsets, determined during the previous iteration.

3. The support of each candidate k-itemset is computed by scanning through all transactions included in the dataset.

4. Candidates with a support lower than the threshold s_{min} are discarded.

5. The algorithm stops if no k-itemset has been generated. Otherwise, set $k = k + 1$ and return to step 2.

To illustrate how the Apriori algorithm works, we will apply it to the transactions in Table 11.1, assuming that the minimum threshold for the support has been fixed at $s_{min} = 0.2$. Step 1, the initialization of the algorithm, determines the relative frequency for each object in $\mathcal{O} = \{a, b, c, d, e\}$, as shown in Table 11.3. All 1-itemsets are frequent since their frequency is higher than the threshold s_{min}. The second iteration then proceeds by generating all candidate 2-itemsets, which can be obtained from the frequent 1-itemsets, and by computing their relative frequency to determine if they are also frequent in turn. Table 11.4 lists the ten 2-itemsets that can be obtained from the frequent 1-itemsets of Table 11.3, showing their relative frequency and status. The next iteration generates the candidate 3-itemsets that can be obtained from the frequent 2-itemsets. These are listed and their relative frequencies shown in Table 11.5: both 3-itemsets are frequent. Since there are no candidate 4-itemsets the procedure stops, having determined a total of $5 + 6 + 2 = 13$ frequent itemsets.

11.3.2   Generation of strong rules

The second phase, generation of the strong rules, receives as input a list of all frequent itemsets, generated during the first phase, each associated with a relative frequency greater than the minimum threshold s_{min}.

Table 11.3   Iteration 1: relative frequency for frequent
1-itemsets

itemset	relative frequency	status
$\{a\}$	$7/10 = 0.7$	frequent
$\{b\}$	$7/10 = 0.7$	frequent
$\{c\}$	$5/10 = 0.5$	frequent
$\{d\}$	$3/10 = 0.3$	frequent
$\{e\}$	$3/10 = 0.3$	frequent

Table 11.4   Iteration 2: relative frequency for frequent
2-itemsets

itemset	relative frequency	status
$\{a, b\}$	$4/10 = 0.4$	frequent
$\{a, c\}$	$4/10 = 0.4$	frequent
$\{a, d\}$	$1/10 = 0.1$	not frequent
$\{a, e\}$	$3/10 = 0.3$	frequent
$\{b, c\}$	$3/10 = 0.3$	frequent
$\{b, d\}$	$3/10 = 0.3$	frequent
$\{b, e\}$	$2/10 = 0.2$	frequent
$\{c, d\}$	$0/10 = 0.0$	not frequent
$\{c, e\}$	$1/10 = 0.1$	not frequent
$\{d, e\}$	$0/10 = 0.0$	not frequent

Table 11.5   Iteration 3: relative frequency for frequent
3-itemsets

itemset	relative frequency	status
$\{a, b, c\}$	$2/10 = 0.2$	frequent
$\{a, b, e\}$	$2/10 = 0.2$	frequent

Given a frequent k-itemset L, it is possible to generate $2^k - 2$ candidate association rules to be extracted as strong rules, leaving out the rules that have an empty set as the body or the head. For each of these the confidence is computed, and this is then compared with the minimum threshold p_{min}. Suppose that the rule $L \Rightarrow H$ has been generated. Observe that the computation of the confidence does not require further scanning of the dataset, since

$$p = \text{conf}\{L \Rightarrow H\} = \frac{f(L \cup H)}{f(L)}, \qquad (11.11)$$

and the frequencies $f(L)$ and $f(L \cup H)$ are known at the end of the first phase of the algorithm, since both L and $L \cup H$ are frequent itemsets.

Procedure 11.2 describes the generation of strong rules for the Apriori algorithm, and Table 11.6 shows its application to the transactions of Table 11.1, with a threshold value $p_{min} = 0.55$.

Procedure 11.2 – Apriori algorithm: generation of strong rules

1. The list of frequent itemsets generated during the first phase is scanned. If the list is empty, the procedure stops. Otherwise, let B be the next itemset to be considered, which is then removed from the list.

2. The set B of objects is subdivided into two non-empty disjoint subsets L and $H = B - L$, according to all possible combinations.

3. For each candidate rule $L \Rightarrow H$, the confidence is computed as

$$p = \text{conf}\{L \Rightarrow H\} = \frac{f(B)}{f(L)}.$$

4. If $p \geq p_{min}$ the rule is included into the list of strong rules, otherwise it is discarded.

The number of operations required by the Apriori algorithm grows exponentially as the number n of objects increases. For example, to generate a frequent 100-itemset, it is necessary to examine at least

$$\sum_{h=1}^{100} \binom{100}{h} = 2^{100} - 1 \approx 10^{30} \qquad (11.12)$$

candidate sets. Several tactics have been suggested, which are briefly described here, to increase the efficiency of the Apriori algorithm; for further information see the references at the end of the chapter. First, it is useful to adopt more advanced data structures for representing the transactions, such as dictionaries and binary trees. Furthermore, it is possible to partition the transactions into disjoint subsets and then separately apply the Apriori algorithm to each partition to identify local frequent itemsets. At a later stage, starting from the latter, the entire set of transactions is considered to obtain global frequent itemsets and the corresponding strong rules. Finally, it is possible to randomly extract a significant sample of transactions and to obtain strong association rules for the extracted sample, with a greater efficiency, even at the expense of possible inaccuracies with respect to the full set of transactions.

Table 11.6   An example of strong rule generation

itemset	rule	confidence	status
$\{a, b\}$	$\{a \Rightarrow b\}$	$p = 4/7 = 0.57$	strong
$\{a, b\}$	$\{b \Rightarrow a\}$	$p = 4/7 = 0.57$	strong
$\{a, c\}$	$\{a \Rightarrow c\}$	$p = 4/7 = 0.57$	strong
$\{a, c\}$	$\{c \Rightarrow a\}$	$p = 4/5 = 0.80$	strong
$\{a, e\}$	$\{a \Rightarrow e\}$	$p = 3/7 = 0.43$	not strong
$\{a, e\}$	$\{e \Rightarrow a\}$	$p = 3/3 = 1.00$	strong
$\{b, c\}$	$\{b \Rightarrow c\}$	$p = 3/7 = 0.43$	not strong
$\{b, c\}$	$\{c \Rightarrow b\}$	$p = 3/5 = 0.60$	strong
$\{b, d\}$	$\{b \Rightarrow d\}$	$p = 3/7 = 0.43$	not strong
$\{b, d\}$	$\{d \Rightarrow b\}$	$p = 3/3 = 1.00$	strong
$\{b, e\}$	$\{b \Rightarrow e\}$	$p = 2/7 = 0.29$	not strong
$\{b, e\}$	$\{e \Rightarrow b\}$	$p = 2/3 = 0.67$	strong
$\{a, b, c\}$	$\{a, b \Rightarrow c\}$	$p = 2/4 = 0.50$	not strong
$\{a, b, c\}$	$\{c \Rightarrow a, b\}$	$p = 2/5 = 0.40$	not strong
$\{a, b, c\}$	$\{a, c \Rightarrow b\}$	$p = 2/4 = 0.50$	not strong
$\{a, b, c\}$	$\{b \Rightarrow a, c\}$	$p = 2/7 = 0.29$	not strong
$\{a, b, c\}$	$\{b, c \Rightarrow a\}$	$p = 2/3 = 0.67$	strong
$\{a, b, c\}$	$\{a \Rightarrow b, c\}$	$p = 2/7 = 0.29$	not strong
$\{a, b, e\}$	$\{a, b \Rightarrow e\}$	$p = 2/4 = 0.50$	not strong
$\{a, b, e\}$	$\{e \Rightarrow a, b\}$	$p = 2/3 = 0.67$	strong
$\{a, b, e\}$	$\{a, e \Rightarrow b\}$	$p = 2/3 = 0.67$	strong
$\{a, b, e\}$	$\{b \Rightarrow a, e\}$	$p = 2/7 = 0.29$	not strong
$\{a, b, e\}$	$\{b, e \Rightarrow a\}$	$p = 2/2 = 1.00$	strong
$\{a, b, e\}$	$\{a \Rightarrow b, e\}$	$p = 2/7 = 0.29$	not strong

11.4   General association rules

The association rules described in the previous sections are *binary*, *asymmetric* and *one-dimensional*. They are *binary* since they refer to the presence or absence of each object in the transactions that make up the dataset, as shown by the rows of the matrix **X** given in Table 11.2 for the example in Table 11.1. Furthermore, the rules are *asymmetric* since we implicitly assumed, as is customary in market basket analysis, that the presence of an object in a single transaction, corresponding to an actual purchase, is far more relevant than its absence, corresponding to a non-purchase. Finally, the rules are *one-dimensional* because they involve only one logical dimension of data from multi-dimensional data warehouses and data marts.

However, it is possible to identify more general association rules that may be useful across a range of applications.

Association rules for symmetric binary attributes. There are situations where we might be interested in binary attributes for which the 0 and 1 values are equally relevant. Suppose, for example, that we wish to analyze the answers given in a registration form filled out on-line by visitors to a website seeking the activation of a newsletter service. As an alternative, the registration forms may be filled out by customers wishing to receive a loyalty card at a point of sale in the retail industry. In these situations, the answers given to questions on the gender of an individual, classified as {male, female}, or on consent to receive advertising material, expressed as {yes, no}, are examples of symmetric binary variables. It is possible to extend the analysis with symmetric attributes to the case of asymmetric binary attributes, in order to apply the algorithms for the generation of association rules described in previous sections. In fact, a simple transformation can be used: for each symmetric binary variable, two asymmetric binary variables are introduced which correspond respectively to the presence of a given value and to the presence of its opposite. If, for example, the symmetric variable *communications* represents the answer provided in a registration form by customers regarding the consent to receive communications from the company, two asymmetric binary variables *consent* and *do not consent*, taking on the values {0,1} in relation to the value assumed by the attribute *communications*, can be introduced for each record.

Association rules for categorical attributes. In other situations the attributes may be categorical but not binary. In the previous registration form examples there may be attributes such as the province of residence or the level of education. In this case too it is possible to introduce, for each categorical attribute, a set of asymmetric binary variables, equal in number to the levels of the categorical attribute. Each binary variable takes the value 1 if in the corresponding record the categorical attribute assumes the level associated with the binary variable, otherwise it takes the value 0.

Association rules for continuous attributes. When dealing with continuous attributes, such as the age or income of those who have filled out a registration form, it is appropriate to proceed in two sequential phases. First, the continuous attribute is transformed into a categorical variable through one of the discretization techniques reviewed in Chapter 6. Then the discretized attribute is also transformed into asymmetric binary variables, as previously described, in order to proceed with the extraction of strong rules.

Multidimensional association rules. As shown in Chapter 3, data are gathered and stored in data warehouses or data marts, and organized according to several logical dimensions, which then direct the analyses and the visualizations. As a consequence, it is reasonable to require the association rules to involve multiple dimensions. A rule in the context of market basket analysis may appear in the form 'if a customer buys a digital camera, is 30–40 years old, and has an annual average expenditure of €300–500 on electronic equipment, then she will also buy a color printer with probability 0.78'. This is a three-dimensional rule, since the body consists of three dimensions (purchased items, age and expenditure amount), while the head refers to a purchase. Observe that the numerical variables that represent the age and the expenditure amount have previously been discretized, as described in the previous paragraph.

Multi-level association rules. In some applications, association rules do not allow strong associations to be extracted due to data rarefaction. It may be, for example, that each of the items for sale can be found in too small a proportion of transactions to be included in the frequent itemsets, thus thwarting the search for association rules. However, the objects making up the transactions usually belong to hierarchies of concepts, which we mentioned in Chapter 3 in connection with the design of a data warehouse. Items are usually grouped by type and in turn by sales department. A possible way to remedy the rarefaction of objects in the transactions is to transfer the analysis to a higher level in the hierarchy of concepts. In this way the number of objects decreases and consequently the number of transactions containing the same object increases. Also the computational complexity significantly decreases, due to the reduced number of objects, so that the algorithms for rule extraction become much more efficient.

Sequential association rules. Often the transactions are recorded according to a specific temporal sequence. For example, the transactions for a loyalty card holder correspond to the sequence of sale receipts. By the same token, the transactions that gather the navigation paths followed by a given web user are associated with the temporal sequence of the sessions. In situations like these, analysts are often interested in extracting association rules that take into account temporal dependencies.

The description of the algorithms used to extract association rules in the different situations reviewed in this section goes beyond the scope of this book, and therefore readers are referred to the references given below. For now readers should bear in mind that the proposed techniques usually consist of extensions of the Apriori algorithm and its variants.

11.5   Notes and readings

The first contributions on association rules can be found in Agrawal (1993a,b) and Agrawal and Srikant (1994). Surveys are presented in Zaki (1999) and Hipp *et al.* (2000). Considerable research effort has been devoted to the efficient generation of rules for massive datasets using parallel algorithms: see Mannila *et al.* (1994), Park *et al.* (1995), Adamo (2001), Agrawal *et al.* (1996), Agarwal *et al.* (2001), Agrawal and Shafer (1998), and Aggarwal and Yu (1998). For the extension to sequential rules see Agrawal and Srikant (1995, 1997) and Silverstein *et al.* (1998) consider generalized rules; Han and Fu (1995) describe multidimensional rules. Methods based on sampling techniques are described in Toivonen (1996). For papers based on the formal analysis of concepts for rules extraction see Pasquier *et al.* (1999), Zaki (2000), Zaki and Ogihara (1998). Pei *et al.* (2000) is focused on web transactions. Finally, for the relationship between association rules and classification see Freitas (2000).

12

Clustering

The second class of models for unsupervised learning is represented by *clustering* methods, which will be covered in this chapter. By defining appropriate metrics and the induced notions of distance and similarity between pairs of observations, the purpose of clustering methods is the identification of homogeneous groups of records called *clusters*. With respect to the specific distance selected, the observations belonging to each cluster must be close to one another and far from those included in other clusters.

We begin this chapter by reviewing the main features of clustering models. We will then illustrate the most popular measures of distance between pairs of observations, in relation to the nature of the attributes contained in the dataset. Partition methods will then be described, focusing in particular on *K*-means and *K*-medoids algorithms. Finally, we will illustrate both agglomerative and divisive hierarchical methods, in connection with the main metrics that express the inhomogeneity among distinct clusters. We will also introduce some indicators of the quality of clustering models.

12.1 Clustering methods

The aim of clustering models is to subdivide the records of a dataset into homogeneous groups of observations, called *clusters*, so that observations belonging to one group are similar to one another and dissimilar from observations included in other groups.

Grouping objects by affinity is a typical reasoning pattern applied by the human brain. Also for this reason clustering models have long been used in various disciplines, such as social sciences, biology, astronomy, statistics, image recognition, processing of digital information, marketing and data mining.

Clustering models are useful for various purposes. In some applications, the clusters generated may provide a meaningful interpretation of the phenomenon of interest. For example, grouping customers based on their purchase behaviors may reveal the existence of a cluster corresponding to a market niche to which it might be appropriate to address specific marketing actions for promotional purposes. Furthermore, subdivision into clusters may be the preliminary phase of a data mining project that will be followed by the application of other methodologies within each cluster. In a retention analysis, a preliminary subdivision into clusters may be followed by the development of distinct classification models, with the aim of identifying with greater accuracy the customers characterized by a high probability of churning. Finally, grouping into clusters may prove useful in the course of exploratory data analysis to highlight outliers and to identify an observation that might represent on its own an entire cluster, in order to reduce the size of the dataset.

Clustering methods must fulfill a few general requirements, as indicated below.

Flexibility. Some clustering methods can be applied to numerical attributes only, for which it is possible to use the Euclidean metrics to calculate the distances between observations. However, a flexible clustering algorithm should also be able to analyze datasets containing categorical attributes. Algorithms based on the Euclidean metrics tend to generate spherical clusters and have difficulty in identifying more complex geometrical forms.

Robustness. The robustness of an algorithm manifests itself through the stability of the clusters generated with respect to small changes in the values of the attributes of each observation. This property ensures that the given clustering method is basically unaffected by the noise possibly existing in the data. Moreover, the clusters generated must be stable with respect to the order of appearance of the observations in the dataset.

Efficiency. In some applications the number of observations is quite large and therefore clustering algorithms must generate clusters efficiently in order to guarantee reasonable computing times for large problems. In the case of massive datasets, one may also resort to the extraction of samples of reduced size in order to generate clusters more efficiently. However, this approach inevitably implies a lower robustness for the clusters so generated. Clustering algorithms must also prove efficient with respect to the number of attributes existing in the dataset.

12.1.1   Taxonomy of clustering methods

Clustering methods can be classified into a few main types based on the logic used for deriving the clusters: *partition* methods, *hierarchical* methods, *density based* methods and *grid* methods.

Partition methods. Partition methods, described in Section 12.2, develop a subdivision of the given dataset into a predetermined number K of non-empty subsets. They are suited to obtaining groupings of a spherical or at most convex shape, and can be applied to datasets of small or medium size.

Hierarchical methods. Hierarchical methods, described in Section 12.3, carry out multiple subdivisions into subsets, based on a tree structure and characterized by different homogeneity thresholds within each cluster and inhomogeneity thresholds between distinct clusters. Unlike partition methods, hierarchical algorithms do not require the number of clusters to be predetermined.

Density-based methods. Whereas the two previous classes of algorithms are founded on the notion of distance between observations and between clusters, density-based methods derive clusters from the number of observations locally falling in a neighborhood of each observation. More precisely, for each record belonging to a specific cluster, a neighborhood with a specified diameter must contain a number of observations which should not be lower than a minimum threshold value. Density-based methods can identify clusters of non-convex shape and effectively isolate any possible outliers.

Grid methods. Grid methods first derive a discretization of the space of the observations, obtaining a grid structure consisting of cells. Subsequent clustering operations are developed with respect to the grid structure and generally achieve reduced computing times, despite a lower accuracy in the clusters generated.

A second distinction is concerned with the methods for assigning the observations to each single cluster. It is possible to include each observation *exclusively* in a single cluster or to place it by *superposition* into multiple clusters. Furthermore, *fuzzy* methods have been developed which assign the observations to the clusters with a weight between 0 (the observation is totally extraneous to the cluster) and 1 (the observation exclusively belongs to the cluster), with the additional condition that the sum of the weights over all clusters be equal to 1. Finally, a distinction should be made between *complete* clustering methods, which assign each observation to at least one cluster, and *partial* methods, which may leave some observations outside the clusters. The latter methods are useful for identifying outliers.

Most clustering methods are heuristic in nature, in the sense that they generate a subdivision into clusters of good quality although not necessarily optimal. Actually, an exhaustive method based on a complete enumeration of the possible subdivisions of m observations into K clusters would require

$$\frac{1}{K!} \sum_{h=1}^{K} \binom{K}{h} h^m. \tag{12.1}$$

combinations to be examined. Therefore, it would be inapplicable even to small datasets, due to the exponential growth in computing time. In terms of computational complexity, clustering problems actually belong to the class of difficult (*NP*-hard) problems whenever $K \geq 3$.

12.1.2   Affinity measures

Clustering models are usually based on a measure of similarity between observations. In many instances this can be obtained by defining an appropriate notion of distance between each pair of observations. In this section we will review the most popular metrics, in connection with the type of attributes analyzed.

Given a dataset $\mathcal{D}$, we can represent the m observations by means of n-dimensional vectors of attributes, so as to formally represent the data using a matrix $\mathbf{X}$ with m rows and n columns, as described in Section 5.2. Let

$$
\mathbf{D} = [d_{ik}] =
\begin{bmatrix}
0 & d_{12} & \cdots & d_{1,m-1} & d_{1m} \\
 & 0 & \cdots & d_{2,m-1} & d_{2m} \\
 & & \cdots & \vdots & \vdots \\
 & & 0 & d_{m-1,m} \\
 & & & & 0
\end{bmatrix}
\tag{12.2}
$$

be the symmetric $m \times m$ matrix of distances between pairs of observations, obtained by setting

$$
d_{ik} = \mathrm{dist}(\mathbf{x}_i, \mathbf{x}_k) = \mathrm{dist}(\mathbf{x}_k, \mathbf{x}_i), \quad i, k \in \mathcal{M},
\tag{12.3}
$$

where $\mathrm{dist}(\mathbf{x}_i, \mathbf{x}_k)$ denotes the distance between observations $\mathbf{x}_i$ and $\mathbf{x}_k$.

Notice that it is possible to transform the distance d_{ik} between two observations into a similarity measure s_{ik}, by alternatively using

$$
s_{ik} = \frac{1}{1 + d_{ik}}, \quad \text{or} \quad s_{ik} = \frac{d_{\max} - d_{ik}}{d_{\max}},
\tag{12.4}
$$

in which $d_{\max} = \max_{i,k} d_{ik}$ denotes the maximum distance between the observations in the dataset $\mathcal{D}$.

As shown in the rest of this section, the definition of an appropriate notion of distance depends on the nature of the attributes that make up the dataset $\mathcal{D}$, which may be numerical, binary, nominal categorical, ordinal categorical or of mixed composition.

Numerical attributes

If all n attributes in a dataset are numerical, we may turn to the *Euclidean distance* between the vectors associated with the pair of observations

$\mathbf{x}_i = (x_{i1}, x_{i2}, \ldots, x_{in})$ and $\mathbf{x}_k = (x_{k1}, x_{k2}, \ldots, x_{kn})$ in n-dimensional space, which is defined as

$$\text{dist}(\mathbf{x}_i, \mathbf{x}_k) = \sqrt{\sum_{j=1}^{n}(x_{ij} - x_{kj})^2}$$

(12.5)

$$= \sqrt{(x_{i1} - x_{k1})^2 + (x_{i2} - x_{k2})^2 + \cdots + (x_{in} - x_{kn})^2}.$$

As an alternative we may consider the *Manhattan distance*

$$\text{dist}(\mathbf{x}_i, \mathbf{x}_k) = \sum_{j=1}^{n}|x_{ij} - x_{kj}|$$

(12.6)

$$= |x_{i1} - x_{k1}| + |x_{i2} - x_{k2}| + \cdots + |x_{in} - x_{kn}|,$$

which is so called because in order to reach one point from another we have to travel along two sides of a rectangle having the two points as its opposite vertices. Figure 12.1 shows the difference between Euclidean and Manhattan metrics, using a two-dimensional example.

A third option, which generalizes both the Euclidean and Manhattan metrics, is the *Minkowski distance*, defined as

$$\text{dist}(\mathbf{x}_i, \mathbf{x}_k) = \sqrt[q]{\sum_{j=1}^{n}|x_{ij} - x_{kj}|^q}$$

(12.7)

$$= \sqrt[q]{|x_{i1} - x_{k1}|^q + |x_{i2} - x_{k2}|^q + \ldots, |x_{in} - x_{kn}|^q},$$

for some positive integer q. The Minkowski distance reduces to the Manhattan distance when $q = 1$, and to the Euclidean distance when $q = 2$.

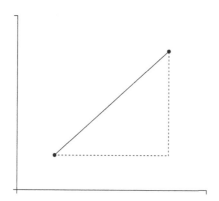

Figure 12.1   Euclidean distance (full line) and Manhattan distance (dotted line) between two points in the plane

A further generalization of the Euclidean distance can be obtained through the *Mahalanobis distance*, defined as

$$\text{dist}(\mathbf{x}_i, \mathbf{x}_k) = \sqrt{(\mathbf{x}_i - \mathbf{x}_k)\mathbf{V}_{ik}^{-1}(\mathbf{x}_i - \mathbf{x}_k)'}, \tag{12.8}$$

where $\mathbf{V}_{ik}^{-1}$ is the inverse of the covariance matrix of the pair of observations $\mathbf{x}_i$ and $\mathbf{x}_k$. If the observations $\mathbf{x}_i$ and $\mathbf{x}_k$ are independent, so that the covariance matrix reduces to the identity matrix, the Mahalanobis distance coincides with the Euclidean distance. The Mahalanobis distance is used when the observations are highly correlated, with different variances and a different range.

The expressions described above for computing the distance between pairs of observations are affected by the possible existence of attributes taking on absolute values significantly greater than the others. In extreme situations, a single prevailing attribute could affect the groupings generated by clustering algorithms.

A few tactics can be adopted in order to avoid such imbalance in the evaluation of distances between observations. On the one hand, one may standardize the values of the numerical attributes, using the techniques described in Chapter 6, so as to obtain new values on a uniform scale, for example in the interval $[0, 1]$ or $[-1, 1]$.

Alternatively, one may assign to each attribute $\mathbf{a}_j$ a weight w_j that is inversely proportional to the greatest value assumed by that same attribute, modifying the previous definitions to obtain weighted distances and thus balancing the importance of each attribute. For example, one can set

$$\text{dist}(\mathbf{x}_i, \mathbf{x}_k) = \sqrt[q]{\sum_{j=1}^{n} w_j |x_{ij} - x_{kj}|^q}. \tag{12.9}$$

A further possibility is to use the *arccosine distance*,[1] which depends on the angle formed by the vectors associated with the observations. Besides being independent of the absolute value of the components of the observations, this measure can be applied not only to numerical attributes but also to categorical and binary attributes. For this metric, the similarity coefficient based on the cosine of the angle between the two vectors $\mathbf{x}_i$ and $\mathbf{x}_k$ should be defined using the expression

$$B_{\cos}(\mathbf{x}_i, \mathbf{x}_k) = \cos(\mathbf{x}_i, \mathbf{x}_k) = \frac{\sum_{j=1}^{n} x_{ij} x_{kj}}{\sqrt{\sum_{j=1}^{n} x_{ij}^2 \sum_{j=1}^{n} x_{kj}^2}}. \tag{12.10}$$

The coefficient $B_{\cos}(\mathbf{x}_i, \mathbf{x}_k)$ lies in the interval $[0, 1]$, and it is closer to 1 when the two vectors $\mathbf{x}_i$ and $\mathbf{x}_k$ are similar to each other (parallel), and closer to 0

---

[1]The arccosine $\arccos(\alpha)$ is the inverse function of the cosine and gives the angle corresponding to the value α of the cosine.

when they are dissimilar (orthogonal). Starting from the coefficient $B_{\cos}(\mathbf{x}_i, \mathbf{x}_k)$, a notion of distance between the two observations can be derived in different ways, for example using the arccosine or the sine of the angle, as in

$$\text{dist}(\mathbf{x}_i, \mathbf{x}_k) = \arccos(\mathbf{x}_i, \mathbf{x}_k),$$
$$\text{dist}(\mathbf{x}_i, \mathbf{x}_k) = \sin(\mathbf{x}_i, \mathbf{x}_k) = \sqrt{1 - \cos^2(\mathbf{x}_i, \mathbf{x}_k)}.$$
(12.11)

Binary attributes

Suppose that a given attribute $\mathbf{a}_j = (x_{1j}, x_{2j}, \ldots, x_{mj})$ is binary, so that it assumes only one of the two values 0 or 1. Even if it is possible to formally calculate the difference $x_{ij} - x_{kj}$ for every two observations of the dataset, it is clear that this quantity does not represent a distance that can be meaningfully associated with the metrics defined for numerical attributes, since the values 0 and 1 are purely conventional, and their meanings could be interchanged. The need thus arises to define an appropriate notion of proximity that can be applied to binary attributes.

Assume for the moment that all n attributes in the dataset $\mathcal{D}$ are binary. In order to define a metric, reference should be made to the contingency table associated with the n attributes at two distinct observations $\mathbf{x}_i$ and $\mathbf{x}_k$, as indicated in Table 12.1. The value p is the number of binary attributes for which both observations $\mathbf{x}_i$ and $\mathbf{x}_k$ assume the value 0. Similarly, q is the number of attributes for which observation $\mathbf{x}_i$ takes the value 0 and observation $\mathbf{x}_k$ takes the value 1. The value u is the number of attributes for which $\mathbf{x}_i$ assumes the value 1 and $\mathbf{x}_k$ assumes the value 0. Finally, there are v attributes for which the value 1 is assumed by both observations.

As remarked in Chapter 11, binary attributes can be *symmetric* or *asymmetric*. In the first case the presence of the value 0 is as interesting as the presence of the value 1. For instance, customer authorization of promotional mailings assumes the values {no, yes}. In the second case we are mostly interested in the presence of the value 1, which can usually be found in a small proportion of the observations. In order to discriminate between those individuals suffering from a medical condition and those who are not affected, it is clearly more

Table 12.1   Contingency table for a binary attribute

		observation $\mathbf{x}_k$		
		0	1	total
observation $\mathbf{x}_i$	0	p	q	$p + q$
	1	u	v	$u + v$
	total	$p + u$	$q + v$	n

significant to analyze the presence of the value 1, assuming that this codes the presence of the condition.

If all the n binary attributes are symmetric, we can define the degree of similarity between pairs of observations through the *coefficient of similarity*, given by

$$\text{dist}(\mathbf{x}_i, \mathbf{x}_k) = \frac{q + u}{p + q + u + v} = \frac{q + u}{n}. \qquad (12.12)$$

Assume instead that all n attributes are binary and asymmetric. As stated before, for a pair of asymmetric attributes it is much more interesting to match *positives*, i.e. records possessing the property relative to each attribute and therefore being coded by the value 1, with respect to matching *negatives*, representing observations coded by the value 0. For binary variables, the *Jaccard coefficient* is therefore used, defined as

$$\text{dist}(\mathbf{x}_i, \mathbf{x}_k) = \frac{q + u}{q + u + v}. \qquad (12.13)$$

Finally, if the dataset includes both symmetric and asymmetric binary attributes, as well as for simultaneous numerical and binary attributes, the methodologies described in the section on mixed composition attributes should be applied.

Nominal categorical attributes

A nominal categorical attribute, defined in Chapter 5, can be interpreted as a generalization of a symmetric binary attribute, for which the number of distinct values is greater than two. As a consequence, for nominal attributes we can also use an extension of the similarity coefficient, previously defined for symmetric binary variables, using the new expression

$$\text{dist}(\mathbf{x}_i, \mathbf{x}_k) = \frac{n - f}{n}, \qquad (12.14)$$

in which f is the number of attributes for which the observations $\mathbf{x}_i$ and $\mathbf{x}_k$ take the same nominal value.

Alternatively, as shown in Section 8.3.3, it is possible to represent each nominal categorical variable by means of a number of asymmetric binary attributes equal to the number of distinct values assumed by the nominal variable. Once the nominal variables in the dataset have been replaced by the categorical attributes, we can therefore use the Jaccard coefficient (12.13) to express the degree of affinity.

Ordinal categorical attributes

Ordinal categorical attributes can be placed on a natural ordering scale, whose numerical values are, however, arbitrary. Therefore, they require a preliminary standardization so that they can be adapted to the metrics defined for numerical attributes.

To clarify this concept, assume that the values of each ordinal categorical attribute are represented through the corresponding position in the natural order. If, for example, the ordinal variable corresponds to the level of education, and can be expressed by the values {elementary school, high school graduate, bachelor's degree, postgraduate}, the elementary level is matched with numerical value 1, the high school graduate level with 2, and so on. Let $\mathcal{H}_j = \{1, 2, \ldots H_j\}$ be the ordered values associated with the ordinal attribute $\mathbf{a}_j$.

To standardize the values assumed by the attribute $\mathbf{a}_j$ to the interval $[0,1]$, the following transformation is used

$$x'_{ij} = \frac{x_{ij} - 1}{H_j - 1}. \tag{12.15}$$

After we have carried out the transformation indicated for all the ordinal variables, we can use the measures of distance previously introduced for numerical attributes.

Mixed composition attributes

Suppose that the n attributes in the dataset $\mathcal{D}$ have a mixed composition, in the sense that some of them are numerical, while others are binary symmetric or binary asymmetric or nominal categorical or ordinal categorical. We wish to define an overall affinity measure that can be used to evaluate the degree of similarity between two observations $\mathbf{x}_i$ and $\mathbf{x}_k$ even if mixed composition attributes are present.

Let us define a binary indicator δ_{ikj} that takes the value 0 if and only if one of the following cases occurs:

- at least one of the two values x_{ij} or x_{kj} is missing in the corresponding records of the dataset;

- the attribute $\mathbf{a}_j$ is binary asymmetric and, moreover, $x_{ij} = x_{kj} = 0$.

For all the remaining cases the indicator δ_{ikj} takes the value 1. Also define the contribution Δ_{ikj} of the variable $\mathbf{a}_j$ to the similarity between $\mathbf{x}_i$ and $\mathbf{x}_k$, based on the nature of the attribute $\mathbf{a}_j$, as follows:

- if the attribute $\mathbf{a}_j$ is binary or nominal, we set $\Delta_{ikj} = 0$ if $x_{ij} = x_{kj}$ and $\Delta_{ikj} = 1$ otherwise;

- if the attribute $\mathbf{a}_j$ is numerical, we set

$$\Delta_{ikj} = \frac{|x_{ij} - x_{kj}|}{\max_l x_{lj}}; \tag{12.16}$$

- if the attribute $\mathbf{a}_j$ is ordinal, its standardized value is computed as previously described and we set Δ_{ikj} as for numerical attributes.

Therefore, we can define the similarity coefficient between the observations $\mathbf{x}_i$ and $\mathbf{x}_k$ as

$$\text{dist}(\mathbf{x}_i, \mathbf{x}_k) = \frac{\sum_{j=1}^{n} \delta_{ikj} \Delta_{ikj}}{\sum_{j=1}^{n} \delta_{ikj}}. \tag{12.17}$$

12.2   Partition methods

Given a dataset $\mathcal{D}$ of m observations, each represented by a vector in n-dimensional space, partition methods construct a subdivision of $\mathcal{D}$ into a collection of non-empty subsets $\mathcal{C} = \{C_1, C_2, \ldots, C_K\}$, where $K \leq m$. In general, the number K of clusters is predetermined and assigned as an input to partition algorithms. The clusters generated by partition methods are usually exhaustive and mutually exclusive, in the sense that each observation belongs to one and only one cluster. There are, however, *fuzzy* partition methods that assign each observation to different clusters according to a specific proportion.

Partition methods start with an initial assignment of the m available observations to the K clusters. Then, they iteratively apply a reallocation technique whose purpose is to place some observations in a different cluster, so that the overall quality of the subdivision is improved. Although alternative measures of the clustering quality can be used, all the various criteria tend to express the degree of homogeneity of the observations belonging to the same cluster and their heterogeneity with respect to the records included in other clusters. Partition algorithms usually stop when during the same iteration no reallocation occurs, and therefore the subdivision appears stable with respect to the evaluation criterion chosen.

Partition methods are therefore of a heuristic nature, in the sense that they are based on a myopic logic typical of the class of so-called greedy methods, and at each step they make the choice that locally appears the most advantageous. Operating in this way, there is no guarantee that a globally optimal clustering will be reached, but only that a good subdivision will be obtained, at least for the majority of the datasets.

The K-*means* method and the K-*medoids* method, which will be described next, are two of the best-known partition algorithms. They are rather efficient clustering methods that are effective in determining clusters of spherical shape.

12.2.1 K-means algorithm

The K-*means* algorithm receives as input a dataset $\mathcal{D}$, a number K of clusters to be generated and a function $\text{dist}(\mathbf{x}_i, \mathbf{x}_k)$ that expresses the inhomogeneity between each pair of observations, or equivalently the matrix $\mathbf{D}$ of distances between observations.

Given a cluster C_h, $h = 1, 2, \ldots K$, the *centroid* of the cluster is defined as the point $\mathbf{z}_h$ having coordinates equal to the mean value of each attribute for the observations belonging to that cluster, that is,

$$z_{hj} = \frac{\sum_{\mathbf{x}_i \in C_h} x_{ij}}{\mathrm{card}\{C_h\}}. \tag{12.18}$$

Procedure 12.1 – K-means algorithm

1. During the initialization phase, K observations are arbitrarily chosen in $\mathcal{D}$ as the centroids of the clusters.

2. Each observation is iteratively assigned to the cluster whose centroid is the most similar to the observation, in the sense that it minimizes the distance from the record.

3. If no observation is assigned to a different cluster with respect to the previous iteration, the algorithm stops.

4. For each cluster, the new centroid is computed as the mean of the values of the observations belonging to the cluster, and then the algorithm returns to step 2.

The algorithm, described in Procedure 12.1, starts by arbitrarily selecting K observations that constitute the initial centroids. For example, the K points can be randomly selected among the m observations in $\mathcal{D}$. At each subsequent iteration, each record is assigned to the cluster whose centroid is the closest, that is, which minimizes the distance from the observation among all centroids. If no observation is reallocated to a cluster different from the one to which it belongs, determined during the previous iteration, the procedure stops, since any subsequent iteration cannot alter the current subdivision in clusters. Otherwise the new centroid for each cluster is computed and a new assignment made.

Figure 12.2 illustrates the application of the K-means algorithm to a set of observations in the Euclidean plane. Suppose that we have decided on $K = 3$ as the desired number of clusters. Figure 12.2(a) shows the points corresponding to the two-dimensional observations that are to be subdivided into three clusters. During the initialization phase, three observations are arbitrarily selected, which are indicated in the figure as large dots and constitute the three initial centroids. At the first iteration, each observation is assigned to the closest centroid, generating the clusters highlighted in Figure 12.2(b), where the

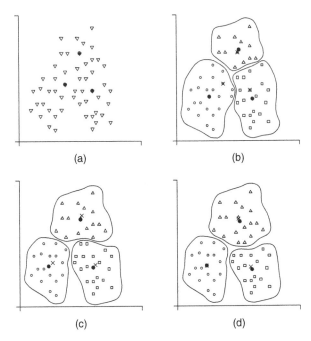

Figure 12.2   An example of application of the K-means algorithm

observations belonging to distinct clusters are indicated by three different graphical symbols. Figure 12.2(b) also shows the new centroids, computed as the mean of the attributes for the observations belonging to each cluster and represented as large dots. The three centroids of the previous iteration are indicated in diagram (b) by the symbol ×. Figures 12.2(c) and (d) respectively show the second and the third iterations of the algorithm. Since at the end of the third iteration no observation is assigned to a different cluster, the algorithm stops.

As stated earlier, if the attributes are numerical or ordinal categorical it is possible to transform them into a standardized representation in n-dimensional Euclidean space. In this case, one may also express the function of the overall heterogeneity between each observation and the point $\mathbf{w}_h$ representing the cluster C_h to which it is assigned, by means of the minimization of the squared error

$$\min_{\mathbf{w}_1,\mathbf{w}_2,\ldots,\mathbf{w}_K} \mathrm{EQ} = \sum_{h=1}^{K} \sum_{\mathbf{x}_i \in C_h} (\mathrm{dist}(\mathbf{x}_i, \mathbf{w}_h))^2$$
$$= \sum_{h=1}^{K} \sum_{\mathbf{x}_i \in C_h} \sum_{j=1}^{n} (x_{ij} - w_{hj})^2,$$

(12.19)

where the function dist$(\cdot, \cdot)$ represents the Euclidean distance between two points.

The function EQ is minimized by setting $\mathbf{w}_h = \mathbf{z}_h$, $h = 1, 2, \ldots, K$. This means that the centroids, calculated as the mean of the observations belonging to each cluster, minimize the sum of squared differences of the observations from the K points that represent the clusters, as these points vary.

If the dataset also includes binary or nominal categorical attributes, it is possible to use other notions of distance, as seen in Section 12.1.2. However, the problem remains of how to replace the mean of the observations, which is not defined for nominal variables, with a corresponding indicator that allows the centroid of a cluster to be identified. The algorithm that uses the mode of each attribute, calculated for the observations belonging to each cluster, in place of its mean, is called the K-modes method.

Several variants and extensions of the K-means algorithm have been proposed. Since the initial assignment tends to strongly influence the final subdivision into clusters, it is useful to generate different initial random assignments and then derive for each of them a different clustering, finally choosing the best one out of those that have been generated. Outliers can also affect the final result, since they take large numerical values in the quadratic objective function EQ. Therefore the K-means method is applied preferably only after the outliers have been identified and removed. Finally, it is possible to apply a posteriori a few tactics to improve the quality of the subdivision generated by the algorithm, for example by separating one or more clusters and increasing in this way the total number K of clusters identified. On the other hand, it may be appropriate to merge two or more clusters into a single one, reducing the number of clusters.

12.2.2 K-medoids algorithm

The K-medoids algorithm, also known as *partitioning around medoids*, is a variant of the K-means method. It is based on the use of *medoids* instead of the means of the observations belonging to each cluster, with the purpose of mitigating the sensitivity of the partitions generated with respect to the extreme values in the dataset.

Given a cluster C_h, a *medoid* $\mathbf{u}_h$ is the most *central* observation, in a sense that will be formally defined below, among those that are assigned to C_h. Once the medoid representing each cluster has been identified, the K-medoids algorithm proceeds like the K-means method, using medoids instead of centroids.

The initialization phase involves the arbitrary selection of K observations representing the medoids at the first iteration. Each of the remaining observations is assigned to the medoid to which it is most similar, so as to minimize the distance measure dist$(\mathbf{x}_i, \mathbf{u}_h)$. During each subsequent iteration, the algorithm tries to substitute each medoid with an observation that does not represent a medoid, provided that the overall quality of the subdivision in clusters is improved. The quality of the subdivision is evaluated through a measure of average inhomogeneity between each observation and the medoid associated with the cluster to which it belongs.

Let U be the set of K medoids at a given iteration. To assess the advantage afforded by an exchange of medoids, the algorithm considers all the $(m - K)K$ pairs $(\mathbf{x}_i, \mathbf{u}_h)$ composed of an observation $\mathbf{x}_i \notin U$ and a medoid $\mathbf{u}_h \in U$. For each pair, every observation $\mathbf{x}_k \notin U$ different from $\mathbf{x}_i$ is considered, and the contribution R_{ihk} made by $\mathbf{x}_k$ to the exchange between $\mathbf{x}_i$ and $\mathbf{u}_h$ is evaluated. The following four cases are then considered:

- $\mathbf{x}_k$ is currently assigned to the cluster corresponding to medoid $\mathbf{u}_h$, and also $\mathrm{dist}(\mathbf{x}_i, \mathbf{x}_k) \leq \min_{\mathbf{u}_e \in U, e \neq h} \mathrm{dist}(\mathbf{u}_e, \mathbf{x}_k)$. In this first case the observation $\mathbf{x}_k$ is assigned to the new medoid $\mathbf{x}_i$ intended to represent cluster C_h, and the contribution to the exchange, which can be positive, zero or negative, is given by

$$R_{ihk} = \mathrm{dist}(\mathbf{x}_i, \mathbf{x}_k) - \mathrm{dist}(\mathbf{u}_h, \mathbf{x}_k). \qquad (12.20)$$

- $\mathbf{x}_k$ is currently assigned to the cluster corresponding to medoid $\mathbf{u}_h$, and also $\mathrm{dist}(\mathbf{x}_i, \mathbf{x}_k) \leq \min_{\mathbf{u}_e \in U, e \neq h} \mathrm{dist}(\mathbf{u}_e, \mathbf{x}_k)$. In this second case the observation $\mathbf{x}_k$ is assigned to a different cluster, and the contribution to the exchange is given by

$$R_{ihk} = \min_{\mathbf{u}_e \in U, e \neq h} \mathrm{dist}(\mathbf{u}_e, \mathbf{x}_k) - \mathrm{dist}(\mathbf{u}_h, \mathbf{x}_k). \qquad (12.21)$$

- $\mathbf{x}_k$ is not currently assigned to the cluster corresponding to medoid $\mathbf{u}_h$, and also $\mathrm{dist}(\mathbf{x}_i, \mathbf{x}_k) \geq \min_{\mathbf{u}_e \in U, e \neq h} \mathrm{dist}(\mathbf{u}_e, \mathbf{x}_k)$. In this third case the observation $\mathbf{x}_k$ remains assigned to a different cluster, and the contribution to the exchange is given by

$$R_{ihk} = 0.$$

- $\mathbf{x}_k$ is not currently assigned to the cluster corresponding to medoid $\mathbf{u}_h$, and also $\mathrm{dist}(\mathbf{x}_i, \mathbf{x}_k) < \min_{\mathbf{u}_e \in U, e \neq h} \mathrm{dist}(\mathbf{u}_e, \mathbf{x}_k)$. In this last case the observation $\mathbf{x}_k$ is assigned to cluster C_h, and the contribution to the exchange is given by

$$R_{ihk} = \mathrm{dist}(\mathbf{x}_i, \mathbf{x}_k) - \min_{\mathbf{u}_e \in U, e \neq h} \mathrm{dist}(\mathbf{u}_e, \mathbf{x}_k). \qquad (12.22)$$

Once the contribution R_{ihk} made by each observation $\mathbf{x}_k \notin U$ to the exchange between $\mathbf{x}_i$ and $\mathbf{u}_h$ has been calculated, the overall contribution can be evaluated as

$$T_{ih} = \sum_{\mathbf{x}_k \notin U} R_{ihk}. \qquad (12.23)$$

If the minimum value among the overall contributions T_{ih} is negative, the pair $(\mathbf{x}_i, \mathbf{u}_h)$ that corresponds to such minimum value is actually exchanged and the algorithm proceeds with a new iteration, stopping when no exchange occurs, that is, when the minimum value of the overall contribution is non-negative.

The K-medoids algorithm requires a large number of iterations and is not suited to deriving clusters for large datasets. A few variants have therefore been proposed, based on the extraction of samples of observations to which the K-medoids method can be applied.

12.3   Hierarchical methods

Hierarchical clustering methods are based on a tree structure. Unlike partition methods, they do not require the number of clusters to be determined in advance. Hence, they receive as input a dataset $\mathcal{D}$ containing m observations and a matrix of distances dist$(\mathbf{x}_i, \mathbf{x}_k)$ between all pairs of observations.

In order to evaluate the distance between two clusters, most hierarchical algorithms resort to one of five alternative measures: minimum distance, maximum distance, mean distance, distance between centroids, and Ward distance.

Suppose that we wish to calculate the distance between two clusters C_h and C_f and let $\mathbf{z}_h$ and $\mathbf{z}_f$ be the corresponding centroids.

Minimum distance. According to the criterion of *minimum distance*, also called the *single linkage* criterion, the dissimilarity between two clusters is given by the minimum distance among all pairs of observations such that one belongs to the first cluster and the other to the second cluster, that is,

$$\text{dist}(C_h, C_f) = \min_{\substack{\mathbf{x}_i \in C_h \\ \mathbf{x}_k \in C_f}} \text{dist}(\mathbf{x}_i, \mathbf{x}_k). \qquad (12.24)$$

Maximum distance. According to the criterion of *maximum distance*, also called the *complete linkage* criterion, the dissimilarity between two clusters is given by the maximum distance among all pairs of observations such that one belongs to the first cluster and the other to the second cluster, that is,

$$\text{dist}(C_h, C_f) = \max_{\substack{\mathbf{x}_i \in C_h \\ \mathbf{x}_k \in C_f}} \text{dist}(\mathbf{x}_i, \mathbf{x}_k). \qquad (12.25)$$

Mean distance. The *mean distance* criterion expresses the dissimilarity between two clusters via the mean of the distances between all pairs of observations belonging to the two clusters, that is,

$$\text{dist}(C_h, C_f) = \frac{\sum_{\mathbf{x}_i \in C_h} \sum_{\mathbf{x}_k \in C_f} \text{dist}(\mathbf{x}_i, \mathbf{x}_k)}{\text{card}\{C_h\}\,\text{card}\{C_f\}}. \qquad (12.26)$$

Distance between centroids. The criterion based on the *distance between centroids* determines the dissimilarity between two clusters through the distance between the centroids representing the two clusters, that is,

$$\text{dist}(C_h, C_f) = \text{dist}(\mathbf{z}_h, \mathbf{z}_f). \qquad (12.27)$$

Ward distance. The criterion of *Ward distance*, based on the analysis of the variance of the Euclidean distances between the observations, is slightly more complex than the criteria described above. Indeed, it requires the algorithm to first calculate the sum of squared distances between all pairs of observations belonging to a cluster. Afterwards, all pairs of clusters that could be merged at the current iteration are considered, and for each pair the total variance is computed as the sum of the two variances between the distances in each cluster, evaluated in the first step. Finally, the pair of clusters associated with the minimum total variance are merged. Methods based on the Ward distance tend to generate a large number of clusters, each containing a few observations.

Hierarchical methods can be subdivided into two main groups: *agglomerative* and *divisive* methods.

12.3.1   Agglomerative hierarchical methods

Agglomerative methods are *bottom-up* techniques in which each single observation initially represents a distinct cluster. These clusters are then aggregated during subsequent iterations, deriving clusters of increasingly larger cardinalities. The algorithm is stopped when a single cluster including all the observations has been reached. It is then necessary for the user to decide on a cut point and thus determine the number of clusters.

Procedure 12.2 describes the general structure of an agglomerative method. At the end of the algorithm it is possible to graphically represent the process of subsequent mergers using a *dendrogram*, indicating on one axis the value of the minimum distance corresponding to each merger and the observations on the other axis.[2]

---

[2]There is no standard convention relative to the placement of the axes in dendrograms, since sometimes the observations are positioned on the horizontal axis and sometimes on the vertical axis.

Procedure 12.2 – Agglomerative algorithm

1. In the initialization phase, each observation constitutes a cluster. The distance between clusters therefore corresponds to the matrix $\mathbf{D}$ of the distances between all pairs of observations.

2. The minimum distance between the clusters is then computed, and the two clusters C_h and C_f with the minimum distance are merged, thus deriving a new cluster C_e. The corresponding minimum distance $\text{dist}(C_h, C_f)$ originating the merger is recorded.

3. The distance between the new cluster C_e, resulting from the merger between C_h and C_f, and the preexisting clusters is computed.

4. If all the observations are included into a single cluster, the procedure stops. Otherwise it is repeated from step 2.

Figure 12.3 shows the dendrograms for the clustering obtained by applying an agglomerative hierarchical algorithm to the *mtcars* dataset, described in Appendix B, using (a) the mean Euclidean distance and (b) the mean Manhattan distance. As we can see, the observations are placed on the horizontal axis and the distance value on the vertical axis. Actually, each dendrogram provides

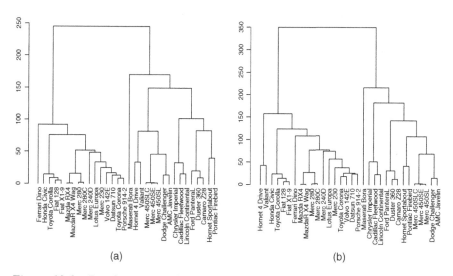

Figure 12.3   Dendrograms for an agglomerative hierarchical algorithm applied to the mtcars dataset with (a) the mean Euclidean distance and (b) the mean Manhattan distance

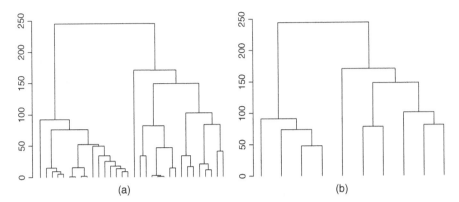

Figure 12.4 Dendrograms for an agglomerative hierarchical algorithm applied to the mtcars dataset with the mean Euclidean distance and two different levels of clusters

a whole hierarchy of clusters, corresponding to different threshold values for the minimum distance between clusters, indicated on the vertical axis. In the dendrogram shown in Figure 12.3(a), four clusters can be obtained if the tree is truncated at the distance 125, while nine clusters are generated by cutting the tree at 70. It can be observed that the dendrograms and the corresponding hierarchies of clusters obtained using the Euclidean and the Manhattan distances differ significantly.

Figure 12.4, in which the labels of each single observation have been removed, illustrates the effect of tree pruning. Figure 12.4(a) shows the same tree as in Figure 12.3(a), while Figure 12.4(b) has been obtained by means of a cut that results in ten clusters. In this way, a more aggregated view of the phenomenon can be achieved that leads to an easier interpretation.

Figures 12.5 and 12.6 show the effect of changes in the metrics used by the agglomerative hierarchical algorithm. Figure 12.5(a) shows the dendrogram obtained with the minimum distance and Figure 12.5(b) the dendrogram generated with the maximum distance, respectively. Figure 12.6(a) shows the dendrogram obtained with the Ward distance and Figure 12.6(b) the dendrogram generated with the distance between the centroids, respectively. Confirming previous remarks, the Ward method produces a tree with several ramifications toward the leaves and therefore a structure made up of many small clusters. More generally, these figures suggest that the type of metric selected has a relevant influence on the clustering generated.

12.3.2   Divisive hierarchical methods

Divisive algorithms are the opposite of agglomerative methods, in that they are based on a *top-down* technique, which initially places all the observations in a single cluster. This is then subdivided into clusters of smaller size, so that the distances between the generated subgroups are minimized. The procedure is repeated until clusters containing a single observation are obtained, or until an analogous stopping condition is met.

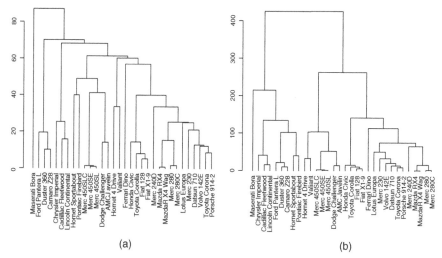

(a)   (b)

Figure 12.5 Dendrograms for an agglomerative hierarchical algorithm applied to the mtcars dataset with (a) the mean Euclidean distance and (b) the maximum Euclidean distance

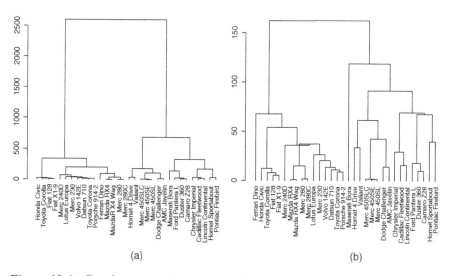

(a)   (b)

Figure 12.6 Dendrograms for an agglomerative hierarchical algorithm applied to the mtcars dataset with (a) the Ward distance and (b) the distance of centroids

Unlike agglomerative methods, in order to keep computing times within reasonable limits, divisive algorithms require a strict limitation on the number of combinations that can be analyzed. In order to understand the motivations for this choice, consider the first step of the algorithm. The only cluster containing all the observations must be subdivided into two subsets, so that the distance between the two resulting clusters is maximized. Since there are 2^{m-2} possible partitions of the whole set into two non-empty disjoint subsets, this results in an exponential number of operations already at the first iteration.

To circumvent this difficulty, at any given iteration divisive hierarchical algorithms usually determine for each cluster the two observations that are furthest from each other, and subdivide the cluster by assigning the remaining records to the one or the other, based on their proximity.

Figure 12.7(a) shows the dendrogram obtained by applying a divisive hierarchical algorithm to the *mtcars* dataset, using the mean Euclidean distance. This can be compared with the corresponding dendrogram shown in Figure 12.3(a), obtained using the same metrics but with an agglomerative hierarchical algorithm. As we can see, for this dataset the type of algorithm seems to exert less influence than the metric adopted.

12.4   Evaluation of clustering models

We have seen in previous chapters that for supervised learning methods, such as classification, regression and time series analysis, the evaluation of the

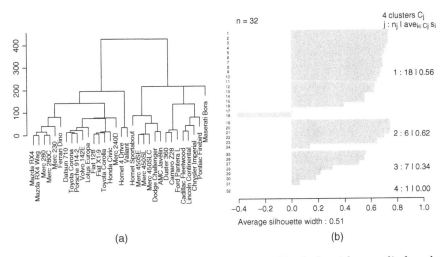

(a)   (b)

Figure 12.7   Dendrograms for a divisive hierarchical algorithm applied to the mtcars dataset (a) with the mean Euclidean distance and (b) the corresponding silhouette with four clusters

predictive accuracy is part of the development process of a model and is based on specific numerical indicators. The same does not apply to clustering methods and, more generally, to unsupervised learning models. Even though the absence of a target attribute makes the evaluation of an unsupervised model less direct and intuitive, it is possible to define reasonable measures of quality and significance for clustering methods.

To evaluate a clustering method it is first necessary to verify that the clusters generated correspond to an actual regular pattern in the data. It is therefore appropriate to apply other clustering algorithms and to compare the results obtained by different methods. In this way it is also possible to evaluate if the number of identified clusters is robust with respect to the different techniques applied.

At a subsequent phase it is recommended to calculate some performance indicators. Let $C = \{C_1, C_2, \ldots, C_K\}$ be the set of K clusters generated. An indicator of homogeneity of the observations within each cluster C_h is given by the *cohesion*, defined as

$$\text{coh}(C_h) = \sum_{\substack{x_i \in C_h \\ x_k \in C_h}} \text{dist}(\mathbf{x}_i, \mathbf{x}_k). \tag{12.28}$$

The overall cohesion of the partition C can therefore be defined as

$$\text{coh}(C) = \sum_{C_h \in C} \text{coh}(C_h). \tag{12.29}$$

One clustering is preferable over another, in terms of homogeneity within each cluster, if it has a smaller overall cohesion.

An indicator of inhomogeneity between a pair of clusters is given by the *separation*, defined as

$$\text{sep}(C_h, C_f) = \sum_{\substack{x_i \in C_h \\ x_k \in C_f}} \text{dist}(\mathbf{x}_i, \mathbf{x}_k). \tag{12.30}$$

Again the overall separation of the partition C can be defined as

$$\text{sep}(C) = \sum_{\substack{C_h \in C \\ C_f \in C}} \text{sep}(C_h, C_f). \tag{12.31}$$

One clustering is preferable over another, in terms of inhomogeneity among all clusters, if it has a greater overall separation.

A further indicator of the clustering quality is given by the *silhouette coefficient*, which involves a combination of cohesion and separation. To calculate the silhouette coefficient for a single observation $\mathbf{x}_i$, three steps should be followed, as detailed in Procedure 12.3

Procedure 12.3 – Calculation of the silhouette coefficient

1. The mean distance u_i of $\mathbf{x}_i$ from all the remaining observations belonging to the same cluster is computed.

2. For each cluster C_f other than the cluster to which $\mathbf{x}_i$ belongs, the mean distance w_{if} between $\mathbf{x}_i$ and all the observations in C_f is calculated. The minimum v_i among the distances w_{if} is determined by varying the cluster C_f.

3. The silhouette coefficient of $\mathbf{x}_i$ is defined as

$$\text{silh}(\mathbf{x}_i) = \frac{v_i - u_i}{\max(u_i, v_i)}.$$

The silhouette coefficient varies between -1 and 1. A negative value indicates that the mean distance u_i of the observation $\mathbf{x}_i$ from the points of its cluster is greater than the minimum value v_i of the mean distances from the observations of the other clusters, and it is therefore undesirable since the membership of $\mathbf{x}_i$ in its cluster is not well characterized. Ideally the silhouette coefficient should be positive and u_i should be as close as possible to 0. Finally, it should be noticed that the overall silhouette coefficient of a clustering may be computed as the mean of the silhouette coefficients for all the observations in the dataset $\mathcal{D}$.

Silhouette coefficients can be illustrated by *silhouette diagrams*, in which the observations are placed on the vertical axis, subdivided by clusters, and the values of the silhouette coefficient for each observation are shown on the horizontal axis. Usually the mean value of the silhouette coefficient for each cluster is also given in a silhouette diagram, as well as the overall mean for the whole dataset.

Figures 12.7 and 12.8 show the silhouette diagrams corresponding to different clusterings. In particular, Figure 12.7(b) shows the silhouette diagram corresponding to a cut of four clusters in the dendrogram shown in Figure 12.7(a), obtained by applying a divisive hierarchical algorithm to the *mtcars* dataset using the mean Euclidean distance. Figure 12.8(a) shows the silhouette diagram corresponding to a cut of four clusters in the dendrogram shown in Figure 12.3(a), obtained by applying an agglomerative hierarchical algorithm to the *mtcars* dataset using the mean Euclidean distance. Finally, Figure 12.8(b) shows the silhouette diagram corresponding to the clustering with $K = 4$ obtained by applying a medoids partitioning algorithm.

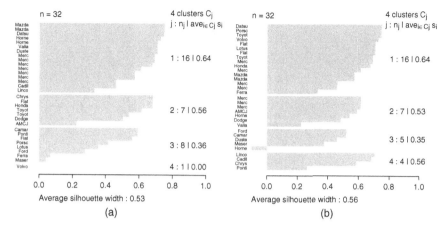

Figure 12.8 Silhouette diagrams with four clusters for an agglomerative hierarchical algorithm, applied to the mtcars dataset (a) with the mean Euclidean distance, and (b) for a medoids partitioning algorithm

12.5 Notes and readings

Among the various texts devoted to clustering, we wish to mention Anderberg (1973), Hartigan (1975), Romesburg (1984), Murtagh (1985), Aldenderfer and Blashfield (1985), Kaufman and Rousseeuw (1990) and Everitt *et al.* (2001). Surveys are also presented in Jain *et al.* (1999) and Mirkin (1996). More specifically, for hierarchical methods the reader is also referred to Fisher (1996) and Jain and Dubes (1988). The evaluation of clustering models is described in Halkidi *et al.* (2002a, b). The interconnections between clustering and classification are discussed in Fraley and Raftery (2002) and Spaeth (1980).

Part III

Business intelligence applications

13

Marketing models

Marketing decision processes are characterized by a high level of complexity due to the simultaneous presence of multiple objectives and countless alternative actions resulting from the combination of the major choice options available to decision makers. Therefore, it should come as no surprise that a large number of mathematical models for marketing have been successfully developed and applied in recent decades. The importance of mathematical models for marketing has been further strengthened by the availability of massive databases of sales transactions that provide accurate information on how customers make use of services or purchase products.

This chapter will primarily focus on two prominent topics in the field of *marketing intelligence*. The first theme is particularly broad and concerns the application of predictive models to support relational marketing strategies, whose purpose is to customize and strengthen the relationship between a company and its customers. After a brief introduction to relational marketing, we will describe the main streams of analysis that can be dealt with in this domain of application, indicating for each of them the classes of predictive models that are best suited to dealing with the problems considered. The subjects discussed in this context can be partly extended to the relationship between citizens and the public administration.

The second theme concerns *salesforce management*. First, we will provide an overview of the major decision-making processes emerging in the organization of a sales staff, highlighting also the role played by response functions. Then, we will illustrate some optimization models which aim to allocate a set of geographical territories to sales agents as well as planning the activities of sales agents. Finally, some business cases consisting of applied marketing models will be discussed.

Business Intelligence: Data Mining and Optimization for Decision Making   C. Vercellis
© 2009 John Wiley & Sons, Ltd

13.1   Relational marketing

In order to fully understand the reasons why enterprises develop relational marketing initiatives, consider the following three examples: an insurance company that wishes to select the most promising market segment to target for a new type of policy; a mobile phone provider that wishes to identify those customers with the highest probability of churning, that is, of discontinuing their service and taking out a new contract with a competitor, in order to develop targeted retention initiatives; a bank issuing credit cards that needs to identify a group of customers to whom a new savings management service should be offered. These situations share some common features: a company owning a massive database which describes the purchasing behavior of its customers and the way they make use of services, wishes to extract from these data useful and accurate knowledge so as to develop targeted and effective marketing campaigns.

The aim of a *relational marketing* strategy is to initiate, strengthen, intensify and preserve over time the relationships between a company and its stakeholders, represented primarily by its customers, and involves the analysis, planning, execution and evaluation of the activities carried out to pursue these objectives.

Relational marketing became popular during the late 1990s as an approach to increasing customer satisfaction in order to achieve a sustainable competitive advantage. So far, most enterprises have taken at least the first steps in this direction, through a process of cultural change which directs greater attention toward customers, considering them as a formidable asset and one of the main sources of competitive advantage. A relational marketing approach has been followed in a first stage by service companies in the financial and telecommunications industries, and has later influenced industries such as consumer goods, finally reaching also manufacturing companies, from automotive and commercial vehicles to agricultural equipments, traditionally more prone to a vision characterized by the centrality of products with respect to customers.

13.1.1   Motivations and objectives

The reasons for the spread of relational marketing strategies are complex and interconnected. Some of them are listed below, although for additional information the reader is referred to the suggested references at the end of the chapter.

- The increasing concentration of companies in large enterprises and the resulting growth in the number of customers have led to greater complexity in the markets.

- Since the 1980s, the innovation–production–obsolescence cycle has progressively shortened, causing a growth in the number of customized options on the part of customers, and an acceleration of marketing activities by enterprises.

- The increased flow of information and the introduction of e-commerce have enabled global comparisons. Customers can use the Internet to compare features, prices and opinions on products and services offered by the various competitors.

- Customer loyalty has become more uncertain, primarily in the service industries, where often filling out an on-line form is all one has to do to change service provider.

- In many industries a progressive commoditization of products and services is taking place, since their quality is perceived by consumers as equivalent, so that differentiation is mainly due to levels of service.

- The systematic gathering of sales transactions, largely automated in most businesses, has made available large amounts of data that can be transformed into knowledge and then into effective and targeted marketing actions.

- The number of competitors using advanced techniques for the analysis of marketing data has increased.

Relational marketing strategies revolve around the choices shown in Figure 13.1, which can be effectively summarized as formulating for each segment, ideally

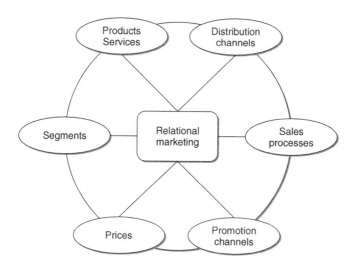

Figure 13.1   Decision-making options for a relational marketing strategy

Figure 13.2   Components of a relational marketing strategy

for each customer, the appropriate offer through the most suitable channel, at the right time and at the best price.

The ability to effectively exploit the information gathered on customers' behavior represents today a powerful competitive weapon for an enterprise. A company capable of gathering, storing, analyzing and understanding the huge amount of data on its customers can base its marketing actions on the knowledge extracted and achieve sustainable competitive advantages. Enterprises may profitably adopt relational marketing strategies to transform occasional contacts with their customers into highly customized long-term relationships. In this way, it is possible to achieve increased customer satisfaction and at the same time increased profits for the company, attaining a win–win relationship.

To obtain the desired advantages, a company should turn to relational marketing strategies by following a correct and careful approach. In particular, it is advisable to stress the distinction between a relational marketing vision and the software tools usually referred to as *customer relationship management* (CRM). As shown in Figure 13.2, relational marketing is not merely a collection of software applications, but rather a coherent project where the various company departments are called upon to cooperate and integrate the managerial culture and human resources, with a high impact on the organizational structures. It is then necessary to create within a company a true *data culture*, with the awareness that customer-related information should be enhanced through the adoption of business intelligence and data mining analytical tools.

Based on the investigation of cases of excellence, it can be said that a successful relational marketing strategy can be achieved through the development of a company-wide vision that puts customers at the center of the whole organization. Of course, this goal cannot be attained by exclusively relying on innovative computer technologies, which at most can be considered a relevant enabling factor.

The overlap between relational marketing strategies and CRM software led to a misunderstanding with several negative consequences. On one hand, the notion that substantial investments in CRM software applications were in themselves sufficient to generate a relational marketing strategy represents a dangerous simplification, which caused many project failures. On the other hand, the high cost of software applications has led many to believe that a viable approach to relational marketing was only possible for large companies in the service industries. This is a deceitful misconception: as a matter of fact, the essential components of relational marketing are a well-designed and correctly fed marketing data mart, a collection of business intelligence and data mining analytical tools, and, most of all, the cultural education of the decision makers. These tools will enable companies to carry out the required analyses and translate the knowledge acquired into targeted marketing actions.

The relationship system of an enterprise is not limited to the dyadic relationship with its customers, represented by individuals and companies that purchase the products and services offered, but also includes other actors, such as the employees, the suppliers and the sales network. For most relationships shown

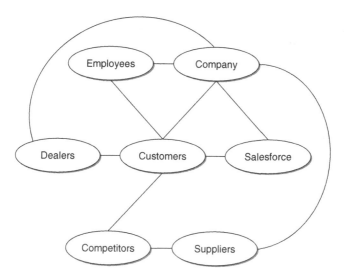

Figure 13.3   Network of relationships involved in a relational marketing strategy

in Figure 13.3, a mutually beneficial exchange occurs between the different subjects involved. More generally, we can widen the boundaries of relational marketing systems to include the stakeholders of an enterprise. The relationship between an enterprise and its customers is sometimes mediated by the sales network, which in some instances can partially obstruct the visibility of the end customers.

Let us take a look at a few examples to better understand the implications of this issue. The manufacturers of consumer goods, available at the points of sale of large and small retailers, do not have direct information on the consumers purchasing their products. The manufacturers of goods covered by guarantees, such as electrical appliances or motor vehicles, have access to personal information on purchasers, even if they rarely also have access to information on the contacts of and promotional actions carried out by the network of dealers. Likewise, a savings management company usually places shares in its investment funds through a network of intermediaries, such as banks or agents, and often knows only the personal data of the subscribers. A pharmaceutical enterprise producing prescription drugs usually ignores the identity of the patients that use its drugs and medicinal products, even though promotional activities to influence consumers are carried out in some countries where the law permits.

It is not always easy for a company to obtain information on its end customers from dealers in the sales network and even from their agents. These may be reluctant to share the wealth of information for fear, rightly or wrongly, of compromising their role. In a relational marketing project specific initiatives should be devised to overcome these cultural and organizational barriers, usually through incentives and training courses.

The number of customers and their characteristics strongly influence the nature and intensity of the relationship with an enterprise, as shown in Figure 13.4. The relationships that might actually be established in a specific economic domain tend to lie on the diagonal shown in the figure. At one extreme, there are highly intense relationships existing between the company and a small number of customers of high individual value. Relationships of this type occur more frequently in *business-to-business* (B2B) activities, although they can also be found in other domains, such as private banking. The high value of each customer justifies the use of dedicated resources, usually consisting of sales agents and key account managers, so as to maintain and strengthen these more intense relationships. In situations of this kind, careful organization and planning of the activities of sales agents is critical. Therefore, optimization models for *salesforce automation* (SFA), described in Section 13.2, can be useful in this context.

At the opposite extreme of the diagonal are the relationships typical of consumer goods and *business-to-consumer* (B2C) activities, for which a high number of low-value customers get in contact with the company in an impersonal

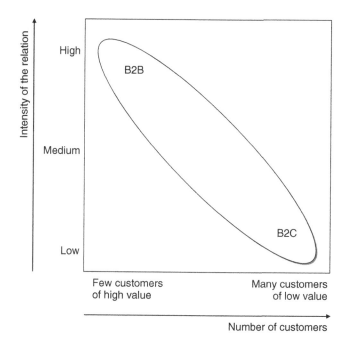

Figure 13.4   Intensity of customer relationships as a function of number of customers

way, through websites, call centers and points of sale. Data mining analyses for segmentation and profiling are particularly valuable especially in this context, characterized by a large number of fragmented contacts and transactions. Relational marketing strategies, which are based on the knowledge extracted through data mining models, enable companies develop a targeted customization and differentiation of their products and/or services, including companies more prone toward a mass-market approach.

Figure 13.5 contrasts the cost of sales actions and the corresponding revenues. Where transactions earn a low revenue per unit, it is necessary to implement low-cost actions, as in the case of mass-marketing activities. Moving down along the diagonal in the figure, more evolved and intense relationships with the customers can be found. The relationships at the end of the diagonal presuppose the action of a direct sales network and for the most part are typical of B2B relational contexts.

Figure 13.6 shows the ideal path that a company should follow so as to be able to offer customized products and services at low cost and in a short time. On the one hand, companies operating in a mass market, well acquainted with fast delivery at low costs, must evolve in the direction of increased customization, by introducing more options and variants of products and services offered

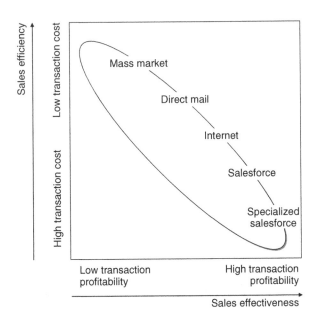

Figure 13.5   Efficiency of sales actions as a function of their effectiveness

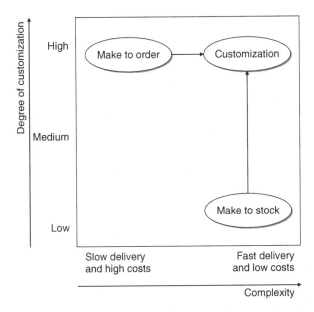

Figure 13.6   Level of customization as a function of complexity of products and services

to the various market segments. Data mining analyses for relational marketing purposes are a powerful tool for identifying the segments to be targeted with customized products. On the other hand, the companies oriented toward make-to-order production must evolve in a direction that fosters reductions in both costs and delivery times, but without reducing the variety and the range of their products.

13.1.2   An environment for relational marketing analysis

Figure 13.7 shows the main elements that make up an environment for relational marketing analysis. Information infrastructures include the company's data warehouse, obtained from the integration of the various internal and external data sources, and a marketing data mart that feeds business intelligence and data mining analyses for profiling potential and actual customers. Using pattern recognition and machine learning models as described in previous chapters, it is possible to derive different segmentations of the customer base, which are then used to design targeted and optimized marketing actions. A classification model can be used, for example, to generate a *scoring* system for customers according to their propensity to buy a service offered by a company, and to direct a cross-selling offer only toward those customers for whom a high probability of acceptance is predicted by the model, thus maximizing the overall *redemption* of the marketing actions.

Effective management of frequent marketing campaign cycles is certainly a complex task that requires planning, for each segment of customers, the content of the actions and the communication channels, using the available

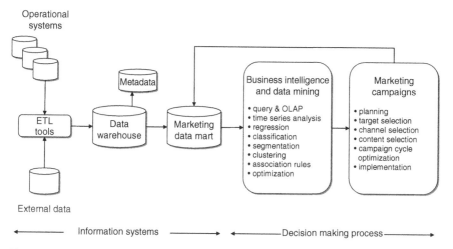

Figure 13.7   Components of an environment for relational marketing analysis

human and financial resources. The corresponding decision-making process can be formally expressed by appropriate optimization models. The cycle of marketing activities terminates with the execution of the planned campaign, with the subsequent gathering of information on the results and the redemption among the recipients. The data collected are then fed into the marketing data mart for use in future data mining analyses. During the execution of each campaign, it is important to set up procedures for controlling and analyzing the results obtained. In order to assess the overall effectiveness of a campaign, it would be advisable to select a *control group* of customers, with characteristics similar to those of the campaign recipients, toward whom no action should be undertaken.

Figure 13.8 describes the main types of data stored in a data mart for relational marketing analyses. A company data warehouse provides demographic and administrative information on each customer and the transactions carried out for purchasing products and using services. The marketing database contains data on initiatives carried out in the past, including previous campaigns and their results, promotions and advertising, and analyses of customer value. A further possible data source is the salesforce database, which provides information on established contacts, calls and applicable sales conditions. Finally, the *contact center* database provides access to data on customers' contacts with the call center, problems reported, sometimes called *trouble tickets*, and

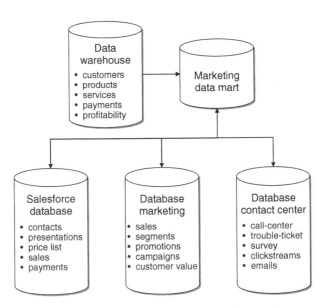

Figure 13.8   Types of data feeding a data mart for relational marketing analysis

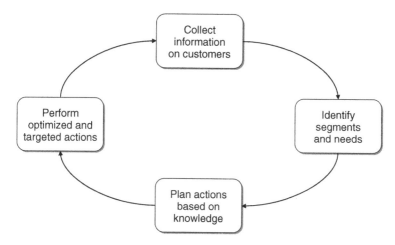

Figure 13.9   Cycle of relational marketing analysis

related outcomes, website navigation paths and forms filled out on-line, and emails exchanged between customers and the support center.

As shown in Figure 13.8, the available data are plentiful, providing an accurate representation of the behaviors and needs of the different customers, through the use of inductive learning models.

The main phases of a relational marketing analysis proceeds as shown in Figure 13.9. The first step is the exploration of the data available for each customer. At a later time, by using inductive learning models, it is possible to extract from those data the insights and the rules that allow market segments characterized by similar behaviors to be identified. Knowledge of customer profiles is used to design marketing actions which are then translated into promotional campaigns and generate in turn new information to be used in the course of subsequent analyses.

13.1.3   Lifetime value

Figure 13.10 shows the main stages during the customer *lifetime*, showing the cumulative value of a customer over time. The figure also illustrates the different actions that can be undertaken toward a customer by an enterprise. In the initial phase, an individual is a *prospect*, or potential customer, who has not yet begun to purchase the products or to use the services of the enterprise. Toward potential customers, *acquisition* actions are carried out, both directly (telephone contacts, emails, talks with sales agents) and indirectly (advertising, notices on the enterprise website). These actions incur a cost that can be assigned to each customer and determine an accumulated loss that lasts until a critical event

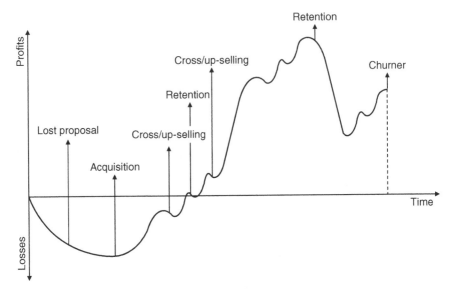

Figure 13.10   Lifetime of a customer

in the relationship with a customer occurs: a prospect becomes a customer. This event may take various forms in different situations: it may consist of a service subscription, the opening of a bank account, the first purchase at a retailer point of sale with the activation of a loyalty card. Before becoming a new customer, a prospect may receive from the enterprise repeated proposals aiming acquiring her custom, shown in the figure as *lost proposals*, which have a negative outcome. From the time of acquisition, each customer generates revenue, which produces a progressive rise along the curve of losses and cumulated profits. This phase, which corresponds to the maturity of the relationship with the enterprise, usually entails alternating *cross-selling*, *up-selling* and *retention* actions, in an effort to extend the duration and the profitability of the relationship so as to maximize the lifetime value of each customer. The last event in a customer lifetime is the interruption of the relationship. This may be *voluntary*, when a customer discontinues the services of an enterprise and switches to those of a competitor, *forced*, when for instance a customer does not comply with payment terms, or *unintentional*, when for example a customer changes her place of residence.

The progress of a customer lifetime highlights the main tasks of relational marketing. First, the purpose is to increase the ability to acquire new customers. Through the analysis of the available information for those customers who in the past have purchased products or services, such as personal socio-demographic characteristics, purchased products, usage of services, previous contacts, and the comparison with the characteristics of those who have not taken up the offers of the enterprise, it is possible to identify the segments with the highest potential. This in turn allows the enterprise to optimize marketing campaigns, to increase the effectiveness of acquisition initiatives and to reduce the waste of resources due to offers addressed to unpromising market segments.

Furthermore, relational marketing strategies can improve the loyalty of customers, extending the duration of their relationship with the enterprise, and thus increasing the profitability. In this case, too, the comparative analysis of the characteristics of those who have remained loyal over time with respect to those who have switched to a competitor leads to predictions of the likelihood of churning for each customer. Retention actions can therefore be directed to the most relevant segments, represented by high-value customers with the highest risk of churning.

Finally, relational marketing analyses can be used to identify customers who are more likely to take up the offer of additional services and products (*cross-selling*), or of alternative services and products of a higher level and with a greater profitability for the enterprise (*up-selling*).

The tasks of acquisition, retention, cross-selling and up-selling, shown in Figure 13.11, are at the heart of relational marketing strategies and their aim is to maximize the profitability of customers during their lifetime. These analysis tasks, which will be described in the next sections, are clearly amenable to classification problems with a binary target class. Notice that attribute selection plays a critical role in this context, since the number of available explanatory variables is usually quite large and it is advisable for learning models to use a

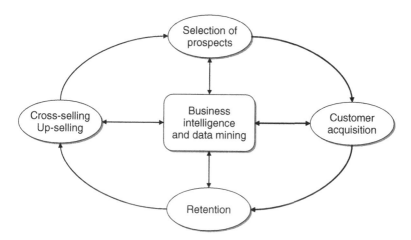

Figure 13.11   Main relational marketing tasks

limited subset of predictive features, in order to generate meaningful and useful classification rules for the accurate segmentation of customers.

13.1.4   The effect of latency in predictive models

Figure 13.12 illustrates the logic of development of a classification model for a relational marketing analysis, also taking into account the temporal dimension. Assume that t is the current time period, and that we wish to derive an inductive learning model for a classification problem. For example, at the beginning of October a mobile telephone provider might want to develop a classification model to predict the probability of churning for its customers. The data mart contains the data for past periods, updated as far as period $t - 1$. In our example, it contains data up to and including September.

Furthermore, suppose that the company wishes to predict the probability of churning h months in advance, since in this way any retention action has a better chance of success. In our example, we wish to predict at the beginning of October the probability of churning in November, using data up to September. Notice that the data for period t cannot be used for the prediction, since they are clearly not available at the beginning of period t.

To develop a classification model we use the value of the target variable for the last known period $t - 1$, corresponding to the customers who churned in the month of September. It should be clear that for training and testing the model the explanatory variables for period $t - 2$ should not be used, since in the training phase it is necessary to reproduce the same situation as will be faced when using the model in the prediction stage. Actually, the target

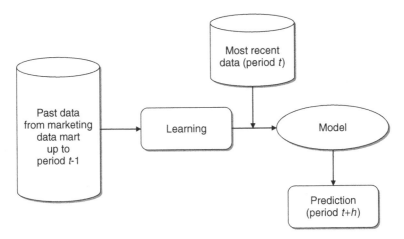

Figure 13.12   Development and application flowchart for a predictive model

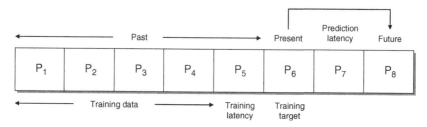

Figure 13.13   Latency of a predictive model

variable must be predicted $h = 2$ periods in advance, and therefore there is an intermediate period of future data that are still unknown at time t (the month of October in our example). To reflect these dynamics, the training phase should be carried out without using August data. In general, the $h - 1$ periods corresponding to data still unknown during the prediction phase, and not used during the training phase, are referred to as the model's *latency*, as shown in Figure 13.13.

13.1.5   Acquisition

Although retention plays a prominent role in relational marketing strategies, for many companies the *acquisition* of new customers also represents a critical factor for growth. The acquisition process requires the identification of new prospects, as they are potential customers who may be totally or partially unaware of the products and services offered by the company, or did not possess in the past the characteristics to become customers, or were customers of competitors. It may also happen that some of the prospects were former customers who switched their custom to competitors, in which case much more information is usually available on them.

Once prospects have been identified, the enterprise should address acquisition campaigns to segments with a high potential profitability and a high probability of acquisition, in order to optimize the marketing resources. Traditional marketing techniques identify interesting segments using predefined profiling criteria, based on market polls and socio-demographic analyses, according to a top-down perspective. This approach can be successfully integrated or even replaced by a top-down segmentation logic which analyzes the data available in the data mart, as shown in Figure 13.8 (demographic information, contacts with prospects, use of products and services of competitors), and derives classification rules that characterize the most promising profiles for acquisition purposes. Also in this case, we are faced with a binary classification problem, which can be analyzed with the techniques described in Chapter 10.

13.1.6   Retention

The maturity stage reached by most products and services, and the subsequent saturation of their markets, have caused more severe competitive conditions. As a consequence, the expansion of the customer base of an enterprise consists more and more of switch mechanisms – the acquisition of customers at the expense of other companies. This phenomenon is particularly apparent in service industries, such as telecommunications, banking, savings management and insurance, although it also occurs in manufacturing, both for consumer goods and industrial products. For this reason, many companies invest significant amounts of resources in analyzing and characterizing the phenomenon of *attrition*, whereby customers switch from their company to a competitor.

There are also economic reasons for devoting substantial efforts to customer *retention*: indeed, it has been empirically observed that the cost of acquiring a new customer, or winning back a lost customer, is usually much higher – of the order of 5 to 9 times higher – than the cost of the marketing actions aimed at retaining customers considered at risk of churning. Furthermore, an action to win back a lost customer runs the risk of being too late and not achieving the desired result. In many instances, winning back a customer requires investments that do not generate a return.

One of the main difficulties in loyalty analysis is actually recognizing a churn. For subscription services there are unmistakable signals, such as a formal notice of withdrawal, while in other cases it is necessary to define adequate indicators that are correlated, a few periods in advance, with the actual churning. A customer who reduces by more than a given percentage her purchases at a selected point of sale using a loyalty card, or a customer who reduces below a given threshold the amount held in her checking account and the number of transactions, represent two examples of disaffection indicators. They also highlight the difficulties involved in correctly defining the appropriate threshold values.

To optimize the marketing resources addressed to retention, it is therefore necessary to target efforts only toward high-value customers considered at risk of churning. To obtain a scoring system corresponding to the probability of churning for each customer, it is necessary to derive a segmentation based on the data on past instances of churning. Predicting the risk of churning requires analysis of records of transactions for each customer and identifying the attributes that are most relevant to accurately explaining the target variable. Again, we are faced with a binary classification problem. Once the customers with the highest risk of churning have been identified, a retention action can be directed toward them. The more accurately such action is targeted, the cheaper it is likely to be.

13.1.7   Cross-selling and up-selling

Data mining models can also be used to support a relational marketing analysis aimed at identifying market segments with a higher propensity to purchase additional services or other products of a company. For example, a bank also offering insurance services may identify among its customers segments interested in purchasing a life insurance policy. In this case, demographic information on customers and their past transactions recorded in a data mart can be used as explanatory attributes to derive a classification model for predicting the target class, consisting in this example of a binary variable that indicates whether the customer accepted the offer or not.

The term *cross-selling* refers to the attempt to sell an additional product or service to an active customer, already involved in a long-lasting commercial relationship with the enterprise. By means of classification models, it is possible to identify the customers characterized by a high probability of accepting a cross-selling offer, starting from the information contained in the available attributes.

In other instances, it is possible to develop an *up-selling* initiative, by persuading a customer to purchase an higher-level product or service, richer in functions for the user and more profitable for the company, and therefore able to increase the lifetime value curve of a customer. For example, a bank issuing credit cards may offer customers holding a standard card an upgrade to a gold card, which is more profitable for the company, but also able to offer a series of complementary services and advantages to interested customers. In this case too, we are dealing with a binary classification problem, which requires construction of a model based on the training data of customers' demographic and operational attributes. The purpose of the model is to identify the most interesting segments, corresponding to customers who have taken up the gold service in the past, and who appear therefore more appreciative of the additional services offered by the gold card. The segments identified in this way represent the target of up-selling actions.

13.1.8   Market basket analysis

The purpose of *market basket analysis* is to gain insight from the purchases made by customers in order to extract useful knowledge to plan marketing actions. It is mostly used to analyze purchases in the retail industry and in e-commerce activities, and is generally amenable to unsupervised learning problems. It may also be applied in other domains to analyze the purchases made using credit cards, the complementary services activated by mobile or fixed telephone customers, the policies or the checking accounts acquired by a same household.

The data used for this purpose mostly refer to purchase transactions, and can be associated with the time dimension if the purchaser can be tracked through a loyalty card or the issue of an invoice. Each transaction consists of a list of purchased items. This list is called a *basket*, just like the baskets available at retail points of sale.

If transactions cannot be connected to one another, say because the purchaser is unknown, one may then apply association rules, described in Chapter 11, to extract interesting correlations between the purchases of groups of items. The rules extracted in this way can then be used to support different decision-making processes, such as assigning the location of the items on the shelves, determining the layout of a point of sale, identifying which items should be included in promotional flyers, advertisements or coupons distributed to customers.

Clustering models, described in Chapter 12, are also useful in determining homogeneous groups of items, once an incidence matrix X has been created for the representation of the dataset, where the rows correspond to the transactions and the columns to the items.

If customers are individually identified and traced, besides the above techniques it is also possible to develop further analyses that take into account the time dimension of the purchases. For instance, one may generate sequential association rules, mentioned at the end of Chapter 11, or apply time series analysis, as described in Chapter 9.

13.1.9   Web mining

The web is a critical channel for the communication and promotion of a company's image. Moreover, e-commerce sites are important sales channels. Hence, it is natural to use *web mining* methods in order to analyze data on the activities carried out by the visitors to a website.

Web mining methods are mostly used for three main purposes, as shown in Figure 13.14: *content mining*, *structure mining* and *usage mining*.

Content mining. Content mining involves the analysis of the content of web pages to extract useful information. Search engines primarily perform content mining activities to provide the links deemed interesting in relation to keywords supplied by users. Content mining methods can be traced back to data mining problems for the analysis of texts, both in free format or HTML and XML formats, images and multimedia content. Each of these problems is in turn dealt with using the learning models described in previous chapters. For example, text mining analyses are usually handled as multicategory classification problems, where the target variable is the subject category to which the text refers, while explanatory variables correspond to the meaningful words contained in the text.

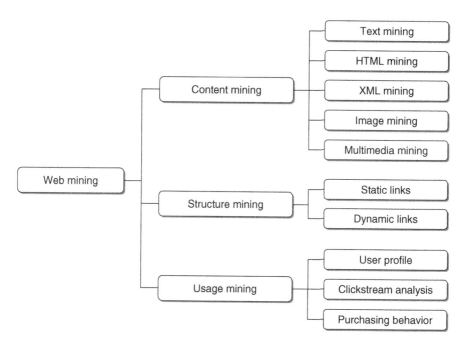

Figure 13.14   Taxonomy of web mining analyses

Once it has been converted into a classification problem, text mining can be approached using the methods described in Chapter 10. Text mining techniques are also useful for analyzing the emails received by a support center. Notice that the input data for content mining analyses are easily retrievable, at least in principle, since they consist of all the pages that can be visited on the Internet.

Structure mining. The aim of this type of analysis is to explore and understand the topological structure of the web. Using the links presented in the various pages, it is possible to create graphs where the nodes correspond to the web pages and the oriented arcs are associated with links to other pages. Results and algorithms from graph theory are used to characterize the structure of the web, that is, to identify areas with a higher density of connections, areas disconnected from others and maximal *cliques*, which are groups of pages with reciprocal links. In this way, it is possible to pinpoint the most popular sites, or to measure the distance between two sites, expressed in terms of the lowest number of arcs along the paths that connect them in the links graph. Besides analyses aimed at exploring the *global* structure of the web, it is also possible to carry out *local* investigations to study how a single website is articulated. In some investigations, the local structure of websites is associated with the time spent by the users on each page, to verify if the organization of the site suffers from inconsistencies that jeopardize its effectiveness. For example, a page whose purpose is to direct navigation on the site should be viewed by each user only briefly. Should this not be the case, the page has a problem due to a possible ambiguity in the articulation of the links offered.

Usage mining. Analyses aimed at *usage mining* are certainly the most relevant from a relational marketing standpoint, since they explore the paths followed by navigators and their behaviors during a visit to a company website. Methods for the extraction of association rules are useful in obtaining correlations between the different pages visited during a session. In some instances, it is possible to identify a visitor and recognize her during subsequent sessions. This happens if an identification key is required to access a web page, or if a cookie-enabling mechanism is used to keep track of the sequence of visits. Sequential association rules or time series models can be used to analyze the data on the use of a site according to a temporal dynamic. Usage mining analysis is mostly concerned with *clickstreams* – the sequences of pages visited during a given session. For e-commerce sites, information on the purchase behavior of a visitor is also available.

13.2   Salesforce management

Most companies have a sales network and therefore rely on a substantial number of people employed in sales activities, who play a critical role in the profitability of the enterprise and in the implementation of a relational marketing strategy. The term *salesforce* is generally taken to mean the whole set of people and roles that are involved, with different tasks and responsibilities, in the sales process. A preliminary taxonomy of salesforces is based on the type of activity carried out, as indicated below.

Residential. *Residential* sales activities take place at one or more sites managed by a company supplying some products or services, where customers go to make their purchases. This category includes sales at retail outlets as well as wholesale trading centers and *cash-and-carry* shops.

Mobile. In *mobile* sales, agents of the supplying company go to the customers' homes or offices to promote their products and services and collect orders. Sales in this category occur mostly within B2B relationships, even though they can also be found in B2C contexts.

Telephone. *Telephone* sales are carried out through a series of contacts by telephone with prospective customers.

There are various problems connected with managing a mobile salesforce management, which will be the main focus of this section. They can be subdivided into a few main categories:

- designing the sales network;

- planning the agents' activities;

- contact management;

- sales opportunity management;

- customer management;

- activity management;

- order management;

- area and territory management;

- support for the configuration of products and services;

- knowledge management with regard to products and services.

Designing the sales network and planning the agents' activities involve decision-making tasks that may take advantage of the use of optimization models, such as those that will be described in the next sections. The remaining activities are operational in nature and may benefit from the use of software tools for *salesforce automation* (SFA), today widely implemented.

13.2.1 Decision processes in salesforce management

The design and management of a salesforce raise several decision-making problems, as shown in Figure 13.15. When successfully solved, they confer multiple advantages: maximization of profitability, increased effectiveness of sales actions, increased efficiency in the use of resources, and greater professional rewards for sales agents.

The decision processes described in Figure 13.15 should take into account the strategic objectives of the company, with respect to other components of the marketing mix, and conform to the role assigned to the salesforce within the broader framework of a relational marketing strategy. The two-way connections indicated in the figure suggest that the different components of the decision-making process interact with each other and with the general objectives of the marketing department.

In particular, the decision-making processes relative to salesforce management can be grouped into three categories: *design*, *planning* and *assessment*.

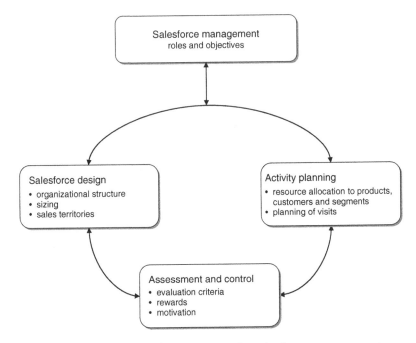

Figure 13.15   Decision processes in salesforce management

Design

Salesforce design is dealt with during the start-up phase of a commercial activity or during subsequent restructuring phases, for example following the merger or acquisition of a group of companies.

As shown in Figure 13.16, the design phase is usually preceded by the creation of market segments through the application of data mining methods and by the articulation of the offer of products and services, which are in turn subdivided into homogeneous classes. Salesforce design includes three types of decisions.

Organizational structure. The *organizational structure* may take different forms, corresponding to hierarchical agglomerations of the agents by group of products, brand or geographical area. In some situations the structure may also be differentiated by markets. In order to determine the organizational structure, it is necessary to analyze the complexity of customers, products and sales activities, and to decide whether and to what extent the agents should be specialized.

Sizing. Sales network *sizing* is a matter of working out the optimal number of agents that should operate within the selected structure, and depends on several factors, such as the number of customers and prospects, the desired level of

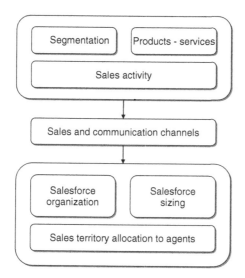

Figure 13.16   Salesforce design process

sales area coverage, the estimated time for each call and the agents' traveling time. One should bear in mind that a reduction in costs due to a decrease in the salesforce size is often followed by a reduction in sales and revenues. A better allocation of the existing salesforce, devised during the planning phase by means of optimization models, is usually more effective than a variation in size.

Sales territories. Designing a *sales territory* means grouping together the geographical areas into which a given region has been divided and assigning each territory to an agent. The design and assignment of sales territories should take into account several factors, such as the sales potential of each geographical area, the time required to travel from one area to another and the total time each agent has available. The purpose of the assignment consists of determining a balanced situation between sales opportunities embedded in each territory, in order to avoid disparities among agents. The assignment of the geographical areas should be periodically reviewed since the sales potential balance in the various territories tends to vary over time.

Decisions concerning the design of the salesforce should take into account decisions about salesforce planning, and this explains the two-way link between the two corresponding blocks in Figure 13.15.

Planning

Decision-making processes for *planning* purposes involve the assignment of sales resources, structured and sized during the design phase, to market entities.

Resources may correspond to the work time of agents or to the budget, while market entities consist of products, market segments, distribution channels and customers.

Allocation takes into account the time spent pitching the sale to each customer, the travel time and cost, and the effectiveness of the action for each product, service or market segment. It is also possible to consider further ancillary activities carried out at the customers' sites, such as making suggestions that are conducive to future sales or explaining the technical and functional features of products and services.

Salesforce planning can greatly benefit from the use of optimization models, as explained below.

Assessment

The purpose of *assessment* and *control* activities is to measure the effectiveness and efficiency of individuals employed in the sales network, in order to design appropriate remuneration and incentive schemes. To measure the efficiency of the sales agents it is necessary to define adequate evaluation criteria that take into account the actual personal contribution of each agent, having removed effects due to the characteristics that may make an area or product more or less advantageous than others. Data envelopment analysis, described in Chapter 15, provides useful models that can be applied to assess agents' performance.

13.2.2   Models for salesforce management

In what follows we will describe some classes of optimization models for designing and planning the salesforce. These models are primarily intended for educational purposes, to familiarize readers with the reasoning behind specific aspects of a sales network, through the formulation of optimization models. For the sake of clarity and conciseness, for each model we have limited the extensions to a single feature. Sales networks simultaneously possess more than one of the distinctive features previously described, and therefore the models developed in real-world applications, just like those described in the last section of the chapter, are more complex and result from a combination of different characteristics.

Before proceeding, it is useful to introduce some notions common to the different models that will be described. Assume that a region is divided into J geographical sales areas, also called *sales coverage units*, and let $\mathcal{J} = \{1, 2, \ldots, J\}$. Areas must be aggregated into disjoint clusters, called *territories*, so that each area belongs to one single territory and is also connected to all the areas belonging to the same territory. The connection property implies that from each area it is possible to reach any other area of the same territory. The time span is divided into T intervals of equal length, which usually correspond to weeks or months, indicated by the index $t \in \mathcal{T} = \{1, 2, \ldots, T\}$.

Each territory is associated with a sales agent, located in one of the areas belonging to the territory, henceforth considered as her area of residence. The choice of the area of residence determines the time and cost of traveling to any other area in the same territory. Let I be the number of territories and therefore the number of agents that form the sales network, and let $\mathcal{I} = \{1, 2, \ldots, I\}$.

In each area there are customers or prospects who can be visited by the agents as part of their promotions and sales activities. In some of the models that will be presented, customers or prospects are aggregated into segments, which are considered homogeneous with respect to the area of residence and possibly to other characteristics, such as value, potential for development and purchasing behaviors. Let H be the number of market entities, which in different models may represent either single customers or segments, and let $\mathcal{H} = \{1, 2, \ldots, H\}$. Let $\mathcal{D}_j$ be the set of customers, or segments of customers where necessary, located in area j.

Finally, assume that a given agent can promote and sell K products and services during the calls she makes on customers or prospects, and let $\mathcal{K} = \{1, 2, \ldots, K\}$.

13.2.3   Response functions

Response functions play a key role in the formulation of models for designing and planning a sales network. In general terms, a response function describes the elasticity of sales in terms of the intensity of the sales actions, and is a formal method to describe the complex relationship existing between sales actions and market reactions.

Sales to which the response function refers are expressed in product units or monetary units, such as revenues or margins. For the sake of uniformity, in the next sections response functions are assumed to be expressed as sales revenues. The intensity of a sales action can be related to different variables, such as the number of calls to a customer in each period, the number of mentions of a product in each period, and the time dedicated to each customer in each period.

In principle, it is possible to consider a response function in relation to each factor that is deemed critical to sales: the characteristics of customers and sales territories; the experience, education and personal skills of the agents; promotions, prices, markdown policies operated by the company and the corresponding features for one or several competitors.

Figures 13.17 and 13.18 show two possible shapes of the response function, obtained by placing the sales of a product or service on the vertical axis and the intensity of the sales action of interest on the horizontal axis. To fix ideas, we

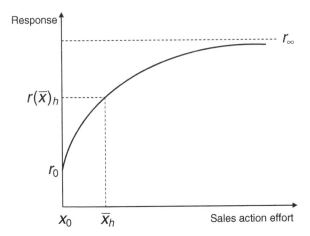

Figure 13.17   A concave response function

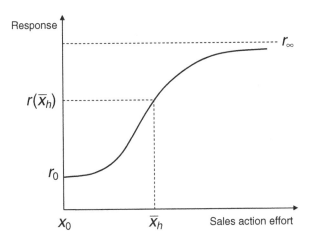

Figure 13.18   A sigmoidal response function

will assume that the number x_h of calls that a specific agent makes to customer h in each period of the planning horizon is placed on the horizontal axis.

The concave response function shown in Figure 13.17 can be interpreted in the following way: as the number of calls increases, revenues grow at a decreasing rate approaching 0, according to the principle of decreasing marginal revenues. In general, a lack of sales actions toward a given customer does not imply a lack of sales, at least for a certain number of periods. This is an effect of the actions executed in previous periods that lasts over time. For this reason the response function is greater than 0 at $x_h = 0$.

The sigmoidal response function in Figure 13.18 reflects a different hypothesis of sales growth as a function of the actions carried out. The assumption made in this case is that the central interval of values on the horizontal axis corresponds to a higher rate of sales growth, while outside that area the growth rate is lower.

It is worth noting that each decision concerning the allocation of sales resources is based on a response function hypothesis, which is implicit and unaware in intuitive decision-making processes, while it is explicit and rigorous in mathematical models such as those presented below.

Response functions can be estimated by considering two types of information. On the one hand, one can use past available data regarding the intensity of the actions carried out and the corresponding sales, to develop a parametric regression model through variants of regression methods. On the other hand, interviews are carried out with agents and sales managers to obtain subjective information which is then incorporated into the procedure for calculating the response function.

We will now show by means of an example how the procedure for estimating the response function works. Let $r_h(x_h)$ be the sales value for customer h associated with a number x_h of calls during a given period. More generally, the variable that determines the response function r expresses the intensity of the sales action that has been carried out. A parametric form should first be selected in order to express the functional dependence. The following function, which may assume both concave and sigmoidal shapes by varying the parameters, can be used:

$$r_h(x_h) = r_0 + (r_\infty - r_0)\frac{x_h^\sigma}{\gamma + x_h^\sigma}. \tag{13.1}$$

The parameters in the expression $r_h(x_h)$ have the following meaning: r_0 represents the sales level that would be obtained at a sales action intensity equal to 0, as a prolonged effect of previous actions; r_∞ represents the maximum sales level, irrespective of the intensity of the sales action; γ and σ are two parameters to be estimated.

To obtain an estimate of the four parameters appearing in the expression for $r_h(x_h)$ it is possible to proceed in two complementary ways. Past sales data can be used to set up a regression model and determine the values through the least squares method. In order to increase the value of the opinions of the sales agents, it is also possible to ask agents and sales managers to estimate the value of the parameters r_0 and r_∞, as well as the values of the expected sales at other three critical points of the response function: $r(\bar{x}_h)$, corresponding to the number of calls carried out at the time of the analysis, $r(\frac{1}{2}\bar{x}_h)$ and $r(\frac{3}{2}\bar{x}_h)$, associated respectively with increasing and decreasing the number of calls by 50% with respect to the current value. Based on a subjective evaluation of the five response values derived through the procedure described above, an estimate by interpolation of the scale parameters γ and σ is then obtained.

13.2.4 Sales territory design

Sales territory design involves allocating sales coverage units to individual agents so as to minimize a weighted sum of two terms, representing respectively the total distance between areas belonging to the same territory and the imbalance of sales opportunities for the agents.

Each region is subdivided into J geographical areas, which should then be clustered into I territories, whose total number has been determined beforehand. A sales agent will be associated with each territory, and she should be located in one of the sales coverage units, to be considered as her area of residence. It is further assumed that travel times within each area are negligible with respect to the corresponding travel times between a pair of distinct areas.

Each area will be identified by the geographical coordinates (e_j, f_j) of one of its points, considered as representative of the entire sales coverage unit. One might, for instance, choose the point whose coordinates are obtained as the average of the coordinates of all points belonging to that area. For each territory, let (e_i, f_i) denote the coordinates of the area where the agent associated with the territory resides. This area will be called *centroid* of territory i.

The parameters in the model are as follows: d_{ij} is the distance between centroid i and area j, given by

$$d_{ij} = \sqrt{(e_i - e_j)^2 + (f_i - f_j)^2};$$ (13.2)

a_j is the opportunity for sales in area j; and β is a relative weight factor between total distance and sales imbalance.

Consider a set of binary decision variables Y_{ij} defined as

$$Y_{ij} = \begin{cases} 1 & \text{if area } j \text{ is assigned to territory } i \\ 0 & \text{otherwise.} \end{cases}$$

Define I additional continuous variables that express the deviations from the average sales opportunity value for each territory:

$$S_i = \text{deviation from the average opportunity value } \frac{1}{I} \sum_{j \in \mathcal{J}} a_j \text{ for territory } i.$$

Hence, the corresponding optimization problem can be formulated as

$$\min \quad \sum_{i \in \mathcal{I}} \sum_{j \in \mathcal{J}} a_j d_{ij}^2 Y_{ij} + \beta \sum_{i \in \mathcal{I}} S_i, \tag{13.3}$$

$$\text{s.to} \quad \sum_{j \in \mathcal{J}} a_j Y_{ij} - \frac{1}{I} \sum_{j \in \mathcal{J}} a_j \leq S_i, \qquad i \in \mathcal{I}, \tag{13.4}$$

$$\sum_{j \in \mathcal{J}} a_j Y_{ij} - \frac{1}{I} \sum_{j \in \mathcal{J}} a_j \geq -S_i, \qquad i \in \mathcal{I}, \tag{13.5}$$

$$\sum_{i \in \mathcal{I}} Y_{ij} = 1, \qquad j \in \mathcal{J}, \tag{13.6}$$

$$S_i \geq 0, \ Y_{ij} \in \{0, 1\}, \qquad i \in \mathcal{I}, j \in \mathcal{J}. \tag{13.7}$$

The purpose of constraints (13.4) and (13.5) is to bound by means of variable S_i the absolute deviation between each territory sales opportunity and the average sales opportunity, to make the assignment to territories more uniform with respect to sales opportunities, hence balancing the sales chances across the agents. Constraints (13.6) represent a multiple choice condition imposed to guarantee that each sales coverage unit is exclusively assigned to one territory, and hence to one and only one agent.

Model (13.3) is a mixed binary optimization problem, which can be solved by a branch-and-bound method, possibly truncated to limit the computing time and to achieve suboptimal solutions. Alternatively, an approximation algorithm can be devised for its *ad hoc* solution.

13.2.5   Calls and product presentations planning

Optimization models for calls and product presentations planning are intended to derive for each agent the optimal sales activity plan.

Calls planning

The aim of the first model described is to identify the optimal number of calls to each customer or prospect (taken together as *market entities* in what follows) located in the territory assigned to a specific agent. The objective function expresses the difference between revenues and transfer costs.

The decision variables are defined as

$$X_h = \text{number of calls to market entity } h,$$
$$W_j = \text{number of trips to market area } j,$$

while the parameters have the following meanings:

$a_h =$ strategic relevance of market entity h,

$c_j =$ transfer cost to area j,

$v_j =$ transfer time to area j,

$t_h =$ time spent with market entity h in each call,

$l_h =$ minimum number of calls to market entity h,

$u_h =$ maximum number of calls to market entity h,

$b =$ total time available to the sales agent.

The corresponding optimization problem can be formulated as

$$\max \quad \sum_{h \in \mathcal{H}} a_h r\left(X_h\right) - \sum_{j \in \mathcal{J}} c_j W_j, \tag{13.8}$$

$$\text{s.to} \quad \sum_{h \in \mathcal{H}} t_h X_h + \sum_{j \in \mathcal{J}} v_j W_j \leq b, \tag{13.9}$$

$$X_h \leq u_h, \qquad X_h \geq l_h, \qquad\qquad h \in \mathcal{H}, \tag{13.10}$$

$$W_j \geq X_h, \qquad\qquad\qquad\qquad j, h \in \mathcal{D}, \tag{13.11}$$

$$X_h, W_j \geq 0 \text{ and integer}, \qquad h \in \mathcal{H}, j \in \mathcal{J}. \tag{13.12}$$

Constraint (13.9) expresses a bound on the total time available to the sales agent within the planning horizon. Constraints (13.10) impose a lower and an upper bound, respectively, on the number of calls to each market entity. Finally, constraints (13.11) establish a logical consistency condition between the decision variables X_h and W_j.

Model (13.8) is a nonlinear mixed integer optimization problem. To obtain a solution, one may proceed as follows. First, the response function is approximated with a piecewise linear function, deriving a set of linear mixed integer optimization problems. These are then solved by using a branch-and-bound method, possibly truncated to limit the computing time and to achieve suboptimal solutions. Alternatively, again, an approximation algorithm can be devised for *ad hoc* solution.

Product presentations planning

The aim of this model is to determine for each period in the planning horizon the optimal number of mentions for each product belonging to the sales portfolio of a given agent. Through an index called *relative exposure* the model also incorporates the dynamic effects determined by the mentions of each product made in past periods.

The decision variables of the model are consequently defined as

X_{kt} = number of calls for product k in period t,

Z_{kt} = cumulated exposure level for product k in period t.

The parameters are

d_{kt} = number of units of product k available in period t,

p = maximum number of mentions for each product,

λ = memoryless parameter.

The quantity $\sigma(X_{kt})$ expresses the relative exposure of product k as a function of the number of times k has been mentioned in period t. The relative exposure formalizes the relationship between the level of cumulative exposure and the number of mentions made in period t through constraints (13.15) in the subsequent optimization model. The response function then depends on the level of cumulated exposure.

The resulting optimization model is formulated as

$$\max \quad \sum_{t \in \mathcal{T}} \sum_{k \in \mathcal{K}} d_{kt} r_k (Z_{kt}), \tag{13.13}$$

$$\text{s.to} \quad \sum_{k \in \mathcal{K}} X_{kt} \leq Kp, \qquad\qquad t \in \mathcal{T}, \tag{13.14}$$

$$Z_{kt} = \lambda \sigma(X_{kt}) + (1 - \lambda)\sigma(X_{kt-1}), \quad k \in \mathcal{K}, t \in \mathcal{T}, \tag{13.15}$$

$$X_{kt}, Z_{kt} \geq 0 \text{ and integer}, \qquad\qquad k \in \mathcal{K}, t \in \mathcal{T}. \tag{13.16}$$

Constraints (13.14) impose a limitation on the maximum number of mentions that can be made in each period. Constraints (13.15) express, through a recursive formula, the relationship between the cumulative exposure level and the number of mentions made in each period.

Model (13.13) is also a nonlinear mixed integer optimization problem, whose solution can be obtained analogously to model (13.8).

Calls and product presentations planning

The aim of the model described in this section is to determine the optimal number of calls to each market entity belonging to a given segment and, for each call, the number of mentions for each product in the sales portfolio. The aggregation of market entities into segments has the purpose of simplifying the estimation of the response function, by limiting its evaluation only to the segments identified within the scope of the analysis.

The decision variables are defined as

X_{kh} = number of mentions of product k to a customer in segment h,

W_h = number of calls to a customer in segment h.

The parameters are

p_h = maximum number of products mentioned in each call

to a customer in segment h,

s_h = number of customers in segment h,

b = maximum number of calls that can be made by each agent.

The resulting optimization problem is formulated as

$$\max \quad \sum_{k \in \mathcal{K}} \sum_{h \in \mathcal{H}} r_{kh}(X_{kh}), \tag{13.17}$$

$$\text{s.to} \quad \sum_{k \in \mathcal{K}} X_{kh} \le p_h W_h, \qquad\qquad h \in \mathcal{H}, \tag{13.18}$$

$$X_{kh} \le W_h, \qquad\qquad k \in \mathcal{K}, h \in \mathcal{H}, \tag{13.19}$$

$$\sum_{h \in \mathcal{H}} s_h W_h \le bI, \tag{13.20}$$

$$X_{kh}, W_h \ge 0 \text{ and integer}, \quad k \in \mathcal{K}, h \in \mathcal{H}. \tag{13.21}$$

Constraints (13.18) express the limitation on the total number of mentions made to each segment. Constraints (13.19) represent a logical consistency condition between decision variables. Finally, constraint (13.20) establishes an overall upper bound on the number of calls that sales agents can make.

Model (13.17) is a nonlinear mixed integer optimization problem, to which remarks similar to those made regarding model (13.8) apply, in particular for its approximate solution.

A general model for sales resources planning

It is possible to provide a somewhat general formulation for salesforce planning problems by adopting a representation framework that involves listing, at least ideally, all tasks that can be assigned to each agent. The resulting model described in this section, like the one discussed in the previous section, derives the optimal plan for the sales agents across multiple time periods, taking into account different shared resources.

For each agent i and for each period t, the set of all possible sales actions, which represent the plan of calls and product presentations to different customers, is identified in advance and denoted by S_{it}. The required resources, denoted by the index $g \in \mathcal{G} = \{1, 2, \ldots, G\}$, represent the overall budget available to implement the sales actions, or other technical factors needed to adopt the different actions.

The required binary decision variables Y_{iut} are defined as

$$Y_{iut} = \begin{cases} 1 & \text{if action } u \in S_{it} \text{ is selected for agent } i \text{ in period } t, \\ 0 & \text{otherwise.} \end{cases}$$

The parameters are

$S_{it} = $ set of feasible actions for agent i in period t,

$w_{giut} = $ quantity of resource g required to implement action $u \in S_{it}$

by agent i in period t,

$V_{gt} = $ quantity of resource g available in period t,

$v_{iut} = $ profit value associated with action $u \in S_{it}$.

The resulting optimization problem is formulated as

$$\max \quad \sum_{t \in T} \sum_{i \in \mathcal{I}} \sum_{u \in S_{it}} v_{iut} Y_{iut}, \tag{13.22}$$

$$\text{s.to} \quad \sum_{i \in \mathcal{I}} \sum_{u \in S_{it}} w_{giut} Y_{iut} \leq V_{gt}, \qquad g \in \mathcal{G}, t \in T, \tag{13.23}$$

$$\sum_{u \in S_{it}} Y_{iut} = 1, \qquad i \in \mathcal{I}, t \in T, \tag{13.24}$$

$$Y_{iut} \in \{0, 1\}, \qquad i \in \mathcal{I}, u \in S_{it}, t \in T. \tag{13.25}$$

Constraints (13.23) express the upper limit on the amount available for each resource in each period. Constraints (13.24) represent a multiple choice condition imposed to guarantee that each agent in each period will perform exactly one action.

Model (13.22) is a binary optimization problem belonging to the class of generalized multiple choice knapsack problems, for whose solution remarks similar to those made for model (13.3) apply. Notice that it is usually advisable to reduce in advance the number of available actions for each agent, by means of a preprocessing phase aimed at discarding those actions that are regarded as less convenient or less profitable.

13.3   Business case studies

In this section we will briefly describe some business case studies that illustrate the application to real-world problems of the methods for marketing analysis presented above. For confidentiality reasons, numerical data for the examples presented will not be given. The purpose of these case studies is to offer readers some ideas on the possible fields of application of business intelligence systems in marketing-related decision-making processes.

13.3.1   Retention in telecommunications

Companies operating in the mobile telephone industry were among the first to use learning models and data mining methods to support relational marketing strategies. One of the main objectives has been customer retention, also known as churn analysis. The effect of market saturation and strong competition have combined to cause instability and disaffection among consumers, who can choose a company based on the rates, services and access methods that they deem most convenient. This phenomenon is particularly critical with regard to prepaid telephone cards, very popular in the mobile phone industry today, as they make changing a telephone service provider quite easy and of little cost. Due to the very nature of the services offered, telephone providers possess a vast array of data on their customers and are in the best position to achieve the maximum benefit from data mining in order to target marketing actions and to optimize the use of resources.

Company and objectives

A mobile phone company wishes to model its customers' *propensity* to churn, that is, a predictive model able to associate each customer with a numerical value (or *score*) that indicates their probability of discontinuing service, based on the value of the available explanatory variables. The model should be able to identify, based on customer characteristics, homogeneous segments relative to the probability of churning, in order to later concentrate on these groups the marketing actions to be carried out for retention, thus reducing attrition and increasing the overall effectiveness. Figure 13.19 shows the possible segments derived using a classification model, using only two predictive attributes in order to create a two-dimensional chart for illustration purposes. The segments with the highest density of churners allow to identify the recipients of the marketing actions.

After an initial exploratory data analysis, the decision is made to develop more than one predictive model, determining a priori some market macro-segments that appear heterogeneous. In this way it is possible to obtain several

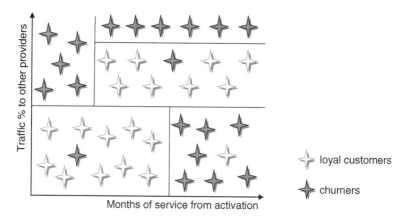

Figure 13.19   An example of segmentation for retention analysis in a mobile telephone company

accurate models instead of a single model related to the entire customer base. The analysis carried out using clustering methods confirms the appropriateness of the segments considered, and leads to the subdivision of customers into groups based on the following dimensions:

- customer type (business or private);
- telephone card type (subscription or prepaid);
- years of service provision, whether above or below a given threshold;
- area of residence.

The marketing data mart provides for each customer a large amount of data:

- personal information (socio-demographic);
- administrative and accounting information;
- incoming and outgoing telephone traffic, subdivided by period (weeks or months) and traffic direction;
- access to additional services, such as fax, voice mail, and special service numbers;
- calls to customer assistance centers;
- notifications of malfunctioning and disservice;
- emails and requests through web pages.

There are approximately 100 explanatory variables available for constructing predictive models.

Analysis and results

Once the dataset for developing the models has been extracted from the data mart, a detailed exploratory analysis can be carried out. On the one hand, it shows a certain number of anomalies, in the form of outliers and missing values, whose removal improves the quality of data. On the other hand, additional variables can be generated through appropriate transformations in order to highlight relevant trends and correlations identified by exploratory data analysis. After applying feature reduction and extraction, the new dataset contains about 150 predictive variables, after the addition of derived variables and removal of some original variables deemed uninfluential on the target. An indicator variable is defined to denote churning by a customer in cases where official notification of service discontinuation is not required, as is the case for prepaid cards. For the different macro-segments, the related *churning signal* is thus defined. For example, if a private customer makes fewer outgoing calls than a preset threshold and receives a number of incoming calls that is below a second threshold value, she is believed to be at a churning stage.

The retention analysis is therefore brought back to a binary classification problem, and models for each macro-segment are then constructed. Different methodologies and different parameterizations are used to obtain several alternative models, the most effective of which can be chosen later based on a comparison that takes into account the indicators of accuracy and the interpretability of the rules generated.

At the end of the development phase, two classes of predictive models are identified. The models based on support vector machines achieve a significantly higher accuracy than other methods, but the corresponding rules they derive are more cumbersome. Classification trees based on axis-parallel splitting rules lead to interpretable rules which are simple and intuitive, but achieve a lower accuracy. The former are preferable for generating the lists of optimal recipients to target marketing campaigns aimed at retention, while the latter are better for investigating loyalty and highlighting relevant market niches. Figure 13.20 shows the cumulative gain curve associated with a classifier based on discrete variants of support vector machines.

13.3.2   Acquisition in the automotive industry

Companies in the automotive industry are striving to develop initiatives aimed at strengthening competitiveness and increasing market share. In this scenario,

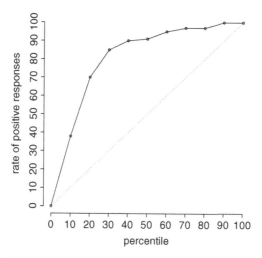

Figure 13.20   Cumulative gains chart for retention analysis in a mobile telephone company

relational marketing projects have been started with the purpose of optimizing the marketing actions and offering products in line with customers' needs, anticipating the evolution of markets and demand. A further element that leads to relational marketing initiatives targeting the sales network is the European directive called the *block exception rule* which introduced new scenarios in the relationships between manufacturing companies and partners in the sales channels, allowing dealers to carry out multi-brand sales activities, and promotion and sales actions across the entire EU territory.

In particular, the use of business intelligence methods is a key factor in strengthening the knowledge of prospects and customers and hence improving new customer acquisition processes and the loyalty of the customer base. It is worth observing that, for some markets in the automotive industry, it is possible to integrate the internal data contained in the marketing data mart with external data sources, which provide a thorough description of the purchases of automobiles and industrial vehicles made by private customers and by companies, easily found at motor vehicles registries. Such an opportunity can be used to enhance segmentation analyses.

Company and objectives

A company manufacturing industrial vehicles wishes to develop a predictive model that can assign to each prospect a score that indicates his propensity to positively respond to a marketing action aimed at acquisition. The main purpose of the model is to provide guidance for promotion actions carried out by the network of dealers. But also, in order to stay one step ahead of competitors, it is required to better understand the trends of the future demand, refining the knowledge of the customer base and of the market scenarios, and identifying the distinctive features that characterize current customers in order to design initiatives directed to stimulate new acquisitions.

Analysis and results

The data available for the acquisition analysis are subdivided into three categories.

Prospects and customers. The first group of data concerns current or potential customers of the company, including customers owning vehicles not necessarily produced by the company. Besides demographic information, the data include 15 explanatory variables that gather meaningful information, such as the top-rated type of vehicle owned by each prospect, the number of new and used vehicles produced by competitors and owned by each prospect, and the number of new and used vehicles bought in the past by each prospect.

Vehicles. The second group of data includes 16 attributes that enable the vehicles owned by each prospect to be identified, among them model, weight class, optional features, fuel type, first and last registration date and status (new or used vehicle). These attributes define for each vehicle the timing of sales transactions, or transfers of ownership.

Works. The third group of data refers to maintenance and repair work undergone by the vehicles during the warranty period and beyond. In particular, these data indicate the type of work carried out, a description of the problem, the vehicle mileage at the time of service, the car shop where the repair was carried out and the date of admission and release of the vehicle to and from the shop.

After exploratory data analysis, which led to the removal of a number of anomalies and outliers, selection of variables was carried out. The analysis led to the addition of new explanatory variables for the time interval that normally elapses between two subsequent purchases, the number of vehicles owned by each prospect, subdivided by weight class, the proportion of vehicles of other brands owned by each potential customer, and the proportion of vehicles of each class included in the vehicle portfolio of each prospect.

Classification models were then developed using different methods. Similar considerations to those in the previous section about retention analysis apply also in this case. Discrete support vector machines turned out to be the most effective method (see Figure 13.22), while classification trees generate rules that can be more easily interpreted (see Figure 13.21).

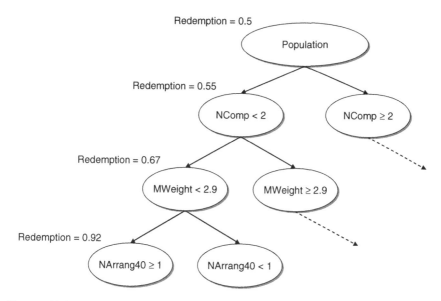

Figure 13.21   Classification tree for acquisition analysis in an industrial vehicle company

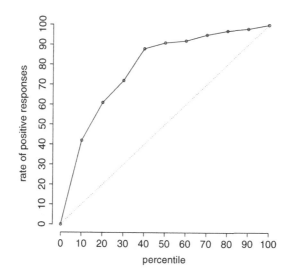

Figure 13.22   Cumulative gains chart for the acquisition analysis in an industrial vehicle company

13.3.3   Cross-selling in the retail industry

The company considered in this section operates in retail consumer electron-ics, and wishes to segment its customer base in order to optimize marketing actions aimed at promoting a specific product or group of products. The goal is therefore to develop a predictive model able to assign to each customer a score that indicates her propensity to respond positively to a cross-selling offer. Besides prediction purposes, the model should be used also to interpret explanatory factors that have a greater effect on the purchase of the product promoted. Finally, the model is to be used for assessing the existence of causal and temporal correlations between the purchase of the product promoted and the purchase of other items.

Analysis and results

The data available for cross-selling analysis are mainly transactional, referring to customers who have signed up for a loyalty card at one of the company's retail stores. These data include the following information:

- personal information (socio-demographic);

- date of signing up for the loyalty card, which can be regarded as the starting date for the relationship between customer and company;

- dates of first and last purchase, marking the boundary of the time interval within which purchases have been made by each individual customer;

- cash slips, indicating which items, and in what quantities, have been purchased by each customer;

- purchases of sale items made by each customer;

- participation in point-earning programs and related prizes won;

- consumer financing requested to make purchases.

Hence, a binary classification problem can be formulated, where the target variable corresponds to the purchase of the specific product to be promoted. Since the prediction should be available a month in advance, classification models should take into account the corresponding latency. Exploratory data analysis enabled the detection and removal of anomalies and missing data. Then a data preparation stage took place at which those variables were removed that showed a low correlation with the target. Finally, some new explanatory variables were generated through transformations of the original attributes, with the purpose of highlighting trends in the temporal sequence of the expenditure amounts.

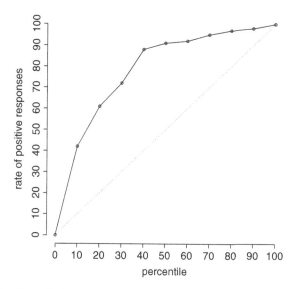

Figure 13.23   Cumulative gains chart for the cross-selling analysis in a retail company

At the end of the exploratory analysis, the dataset used to develop the classification models included approximately 120 explanatory attributes. Although the methods based on variants of support vector machines turned out to be more accurate, a model based on classification trees was deemed more appropriate, in consideration of the interpretability that was indicated among the primary objectives of the analysis. The cumulative gains curve shown in Figure 13.23 corresponds to a model based on support vector machines, able to reach a lift of 8 at a quantile of 0.05, corresponding to 5% of the customers considered by the model to have the highest propensity to buy the promoted product.

Attention then turned to the hypothesis of a causal correlation between the purchase of the product promoted and some other items, identified by marketing analysts on a subjective basis. We proceeded by evaluating the association rules which relate the product promoted to each of the other items, in a series of dyadic relations. The analysis showed that the presence of the product promoted in the head or body of the rule provides similar support and confidence values. This led to rejection of the hypothesis of a significant correlation between purchases. To further confirm this conclusion, sequential rules were analyzed to assess whether the purchase of the item promoted was preceded by the purchase of one of the other items considered, or whether the former was preceded by the latter. Also in this case, the analysis allowed the existence of causal relationships between purchases to be ruled out.

13.4 Notes and readings

There are several books devoted to relational marketing, among which we recommend Peppers and Rogers (1996, 2004), Dyche (2001), Egan (2001), and Bruhn (2002). In particular, for the role of data mining methods and learning models in relational marketing see Berson and Smith (1997), Berson *et al.* (1999), Parr Rud (2000), and Berry and Linoff (2004). Text mining methods are discussed in Drucker *et al.* (1999), Joachims (2002), and Nigam *et al.* (2000). For market basket analysis see Silverstein *et al.* (1998) and Lawrence *et al.* (2001). E-commerce applications are considered in Schafer *et al.* (2001), while Pei *et al.* (2000) and Cadez *et al.* (2003) analyze transactions on the web. The themes of salesforce automation are described in most books generally devoted to relational marketing that have been mentioned above. A more specific book on the subject is Zoltners *et al.* (2001). Optimization models for salesforce management are also considered in Eliashberg and Lilien (1993).

14

Logistic and production models

In Chapter 13 we saw how the combination of relational marketing strategies with business intelligence and data mining models makes it possible to simultaneously increase revenues and reduce the costs of marketing actions, with an overall benefit for the profitability of an enterprise.

Besides acting on the marketing control levers, a manufacturing company can achieve further reductions in costs by improving its processes in another area that has received increasing attention in recent years: an effective *supply chain* management, understood as the logistic and production processes of a single enterprise as well as the network of companies composing the production chain of a given industry.

In this chapter we will focus on optimization models aimed at the integrated planning of the logistic chain from the perspective of a single company. In particular, we will begin with a qualitative description of the relevant processes within a logistic production system, by highlighting the major decisions that logistics managers have to face. The discussion will be confined to medium-term planning processes, which are concerned with some critical choices in the organization of the supply chain and can bring about substantial savings if appropriately optimized. We will then introduce some classes of optimization models, showing how the different features of logistic production systems can be formally represented. Finally, we will discuss a few business case studies, with particular emphasis on a decision support system for supply chain optimization developed for a company in the food industry.

Business Intelligence: Data Mining and Optimization for Decision Making   C. Vercellis
© 2009 John Wiley & Sons, Ltd

14.1   Supply chain optimization

In a broad sense, a *supply chain* may be defined as a network of connected and interdependent organizational units that operate in a coordinated way to manage, control and improve the flow of materials and information originating from the suppliers and reaching the end customers, after going through the procurement, processing and distribution subsystems of a company, as shown in Figure 14.1.

The aim of the integrated planning and operations of the supply chain is to combine and evaluate from a systemic perspective the decisions made and the actions undertaken within the various subprocesses that compose the logistic system of a company.

Many manufacturing companies, such as those operating in the consumer goods industry, have concentrated their efforts on the integrated operations of the supply chain, even to the point of incorporating parts of the logistic chain that are outside the company, both upstream and downstream.

The major purpose of an integrated logistic process is to minimize a function expressing the total cost, which comprises processing costs, transportation costs for procurement and distribution, inventory costs and equipment costs. Note that

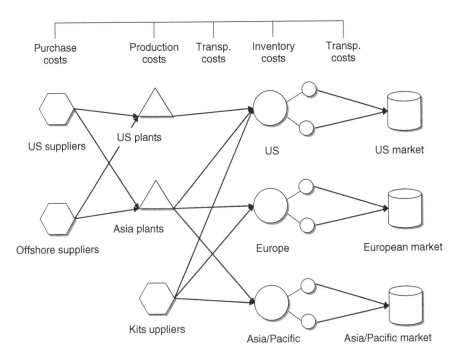

Figure 14.1   An example of global supply chain

the optimization of the costs for each single phase does not generally imply that the minimum total cost of the entire logistic process has been achieved, so that a holistic perspective is required to attain a really optimized supply chain.

The need to optimize the logistic chain, and therefore to have models and computerized tools for medium-term planning and for capacity analysis, is particularly critical in the face of the high complexity of current logistic systems, which operate in a dynamic and truly competitive environment. We are referring here to manufacturing companies that produce a vast array of products and that usually rely on a multicentric logistic system, distributed over several plants and markets, characterized by large investments in highly automated technology, by an intensive usage of the available production capacity and by short-order processing cycles. The features of the logistic system we have described reflect the profile of many enterprises operating in the consumer goods industry.

In the perspective outlined above, the aim of a medium-term planning process is therefore to devise an optimal logistic production plan, that is, a plan that is able to minimize the total cost, understood as the sum of procurement, processing, storage, distribution costs and the penalty costs associated with the failure to achieve the predefined service level. However, to be implemented in practice, an optimal logistic production plan should also be feasible, that is, it should be able to meet the physical and logical constraints imposed by limits on the available production capacity, specific technological conditions, the structure of the bill of materials, the configuration of the logistic network, minimum production lots, as well as any other condition imposed by the decision makers in charge of the planning process.

Optimization models represent a powerful and versatile conceptual paradigm for analyzing and solving problems arising within integrated supply chain planning, and for developing the necessary software. Due to the complex interactions occurring between the different components of a logistic production system, other methods and tools intended to support the planning activity seem today inadequate, such as electronic spreadsheets, simulation systems and planning modules at infinite capacity included in *enterprise resource planning* software. Conversely, optimization models enable the development of realistic mathematical representations of a logistic production system, able to describe with reasonable accuracy the complex relationships among critical components of the logistic system, such as capacity, resources, plans, inventory, batch sizes, lead times and logistic flows, taking into account the various costs. Moreover, the evolution of information technologies and the latest developments in optimization algorithms mean that decision support systems based on optimization models for logistics planning can be efficiently developed.

14.2   Optimization models for logistics planning

In this section we will describe some optimization models that may be used to represent the most relevant features of logistic production systems. As already observed when introducing salesforce planning models in Chapter 13, for the sake of simplicity we have chosen to illustrate for each model a single feature of a logistic system. Readers should keep in mind that real-world logistic production systems feature simultaneously more than one of the elements considered, so that the models developed in applications, such as the business case studies presented in Section 14.4, will be substantially more complex as they result from the combination of the different features.

Before proceeding with the description of specific models, it is useful to introduce some notation common to most models presented in this section. The logistic system includes I products, which will be denoted by the index $i \in \mathcal{I} = \{1, 2, \ldots, I\}$. The planning horizon is subdivided into T time intervals $t \in \mathcal{T} = \{1, 2, \ldots, T\}$, generally of equal length and usually corresponding to weeks or months.

The manufacturing process has at its disposal a set of critical resources shared among the different products and available in limited quantities. These resources may consist of production and assembly lines, to manpower, to specific fixtures and tools required by manufacturing. The R critical resources considered in the logistic production system will be denoted by the index $r \in \mathcal{R} = \{1, 2, \ldots, R\}$. Whenever a single resource is relevant to the manufacturing process, the index r will be omitted for sake of simplicity.

14.2.1   Tactical planning

In its simplest form, the aim of tactical planning is to determine the production volumes for each product over the T periods included in the medium-term planning horizon in such a way as to satisfy the given demand and capacity limits for a single resource, and also to minimize the total cost, defined as the sum of manufacturing production costs and inventory costs.

We therefore consider the decision variables

$P_{it} =$ units of product i to be manufactured in period t,

$I_{it} =$ units of product i in inventory at the end of period t,

and the parameters

$d_{it} =$ demand for product i in period t,

$c_{it} =$ unit manufacturing cost for product i in period t,

h_{it} = unit inventory cost for product i in period t,

e_i = capacity absorption to manufacture a unit of product i,

b_t = capacity available in period t.

The resulting optimization problem is formulated as follows:

$$\min \quad \sum_{t\in T}\sum_{i\in I}(c_{it}P_{it}+h_{it}I_{it}) \tag{14.1}$$

$$\text{s.to} \quad P_{it}+I_{i,t-1}-I_{it}=d_{it}, \quad i\in I, t\in T, \tag{14.2}$$

$$\sum_{i\in I}e_i P_{it}\le b_t, \qquad t\in T, \tag{14.3}$$

$$P_{it}, I_{it}\ge 0, \qquad i\in I, t\in T. \tag{14.4}$$

Constraints (14.2) express the balance conditions among production, inventory and demand, by establishing a connection between successive periods along the planning horizon. Inequalities (14.3) constrain the absorbed capacity not to exceed the available capacity for each period.

Model (14.1) is a linear optimization problem which can be therefore solved efficiently even with a very large number of variables and constraints, of the order of a few million, by means of current state-of-art algorithms and computer technologies.

14.2.2 Extra capacity

A first extension of the basic model (14.1) deals with the possibility of resorting to *extra capacity*, perhaps in the form of overtime, part-time or third-party capacity. In addition to the decision variables already included in model (14.1), we define the variables

O_t = extra capacity used in period t,

and the parameters

q_t = unit cost of extra capacity in period t.

The optimization problem now becomes

$$\min \quad \sum_{t\in T}\sum_{i\in I}(c_{it}P_{it}+h_{it}I_{it})+\sum_{t\in T}q_t O_t \tag{14.5}$$

$$\text{s.to} \quad P_{it}+I_{i,t-1}-I_{it}=d_{it}, \qquad i\in I, t\in T, \tag{14.6}$$

$$\sum_{i\in I}e_i P_{it}\le b_t+O_t, \qquad t\in T, \tag{14.7}$$

$$P_{it}, I_{it}, O_t\ge 0, \qquad i\in I, t\in T. \tag{14.8}$$

Constraints (14.7) have been modified to include the available extra capacity. The extended model (14.5) is still a linear optimization problem which can be therefore solved efficiently.

14.2.3 Multiple resources

If the manufacturing system requires R critical resources, a further extension of model (14.1) can be devised by considering multiple capacity constraints. The decision variables already included in model (14.1) remain unchanged, though it is necessary to consider the additional parameters

$b_{rt} = $ quantity of resource r available in period t,

$e_{ir} = $ quantity of resource r absorbed to manufacture one unit of product i.

The resulting optimization problem is given by

$$\min \quad \sum_{t \in \mathcal{T}} \sum_{i \in \mathcal{I}} (c_{it} P_{it} + h_{it} I_{it}) \qquad (14.9)$$

$$\text{s.to} \quad P_{it} + I_{i,t-1} - I_{it} = d_{it}, \quad i \in \mathcal{I}, t \in \mathcal{T}, \qquad (14.10)$$

$$\sum_{i \in \mathcal{I}} e_{ir} P_{it} \le b_{rt}, \qquad r \in \mathcal{R}, t \in \mathcal{T}, \qquad (14.11)$$

$$P_{it}, I_{it} \ge 0, \qquad i \in \mathcal{I}, t \in \mathcal{T}. \qquad (14.12)$$

Constraints (14.11) have been modified to take into account the upper limits on the capacity of the R resources in the system. Model (14.9) remains a linear optimization problem which can be solved efficiently.

14.2.4 Backlogging

Another feature that needs to be modeled in some logistic systems is *backlogging*. The term *backlog* refers to the possibility that a portion of the demand due in a given period may be satisfied in a subsequent period, incurring an additional penalty cost. Backlogs are a feature of production systems more likely to occur in B2B or make-to-order manufacturing contexts. In B2C industries, such as mass production consumer goods, on the other hand, one is more likely to find a variant of the backlog, known as *lost sales*, in which unfulfilled demand in a period cannot be transferred to a subsequent period and is lost.

To model backlogging, it is necessary to introduce new decision variables

$B_{it} = $ units of demand for product i delayed in period t,

and the parameters

$g_{it} = $ unit cost of delaying the demand for product i in period t.

The resulting optimization problem is

$$\min \quad \sum_{t \in \mathcal{T}} \sum_{i \in \mathcal{I}} (c_{it} P_{it} + h_{it} I_{it} + g_{it} B_{it}) \qquad (14.13)$$

$$\text{s.to} \quad P_{it} + I_{i,t-1} - I_{it} + B_{it} - B_{i,t-1} = d_{it}, \quad i \in \mathcal{I}, t \in \mathcal{T}, \qquad (14.14)$$

$$\sum_{i \in \mathcal{I}} e_i P_{it} \le b_t, \qquad\qquad\qquad t \in \mathcal{T}, \qquad (14.15)$$

$$P_{it}, I_{it}, B_{it} \ge 0, \qquad\qquad\qquad i \in \mathcal{I}, t \in \mathcal{T}. \qquad (14.16)$$

The balance constraints (14.14) have been modified to take backlog variables into account. Specifically, in each period t one is allowed to delay a portion of the demand d_{it}, given precisely by the backlog variable B_{it}, whereas the demand d_{it} itself is increased by the units held as backlog in the previous period. Model (14.13) is again a linear optimization problem which can be therefore solved efficiently.

An alternative way to model backlog is to suppose that the demand d_{it} is made up of separate orders k, each characterized by a request for w_{ik} units of product i and a due delivery date t_{ik}:

$$d_{it} = \sum_{k \in \mathcal{K}_{it}} w_{ik}, \qquad \mathcal{K}_{it} = \{k | t_{ik} = t\}. \qquad (14.17)$$

We further assume that each order should be completed within at most two periods from its due delivery date.

To model this second form of backlogging, define the following binary decision variables:

$$\beta_{ik0} = \begin{cases} 1 & \text{if order } k \text{ for product } i \text{ is delivered on time,} \\ 0 & \text{otherwise;} \end{cases}$$

$$\beta_{ik1} = \begin{cases} 1 & \text{if order } k \text{ for product } i \text{ is delayed by one period,} \\ 0 & \text{otherwise;} \end{cases}$$

$$\beta_{ik2} = \begin{cases} 1 & \text{if order } k \text{ for product } i \text{ is delayed by two periods,} \\ 0 & \text{otherwise.} \end{cases}$$

Consider also the parameters

$g_{ik1} = $ cost of delivering order k for product i delayed by one period,

$g_{ik2} = $ cost of delivering order k for product i delayed by two periods.

The resulting optimization problem is formulated as

$$\min \quad \sum_{t\in T}\sum_{i\in \mathcal{I}}(c_{it}P_{it} + h_{it}I_{it}) +$$

$$\sum_{t\in T}\sum_{i\in \mathcal{I}}\left(\sum_{k\in \mathcal{K}_{i,t-1}} g_{ik1}\beta_{ik1} + \sum_{k\in \mathcal{K}_{i,t-2}} g_{ik2}\beta_{ik2}\right) \qquad (14.18)$$

$$\text{s.to} \quad P_{it} + I_{i,t-1} - I_{it} = \sum_{k\in \mathcal{K}_{it}} w_{ik}\beta_{ik0}$$

$$+ \sum_{k\in \mathcal{K}_{i,t-1}} w_{ik}\beta_{ik1} + \sum_{k\in \mathcal{K}_{i,t-2}} w_{ik}\beta_{ik2}, \qquad i\in \mathcal{I}, t\in T, \quad (14.19)$$

$$\sum_{i\in \mathcal{I}} e_i P_{it} \le b_t, \qquad t\in T, \quad (14.20)$$

$$\beta_{ik0} + \beta_{ik1} + \beta_{ik2} = 1, \qquad i\in \mathcal{I}, k\in \mathcal{K}, \quad (14.21)$$

$$P_{it}, I_{it} \ge 0, \quad \beta_{ik0}, \beta_{ik1}, \beta_{ik2} \in \{0,1\}, \qquad i\in \mathcal{I}, t\in T. \quad (14.22)$$

The balance constraints (14.19) have been modified to take the binary backlog variables into account. Specifically, in each period t are fulfilled the orders k due for period t for which there is no delayed delivery (i.e. such that $\beta_{ik0} = 1$), the orders due for the previous period $t-1$ for which the delivery is delayed by one period (i.e. such that $\beta_{ik1} = 1$), and finally the orders due for period $t-2$ for which the delivery is delayed by two periods (i.e. such that $\beta_{ik2} = 1$). The multiple choice constraints (14.21) establish that each order be fulfilled in exactly one of the three alternative ways corresponding to variables $\beta_{ik0}, \beta_{ik1}, \beta_{ik2}$.

Compared to previous model (14.13), formulation (14.18) allows us to attach a different penalty cost to the K orders, therefore assigning preferences to some customers, for example those considered of strategic importance. In B2B customer–supplier relationships this possibility can be of considerable value.

Unlike the models previously considered, model (14.18) is a mixed binary linear optimization problem, whose solution requires computation times that grow exponentially fast with the number of variables and constraints. However, when the problem size is too large to yield an optimal solution in a reasonable time, for instance through a general purpose exact algorithm such as branch-and-bound, it is usually possible to devise an approximation algorithm achieving suboptimal solutions. This can be done by truncating a branch-and-bound procedure, or by designing *ad hoc* approximation algorithms.

Finally, notice that a multi-objective optimization model can also be formulated by requiring, for instance, that a proportion of at least n_1 of the orders be delivered on time and that a proportion of at most $1 - n_2$ be delayed by two periods, by introducing into model (14.18) the additional constraints

$$\sum_{k=1}^{K_i} \beta_{ik0} \geq n_1 K_i, \quad \sum_{k=1}^{K_i} \beta_{ik0} + \beta_{ik1} \geq n_2 K_i, \tag{14.23}$$

where K_i is the total number of orders referring to product i.

14.2.5   Minimum lots and fixed costs

A further feature often appearing in manufacturing systems is represented by *minimum lot* conditions: for technical or scale economy reasons, it is sometimes necessary that the production volume for one or more products be either equal to 0 (i.e. the product is not manufactured in a specific period) or not less than a given threshold value, the minimum lot.

To incorporate minimum lot conditions into the model, we define the binary decision variables

$$Y_{it} = \begin{cases} 1 & \text{if } P_{it} > 0, \\ 0 & \text{otherwise,} \end{cases} \tag{14.24}$$

and the parameters

$l_i =$ minimum lot for product i,

$\gamma =$ constant value larger than any producible volume for i.

The optimization problem is now

$$\min \sum_{t \in T} \sum_{i \in I} (c_{it} P_{it} + h_{it} I_{it}) \tag{14.25}$$

$$\text{s.to} \quad P_{it} + I_{i,t-1} - I_{it} = d_{it}, \quad i \in I, t \in T, \tag{14.26}$$

$$\sum_{i \in I} e_i P_{it} \leq b_t, \quad\quad t \in T, \tag{14.27}$$

$$P_{it} \geq l_i Y_{it}, \quad\quad i \in I, t \in T, \tag{14.28}$$

$$P_{it} \leq \gamma Y_{it}, \quad\quad i \in I, t \in T, \tag{14.29}$$

$$P_{it}, I_{it} \geq 0, Y_{it} \in \{0, 1\}, \quad i \in I, t \in T. \tag{14.30}$$

Constraints (14.28) express the minimum lot conditions. Constraints (14.29) are logical consistency conditions between the variables Y_{it} and P_{it} needed to force the binary variable Y_{it} to take the value 1 whenever the corresponding production volume P_{it} is greater than 0. The constant γ in (14.29) must be chosen sufficiently large that the condition does not constitute an actual upper bound on the producible volume for product i. Indeed, it should be bounded above only by the available capacity and by the assigned demand.

In all previous model formulations we have implicitly assumed that production costs are proportional to production volumes. For some logistic systems, however, in order to manufacture a product it may be necessary to set up a machine and incur a *setup cost*. However, such costs are required only if the production volume is strictly greater than 0, that is, if production of the product concerned is actually accomplished. A further parameter,

$$f_{it} = \text{unit setup cost for product } i \text{ in period } t,$$

is then assigned and the optimization problem becomes

$$\min \quad \sum_{t \in T} \sum_{i \in I} (c_{it} P_{it} + h_{it} I_{it} + f_{it} Y_{it}) \tag{14.31}$$

$$\text{s.to} \quad P_{it} + I_{i,t-1} - I_{it} = d_{it}, \qquad i \in I, t \in T, \tag{14.32}$$

$$\sum_{i \in I} e_i P_{it} \le b_t, \qquad t \in T, \tag{14.33}$$

$$P_{it} \le \gamma Y_{it}, \qquad i \in I, t \in T, \tag{14.34}$$

$$P_{it}, I_{it} \ge 0, Y_{it} \in \{0, 1\}, \qquad i \in I, t \in T. \tag{14.35}$$

Constraints (14.34) represent the logical consistency conditions between variables Y_{it} and P_{it}, as already observed for model (14.25).

Models (14.25) and (14.31) are mixed binary linear optimization problems, for whose solution the same remarks as for model (14.13) apply.

14.2.6 Bill of materials

A further extension of the basic planning model deals with the representation of products with a complex structure, described via the so-called *bill of materials*, where end-items are made by components that in turn may include other components.

Formally, the following parameters are defined to describe the structure of the bill of materials:

$$a_{ij} = \text{units of product } i \text{ directly required by one unit of product } j,$$

where the term *product* refers here to both end-items and components at various levels of the bill of materials. For each product i we assign an *external* demand d_{it} and an *internal* demand, the latter induced by the requirements of product i needed to manufacture the components or the end-items for which i represents a direct component. The external demand for components may originate from other plants of the same manufacturing company or from outside customers that also buy components.

The resulting optimization problem is formulated as

$$\min \quad \sum_{t \in T} \sum_{i \in I} (c_{it} P_{it} + h_{it} I_{it}) \tag{14.36}$$

$$\text{s.to} \quad P_{it} + I_{i,t-1} - I_{it} = d_{it} + \sum_{j \in I, j \neq i} a_{ij} P_{jt}, \quad i \in I, t \in T, \tag{14.37}$$

$$\sum_{i \in I} e_i P_{it} \leq b_t, \qquad\qquad\qquad t \in T, \tag{14.38}$$

$$P_{it}, I_{it} \geq 0, \qquad\qquad\qquad i \in I, t \in T. \tag{14.39}$$

The balance constraints (14.37) have been modified to take into account the demand internally generated. Model (14.36) is a linear optimization problem which can be therefore solved efficiently.

14.2.7 Multiple plants

In this section it is assumed that a manufacturing company has a network of M production plants, located in geographically distinct sites, that manufacture a single product. The logistic system is responsible for supplying N peripheral depots, located in turn at distinct sites. Each production plant $m \in \mathcal{M} = \{1, 2, \ldots, M\}$ is characterized by a maximum availability of product, denoted by s_m, while each plant $n \in \mathcal{N} = \{1, 2, \ldots, N\}$ has a demand d_n. We further assume that a transportation cost c_{mn} is incurred by sending a unit of product from plant m to depot n, for each pair (m, n) of origins and destinations in the logistic network. The objective of the company is to determine an optimal logistic plan that satisfies at minimum cost the requests of the depots, without violating the maximum availability at the plants. It should be clear that the problem described arises frequently in logistic systems, at different levels in the logistic network (e.g. from suppliers to plants, from plants to warehouses or from warehouses to customers).

The decision variables needed to model the problem described represent the quantity to be transported for each plant–depot pair,

$$x_{mn} = \text{unit of product to be transported from } m \text{ to } n.$$

The resulting optimization problem is

$$\min \quad \sum_{m \in \mathcal{M}} \sum_{n \in \mathcal{N}} c_{mn} x_{mn} \tag{14.40}$$

$$\text{s.to} \quad \sum_{n \in \mathcal{N}} x_{mn} \leq s_m, \qquad m \in \mathcal{M}, \tag{14.41}$$

$$\sum_{m \in \mathcal{M}} x_{mn} \geq d_n, \qquad n \in \mathcal{N}, \tag{14.42}$$

$$x_{mn} \geq 0, \qquad m \in \mathcal{M}, n \in \mathcal{N}. \tag{14.43}$$

Constraints (14.41) ensure that the availability of each plant is not exceeded, whereas constraints (14.42) establish that the demand of each depot be satisfied. Model (14.40) is a linear optimization problem, and can be therefore solved efficiently.

14.3   Revenue management systems

Revenue management is a managerial policy whose purpose is to maximize profits through an optimal balance between demand and supply. It is mainly intended for marketing as well as logistic activities and has found growing interest in the service industry, particularly in the air transportation, tourism and hotel sectors. More recently these methods have also begun to spread within the manufacturing and distribution industries.

The strong interest shown by such enterprises in the themes considered by revenue management should come as no surprise, if we consider the complexity and strategic relevance of decision-making processes concerning demand management, which are addressed by marketing and logistics managers. Consider, for example, the complex interactions among decisions on pricing, sales promotions, markdowns, mix definition and allocation to points of sale, in a highly dynamic and competitive context characterized by multiple sales channels and several alternative ways of contacting customers.

Despite the potential advantages that revenue management initiatives may offer for enterprises, there are certain difficulties that hamper the actual implementation of practical projects and actions aimed at adopting revenue management methodologies and tools. We can identify several explanations for the gap between intentions and initiatives actually undertaken. Certainly the fear of implementation costs and uncertainty over the results that can be achieved play an important role, as happens for many innovation projects. Empirical investigations show, however, that the primary reason for prudence in adopting revenue management should be sought in the prerequisite conditions necessary to successfully start a revenue management project. There is a high level of interaction between revenue management and two other themes that we described earlier – optimization of the supply chain and relational marketing. On the one hand, in order to apply revenue management methods and tools it is necessary to have an integrated and optimized logistic chain that guarantees the efficiency and responsiveness of the logistic flows. On the other hand, it is also necessary to possess a deep knowledge of the customers and an accurate micro-segmentation of the market, achieved through data mining analytical models and therefore based on the analysis of the actual purchasing behaviors regularly recorded in the marketing data mart. Hence, to profitably adopt revenue management a company should be able to enhance and transform into knowledge, through the use of business intelligence methodologies, the huge amount of information collected by means of automatic data gathering technologies.

14.3.1 Decision processes in revenue management

Revenue management involves the application of mathematical models to predict the behavior of customers at a micro-segmentation level and to optimize the availability and price of products in order to maximize profits. In this respect, we can use the same definition introduced in Chapter 13 to summarize relational marketing objectives: to formulate for each segment, ideally for each customer, the appropriate offer through the most suitable channel, at the right time and at the best price.

The purpose of revenue management is therefore to maximize profits, aligning the offer of products and services to the expected demand, using both the major levers of the marketing mix (e.g. prices, promotions, assortment) and the levers of logistics (e.g. efficiency and timeliness). Specific and innovative features of revenue management strategies are a closer focus on demand than supply and a greater emphasis on costs than revenues; such features are often absent from the managerial policies adopted by most enterprises.

As already observed, in recent years revenue management has been applied with more and more success by many companies operating in the service industry. Among the pioneers in this field are airlines, hotel chains, automobile rental companies, theme parks, theaters and other entertainment-related enterprises. The common characteristics of these fields are well apparent: a highly perishable product, a fairly low marginal sales cost and the possibility of applying dynamic pricing policies and exploiting multiple sales channels.

Revenue management affects some highly complex decision-making processes of strategic relevance, as shown in Figure 14.2:

- market segmentation, by product, distribution channel, consumer type and geographic area, performed using data mining models;

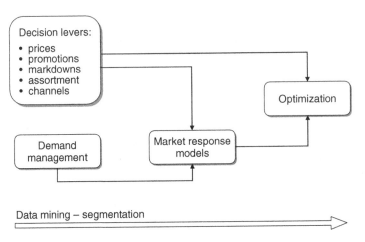

Figure 14.2   Decision processes in revenue management

- prediction of future demand, using time series and regression models;

- identification of the optimal assortment, i.e. the mix of products to be allocated to each point of sale;

- definition of the market response function, obtained by identifying models and rules that explain the demand based on company actions, the initiatives of competitors and other exogenous contextual events;

- management of activities aimed at determining the price of each product (*pricing*) as well as the timing and the amount of markdowns;

- planning, management and monitoring of sales promotions, and assessment of their effectiveness;

- sales analysis and control, and use of the information gathered to evaluate market trends;

- material procurement and stock management policies, such as control policy, frequency of issued orders, reorder quantities;

- integrated management of the different sales and distribution channels.

Revenue management relies on the following basic principles:

- To address sales to micro-segments: segmentation carried out by means of business intelligence and data mining models is critical to achieve an adequate knowledge of the market.

- To exploit the product value cycle: to generate the highest revenues, it is required to grasp the value cycle of products and services, in order to optimally synchronize their availability over time and to determine the price for each market micro-segment. Notice that the value cycle also depends on the sensitivity of micro-segments to price variations.

- To have a price-oriented rather than cost-oriented approach in balancing supply and demand: when supply and demand are out of balance, most enterprises tend to react by increasing or decreasing capacity. In many instances it might, however, be more convenient to adopt price variations, avoiding repeated variations in capacity.

- To make informed and knowledge-based decisions: a consistent use of prediction models tends to mean that decisions rest on a more robust knowledge basis. In particular, a correct prediction of consumer purchasing behaviors is essential to evaluate elasticity and reactions to price variations.

- To regularly examine new opportunities to increase revenues and profits: the possibility of timely access to the available information, combined with the possibility of considering alternative scenarios, strengthens the competencies of marketing analysts and increases the effectiveness of their activity.

The adoption of revenue management methods and tools requires a few prerequisite conditions to be satisfied within a company, since without them the expected results are unlikely to be achieved. As with any innovation project, it is the people and the organization that constitute a key success factor rather than the use of specific software tools. In this case too, the culture and the structure of the processes within an organization must be prepared to adopt powerful tools that may turn out to be unsafe and disrupting if improperly used. It is therefore necessary to develop within the enterprise an information culture, particularly among those knowledge workers who operate in the marketing and logistics departments, more directly involved with the application of revenue management strategies. This means that all marketing data must be systematically gathered, controlled, normalized, integrated and stored in a data mart. To segment the market and to create micro-segments, business intelligence methods and analytical models should be used. It is therefore advisable for an enterprise turning to revenue management to have already developed relational marketing initiatives or at least to be able to carry out data mining analyses.

On the other hand, the decisions involved in revenue management strategies share many aspects with the logistics department, and in particular with the management of flows in the supply chain. In this case too, particularly for manufacturing companies, it is advisable for an enterprise considering revenue management to have previously embarked on supply chain integration and rationalization projects, in order to guarantee an adequate cost reduction that, combined with the increased revenues obtained through revenue management, may lead to a significant increase in profits. Moreover, effective supply chain management is also required to guarantee timely restocking.

14.4   Business case studies

This section describes two examples of real-world applications of optimization models for logistic and production planning. The first is concerned with an enterprise operating in the food industry, while the second refers to a company that manufactures integrated solutions for liquid food product packaging.

14.4.1   Logistics planning in the food industry

The logistic system of the food manufacturing company consists of a network whose nodes represent suppliers of raw materials, production plants and central and peripheral warehouses, as shown in Figure 14.3. Retail and wholesale distribution to the points of sale, placed downstream of the warehouses, is regarded as external to the logistic subsystem considered here.

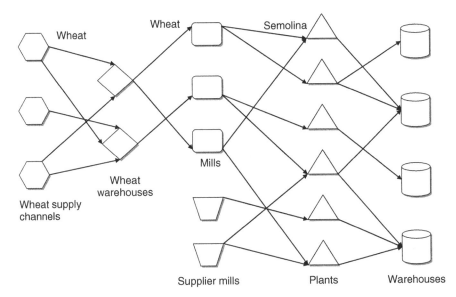

Figure 14.3   Structure of the logistic production system for a company in the food industry

Logistics planning is driven by the demand plan, which in turn depends on the sales forecasts for individual items, disaggregated by warehouse and by period. Supply, production and primary distribution plans are therefore meant to feed end items into central and peripheral warehouses.

Industrial processing manufactures products of low complexity on *transfer lines*, that is, highly automated production lines designed to achieve high output rates even with a limited plant flexibility. The processing cycle basically includes two major phases: during the *production* phase, raw materials are transformed into semi-finished goods, which during the *packaging* phase are then turned into end items. Both production and packaging lines operate in a batch processing mode. Generally, each item can be produced at several plants and on several lines within a given plant. Each plant includes both production and packaging lines, while the transfer of semi-finished goods from one plant to another is not allowed. Based on the demand profile, and on the related production capacity requirements, each of the two phases of the production cycle may represent a bottleneck of the manufacturing system during the periods included in the planning horizon. Since the enterprise targets consumer goods markets, if the demand for an item for a specific warehouse cannot be satisfied in a given period, it is assumed that this translates into lost sales. Therefore, the company does not allow any demand backlog to be met during following periods.

The logistic system is made up of dozens of plants housing hundreds of production and packaging lines. More than a thousand end items are produced and later stored in central and peripheral warehouses. Each production or packaging line may produce up to dozens of semi-finished goods or end products. There are plants that manufacture a small number of items (two or three) and plants that are able to produce a considerable portion of the product range.

Several decisions depend on the company's logistic plan, and would benefit from a production and distribution plan that is more efficient and stable. Some of these decisions are:

- the optimal sizing of the stocks, so as to guarantee the required service level;

- the plan of allocation to the plants of the demand originating at each market warehouse, in order to achieve an optimal balance between production, distribution and supply costs;

- the plan of allocation of the production capacity, with the possible addition of extra capacity, by using extra work shifts and part-time work or by subcontracting some of the production to external partners (*co-packers*); moreover, the company needs to set out medium-term contractual obligations with co-packers, which are significantly affected by the stability and reliability of the logistic plan devised;

- the supply plan, and the corresponding medium-term contracts with suppliers for the procurement of raw materials and packaging.

Before the new planning system based on an optimization model was introduced, it was customary to work out the logistic plan in an aggregate way, by families of items and with infinite capacity, without duly considering costs and therefore with no attempt at optimization. Subsequent simulations were used to determine the corresponding level of production capacity engagement and the portions of the logistic plan that turned out not to be feasible. In some instances, such preliminary processing activities were followed by more focused analyses developed by the planners using simple spreadsheets. This development process proved largely unable to effectively manage such a highly complex supply chain; furthermore, it was also inefficient, since operators were required to continuously apply corrections and assessments.

The optimization system for logistics planning

The reengineering of the supply chain management process has led to the development of a logistic decision support system whose *intelligence* is represented by an integrated optimization model of the entire logistic production system. Figure 14.4 shows a sketch of the system architecture.

The system uses a logistic data mart that constitutes a local database, also to increase the efficiency in the generation and solution of the optimization model by means of the algorithmic engine. The information contained in the data

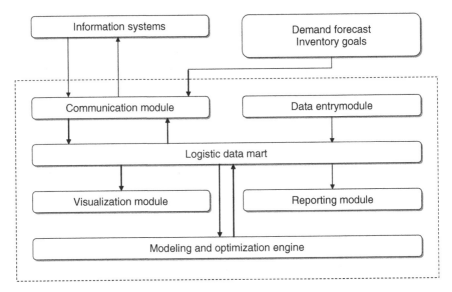

Figure 14.4   Architecture of the logistic production optimization system

mart is entered and updated in two different ways: on the one hand, through the automatic acquisition of data from the company information system; and on the other hand, through direct data entry by the users for those data and parameters required to devise the optimal logistic plan and not contained in the company information system.

Primary input data acquired from the information system are represented by the demand plan, formulated by item, warehouse and period. The demand plan is determined based on sales forecasts, which are available at the highest level of accuracy. Sales forecasts for each point of sale are therefore aggregated, allocating the demand to each central and peripheral warehouse. Finally, to obtain the demand plan, the aggregated sales forecasts are corrected to take into account the desired level of stocks, defined for the purpose of guaranteeing a preset level of service to customers.

The logistic plan, updated every week on a *rolling* basis, is usually developed over a time span of a year, divided into weeks. Downstream, the planning system is integrated with other modules that allow raw and packaging material requirements to be determined via *manufacturing resource planning*.

At the heart of the planning decision support system is the modeling and algorithmic module, which in turn is based on a finite-capacity optimization model of the entire logistic system. In particular, this is a discrete-time deterministic model that uses a network representation of the logistic system. The nodes of the network represent the supply centers, plants and warehouses where the demand is placed, as indicated in Figure 14.3. For each plant, the model considers the two phases of the manufacturing process – both production and packaging lines. The two phases are planned in a coordinated way, by properly adjusting the volume of products processed at each phase during each period of the planning horizon.

Each production and packaging line is assigned the available capacity, expressed in hours, for each period of the planning horizon. The available manpower capacity at the packaging lines is also assigned, plant by plant, since the human resources employed in packaging are multi-skilled workers and can be shared among different lines of the same plant. Finally, one assigns the potential manpower availability in terms of overtime, holidays or seasonal work, which can be used when needed at additional cost.

Regarding stock management, the model input data specify both a safe level of stock, defined on the basis of the demand profile and the desired service level, and an overstock value not to be exceeded. The logistic plan must guarantee at each period that stocks are above the safe level and below the overstock threshold. For each violation of these conditions, a penalty cost assigned as input is applied.

The size of the minimum lots is also specified, both for production and packaging lines, depending on the specific semi-finished goods and end items. Finally, the model deals with secondary conditions, to be assigned at the discretion of the planners, such as enforcing minimum and maximum volumes for selected end items, or specifying a set of semi-finished goods and products allowed to be manufactured at the beginning or end of each period.

The model also determines the supply of semolina to the plants. This can be procured through two channels: it can be purchased from external suppliers; or it can be obtained from production plants (mills) owned by the company, which in turn have to obtain stocks of wheat from supply markets. In brief, the input data to the system are:

- the demand plan, specified for each warehouse, item and period;

- the initial inventory for each warehouse;

- the bill of materials;

- the technological maps, which list all possible combinations and the corresponding yields between semi-finished goods and production lines, as well as between end items and packaging lines;

- the capacity of production and packaging lines over the periods of the planning horizon;

- the time availability of labor, in terms of regular working hours, extra shifts, holidays and seasonal working time;

- the minimum lots and the forced excess of production and packaging;

- the availability of the different types of wheat on the supply markets;

- the capacity at the mills to transform wheat into semolina;

- the availability of semolina at third-party mills;

- the blending quality requirements for the transformation of wheat into semolina.

The objective function of the model takes the following cost factors into account:

- the transportation costs of transferring end items from each plant to each warehouse;

- the production costs for production lines, net of labor costs;

- the production costs for packaging lines, net of labor cost;

- the penalty cost of failing to reach the desired stock level;

- the penalty cost of overstocking;

- the cost determined by lost sales due to the required product being unavailable in the appropriate period;

- the unit cost of labor, regular working hours, extra shifts, holidays and seasonal working;

- the cost of different wheat types available on the supply markets;

- the milling cost;

- the cost of semolina bought at third-party mills.

The representation of the logistic system described leads to a large-scale mixed binary optimization model, due to the presence of binary decision variables. However, despite its high level of complexity, the model requires no more than a few minutes' computation time, due to the existence of an *ad hoc* algorithm which yields an approximate solution.

At the end of the computation, the model provides an optimal integrated logistic plan, providing the following information:

- the allocation of the demand to the plants;

- the distribution plan, expressing the volumes of end products shipped weekly from each plant to the central and peripheral warehouses;

- the production and packaging plans for each plant, indicating the volumes of semi-finished goods and end products processed weekly by each production and packaging line, with the possibility of processing ahead of time with respect to the due dates, whenever needed or advantageous;

- the inventory plan, specifying the optimal levels of weekly stock for the end products at each warehouse;

- the possibly unmet demand for end items for each warehouse;

- the production capacity engagement plan, for both the lines and the labor;

- the supply of raw materials and packaging;

- the employment of additional labor, in the form of extra shifts, holiday work or seasonal work;

- the purchase from the supply markets and storage cost of wheat, based on the foreseen availability;

- the transportation of wheat from supply markets to the company's own mills;

- the blending and processing of wheat into semolina at the company's own mills;

- the purchase of semolina from the mills of third-party suppliers;

- the transportation of semolina from internal mills and from supplier mills to plants.

Main advantages

The use of the decision support system for logistics planning affords several advantages, both in terms of efficiency in devising the plan and effectiveness of the plans generated.

It requires less effort on the part of planners, while at the same time contributing to a substantial job enrichment, since they are required to play a decision-making role that is more gratifying on a personal level and of greater value to the company. Indeed, the system enables users to perform and manage different scenario analyses by experimenting with the input data and the parameters of the model. In this way, planners may simulate and evaluate the effect of different conditions and assumptions, carrying out what-if analyses in order to achieve the most effective logistic plan. For example, they can modify some cost parameters, or the level of the required stocks, or the labor availability, in order to assess the consequences of such changes. Particularly noteworthy and of great practical advantage is the possibility of assigning predetermined quantities for the production volumes, by single item or group of items, as well as by single period or group of periods.

The planners responsible for wheat procurement can also use the system to easily carry out what-if analyses of different alternative scenarios, formulated by varying model parameters such as transportation costs, wheat purchase costs, currency exchange rates, and the processing capacity of the company's own mills.

In addition, the system encourages closer integration between the sales planning department and the supply chain management department, by reducing interdepartmental conflicts and improving the quality of decisions regarding the marketing mix, intended to maximize the overall economic benefits for the company.

Other advantages that are worth mentioning concern the vast array of managerial decisions that benefit from the introduction of an optimized logistic plan, among which are:

- the possibility of assessing the feasibility of the restocking plan devised by the sales department;

- the definition of contracts with the suppliers of raw materials and packaging, as well as with third-party co-packers;

- the definition of the budget, formulated on a rolling basis and with a planning horizon of 18 months;

- the hiring and training plans for seasonal labor;

- the possibility of assessing the impact of expansion plans for the production capacity, obtainable through the expansion of production and packaging lines within existing plants, the construction of new manufacturing plants and the activation of relationships with new co-packers;

- the optimal allocation of the demand to the plants consistent with the plan of distribution to the markets.

14.4.2   Logistics planning in the packaging industry

The second case study considered here refers to an enterprise that produces integrated solutions for processing, packaging and handling liquid food products. The highly complex logistic and production network includes approximately 60 plants, 1000 different end products and 200 market areas assigned to more than 100 sales divisions, called *market companies*.

The organizational structure entails considerable independence in decision making on the part of local market companies, which leads them to assign production to their preferred plant, without taking into account global optimization goals at the enterprise level. The policy of independence in decision making also means a lack of homogeneity of costs and service performance among the various production sites. Delivery costs and times depend on the number of distinct products manufactured at a given plant, and the lack of homogeneity in the allocations is considerable, since the number of products assigned to the various production plants may vary from 10 to 100.

A decision support system for medium-term logistics planning was developed, with functionalities and features not much dissimilar from those described in the previous section. Its primary objective was to allocate the demand to the plants so as to minimize the overall logistic and production costs.

The optimization model representing the *intelligence* component of the system considers for each plant the two stages of production required by the processing technology of the company. For each stage, the model takes into account:

- the production costs by line and by plant;

- the cost of procurement of raw materials from suppliers;

- the transportation costs for raw materials;

- the transportation costs from plants to markets for end products;

- the limitations imposed by the capacity of the production lines;

- the limited availability of some technological components required by the production process;

- the technological maps that describe all possible combinations of lines and products, along with the relative processing times.

The optimization model also determines transfers from one plant to another for some technological process components, whose total number is limited by cost considerations. The solution of the resulting model leads to an optimal logistic and production plan, which includes the following choices:

- the supply plan of raw materials from suppliers, for each plant and for each period;

- the production plan for each line, plant, product and period, and therefore the optimal allocation of the demand to plants;

- the distribution plan for each product, plant, market area and period;

- the allocation and possible transfer between plants of critical technological process components in each period.

14.5   Notes and readings

There are several general texts that take a qualitative approach to supply chain management, among them Poirier and Reiter (1999) and Chopra and Meindl (2003). For a discussion closer to optimization themes we recommend Shapiro (2000), Simchi-Levi *et al.* (2002, 2004a,b), Graves *et al.* (2002), and Graves and De Kok (2003). For an application of logistic and production optimization in the textile industry see Dumoulin and Vercellis (2000), and for an application to tyres production see Fumero and Vercellis (1997, 1999); see also Battistini *et al.* (1999) for food production.

15

Data envelopment analysis

The purpose of *data envelopment analysis* (DEA) is to compare the operating performance of a set of units such as companies, university departments, hospitals, bank branch offices, production plants, or transportation systems. In order for the comparison to be meaningful, the units being investigated must be homogeneous.

The performance of a unit can be measured on several dimensions. For example, to evaluate the activity of a production plant one may use quality indicators, which estimate the rate of rejects resulting from manufacturing a set of products, and also flexibility indicators, which measure the ability of a system to react to changes in the requirements with quick response times and low costs.

Data envelopment analysis relies on a productivity indicator that provides a measure of the efficiency that characterizes the operating activity of the units being compared. This measure is based on the results obtained by each unit, which will be referred to as *outputs*, and on the resources utilized to achieve these results, which will be generically designated as *inputs* or *production factors*. If the units represent bank branches, the outputs may consist of the number of active bank accounts, checks cashed or loans raised; the inputs may be the number of cashiers, managers or rooms used at each branch. If the units are university departments, it is possible to consider as outputs the number of active teaching courses and scientific publications produced by the members of each department; the inputs may include the amount of financing received by each department, the cost of teaching, the administrative staff and the availability of offices and laboratories.

Business Intelligence: Data Mining and Optimization for Decision Making   C. Vercellis
© 2009 John Wiley & Sons, Ltd

15.1   Efficiency measures

In data envelopment analysis the units being compared are called *decision-making units* (DMUs), since they enjoy a certain decisional autonomy. Assuming that we wish to evaluate the efficiency of n units, let $\mathcal{N} = \{1, 2, \ldots, n\}$ denote the set of units being compared.

If the units produce a single output using a single input only, the *efficiency* of the jth decision-making unit DMU_j, $j \in \mathcal{N}$, is defined as

$$\theta_j = \frac{y_j}{x_j}, \tag{15.1}$$

in which y_j is the output value produced by DMU_j and x_j the input value used.

If the units produce multiple outputs using various input factors, the efficiency of DMU_j is defined as the ratio between a weighted sum of the outputs and a weighted sum of the inputs. Denote by $\mathcal{H} = \{1, 2, \ldots, s\}$ the set of production factors and by $\mathcal{K} = \{1, 2, \ldots, m\}$ the corresponding set of outputs. If x_{ij}, $i \in \mathcal{H}$, denotes the quantity of input i used by DMU_j and y_{rj}, $r \in \mathcal{K}$, the quantity of output r obtained, the efficiency of DMU_j is defined as

$$\theta_j = \frac{u_1 y_{1j} + u_2 y_{2j} + \ldots + u_m y_{mj}}{v_1 x_{1j} + v_2 x_{2j} + \ldots + v_s x_{sj}} = \frac{\sum_{r \in \mathcal{K}} u_r y_{rj}}{\sum_{i \in \mathcal{H}} v_i x_{ij}}, \tag{15.2}$$

for weights $u_1, u_2, \ldots, u_m$ associated with the outputs and $v_1, v_2, \ldots, v_s$ assigned to the inputs.

In this second case, the efficiency of DMU_j depends strongly on the system of weights introduced. At different weights, the efficiency value may undergo relevant variations and it becomes difficult to fix a single structure of weights that might be shared and accepted by all the evaluated units. In order to avoid possible objections raised by the units to a preset system of weights, which may privilege certain DMUs rather than others, data envelopment analysis evaluates the efficiency of each unit through the weights system that is best for the DMU itself – that is, the system that allows its efficiency value to be maximized. Subsequently, by means of additional analyses, the purpose of data envelopment analysis is to identify the units that are efficient in absolute terms and those whose efficiency value depends largely on the system of weights adopted.

15.2   Efficient frontier

The *efficient frontier*, also known as *production function*, expresses the relationship between the inputs utilized and the outputs produced. It indicates the maximum quantity of outputs that can be obtained from a given combination of inputs. At the same time, it also expresses the minimum quantity of inputs that must be used to achieve a given output level. Hence, the efficient frontier corresponds to *technically efficient* operating methods.

The efficient frontier may be empirically obtained based on a set of observations that express the output level obtained by applying a specific combination of input production factors. In the context of data envelopment analysis, the observations correspond to the units being evaluated. Most statistical methods of parametric nature, which are based for instance on the calculation of a regression curve, formulate some prior hypotheses on the shape of the production function. Data envelopment analysis, on the other hand, forgoes any assumptions on the functional form of the efficient frontier, and is therefore nonparametric in character. It only requires that the units being compared are not placed above the production function, depending on their efficiency value. To further clarify the notion of efficient frontier consider Example 15.1.

Example 15.1 – Evaluation of the efficiency of bank branches. A bank wishes to compare the operational efficiency of its nine branches, in terms of staff size and total value of savings in active accounts. Table 15.1 shows for each branch the total value of accounts, expressed in hundreds of thousands of euros, and the number of staff employed, with the corresponding efficiency values calculated based on definition (15.1). The graph shown in Figure 15.1 shows for each branch the number of employees on the horizontal axis and the value of accounts on the vertical axis. The slope of the line connecting each point to the origin represents the efficiency value associated with the corresponding branch. The line with the maximum slope, represented in Figure 15.1 by a solid line, is the efficient frontier for all branches being analyzed. The branches that are on this line correspond to efficient units, while the branches that are below the efficient frontier are inefficient units. The area between the efficient frontier and the positive horizontal semi-axis is called the *production possibility set*.

A possible alternative to the efficient frontier is the regression line that can be obtained based on the available observations, indicated in Figure 15.1 by a dashed line. In this case, the units that fall above the regression line may be deemed excellent, and the degree of excellence of each unit could be expressed by its distance from the line. However, it is appropriate to underline the difference that exists between the prediction line obtained using a regression model and the efficient frontier obtained using data envelopment analysis. The

Table 15.1   Input and output values for the bank branches in Example 15.1

bank branch	staff size	accounts value	efficiency
A	3	2.5	0.733
B	2	1.0	0.500
C	5	2.7	0.540
D	3	3.0	1.000
E	7	5.0	0.714
F	5	2.3	0.460
G	4	3.2	0.700
H	5	4.5	0.900
I	6	4.5	0.633

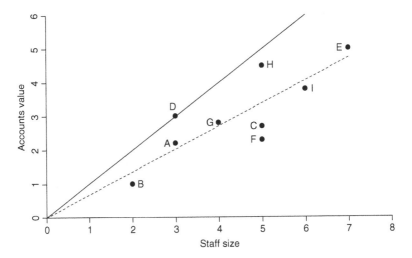

Figure 15.1   Evaluation of efficiency of bank branches

regression line reflects the average behavior of the units being compared, while the efficient frontier identifies the best behavior, and measures the inefficiency of a unit based on the distance from the frontier itself.

Notice also that the efficient frontier provides some indications for improving the performance of inefficient units. Indeed, it identifies for each input level the output level that can be achieved in conditions of efficiency. By the same token, it identifies for each output level the minimum level of input that should be used in conditions of efficiency. In particular, for each DMU$_j$, $j \in \mathcal{N}$, the *input-oriented* efficiency θ_j^I can be defined as the ratio between the ideal input quantity x^* that should be used by the unit if it were efficient and the actually used quantity x_j:

$$\theta_j^I = \frac{x^*}{x_j}. \qquad (15.3)$$

Similarly, the *output-oriented* efficiency θ_j^O is defined as the ratio between the quantity of output y_j actually produced by the unit and the ideal quantity y^* that it should produce in conditions of efficiency:

$$\theta_j^O = \frac{y_j}{y^*}. \tag{15.4}$$

The problem of making an inefficient unit efficient is then turned into one of devising a way by which the inefficient unit can be brought close to the efficient frontier.

If the unit produces a single output only by using two inputs, the efficient frontier assumes the shape shown in Figure 15.2. In this case, the inefficiency of a given unit is evaluated by the length of the segment connecting the unit to the efficient frontier along the line passing through the origin of the axes. For the example illustrated in Figure 15.2, the efficiency value of DMU$_A$ is given by

$$\theta_A = \frac{\overline{OP}}{\overline{OA}}, \tag{15.5}$$

where $\overline{OP}$ and $\overline{OA}$ represent the lengths of segments OP and OA, respectively. The inefficient unit may be made efficient by a displacement along segment OA that moves it onto the efficient frontier. Such displacement is tantamount to progressively decreasing the quantity of both inputs while keeping unchanged the quantity of output. In this case, the production possibility set is defined as the region delimited by the efficient frontier where the observed units being compared are found.

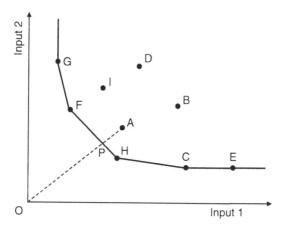

Figure 15.2   Efficient frontier with two inputs and one output

15.3   The CCR model

Using data envelopment analysis, the choice of the optimal system of weights for a generic DMU_j involves solving a mathematical optimization model whose decision variables are represented by the weights u_r, $r \in \mathcal{K}$, and v_i, $i \in \mathcal{H}$, associated with each output and input. Various formulations have been proposed, the best-known of which is probably the Charnes–Cooper–Rhodes (CCR) model. The CCR model formulated for DMU_j takes the form

$$\max \quad \vartheta = \frac{\sum_{r \in \mathcal{K}} u_r y_{rj}}{\sum_{i \in \mathcal{H}} v_i x_{ij}}, \tag{15.6}$$

$$\text{s.to} \quad \frac{\sum_{r \in \mathcal{K}} u_r y_{rj}}{\sum_{i \in \mathcal{H}} v_i x_{ij}} \leq 1, \qquad j \in \mathcal{N}, \tag{15.7}$$

$$u_r, v_i \geq 0, \qquad r \in \mathcal{K}, i \in \mathcal{H}. \tag{15.8}$$

The objective function involves the maximization of the efficiency measure for DMU_j. Constraints (15.7) require that the efficiency values of all the units, calculated by means of the weights system for the unit being examined, be lower than one. Finally, conditions (15.8) guarantee that the weights associated with the inputs and the outputs are non-negative. In place of these conditions, sometimes the constraints $u_r, v_i \geq \delta$, $r \in \mathcal{K}$, $i \in \mathcal{H}$ may be applied, where $\delta > 0$, preventing the unit from assigning a null weight to an input or output.

Model (15.6) can be linearized by requiring the weighted sum of the inputs to take a constant value, for example 1. This condition leads to an alternative optimization problem, the *input-oriented* CCR model, where the objective function consists of the maximization of the weighted sum of the outputs

$$\max \quad \vartheta = \sum_{r \in \mathcal{K}} u_r y_{rj}, \tag{15.9}$$

$$\text{s.to} \quad \sum_{i \in \mathcal{H}} v_i x_{ij} = 1, \tag{15.10}$$

$$\sum_{r \in \mathcal{K}} u_r y_{rj} - \sum_{i \in \mathcal{H}} v_i x_{ij} \leq 0, \qquad j \in \mathcal{N}, \tag{15.11}$$

$$u_r, v_i \geq 0, \qquad r \in \mathcal{K}, i \in \mathcal{H}. \tag{15.12}$$

Let ϑ^* be the optimum value of the objective function corresponding to the optimal solution $(\mathbf{v}^*, \mathbf{u}^*)$ of problem (15.9). DMU_j is said to be *efficient* if $\vartheta^* = 1$ and if there exists at least one optimal solution $(\mathbf{v}^*, \mathbf{u}^*)$ such that $\mathbf{v}^* > \mathbf{0}$ and $\mathbf{u}^* > \mathbf{0}$.

By solving a similar optimization model for each of the n units being compared, one obtains n systems of weights. The flexibility enjoyed by the units in choosing the weights represents an undisputed advantage, in that if a unit turns out to be inefficient based on the most favorable system of weights, its inefficiency cannot be traced back to an inappropriate evaluation process. However, given a unit that scores $\vartheta^* = 1$, it is important to determine whether its efficiency value should be attributed to an actual high-level performance or simply to an optimal selection of the weights structure.

Dual of the CCR model

For the input-oriented CCR model, the following dual problem, which lends itself to an interesting interpretation, can be formulated:

$$\text{min} \quad \vartheta, \tag{15.13}$$

$$\text{s.to} \quad \sum_{j \in \mathcal{N}} \lambda_j x_{ij} - \vartheta x_{ij} \leq 0, \quad i \in \mathcal{H}, \tag{15.14}$$

$$\sum_{j \in \mathcal{N}} \lambda_j y_{rj} - y_{rj} \geq 0, \quad r \in \mathcal{K}, \tag{15.15}$$

$$\lambda_j \geq 0, \quad\quad\quad\quad j \in \mathcal{N}. \tag{15.16}$$

Based on the optimum value of the variables λ_j^*, $j \in \mathcal{N}$, the aim of model (15.13) is to identify an ideal unit that lies on the efficient frontier and represents a term of comparison for DMU$_j$. Constraints (15.14) and (15.15) of the model require that this unit produces an output at least equal to the output produced by DMU$_j$, and uses a quantity of inputs equal to a fraction of the quantity used by the unit examined. The ratio between the input used by the ideal unit and the input absorbed by DMU$_j$ is defined as the optimum value ϑ^* of the dual variable ϑ. If $\vartheta^* < 1$, DMU$_j$ lies below the efficient frontier. In order to be efficient, this unit should employ $\vartheta^* x_{ij}$, $i \in \mathcal{H}$, of each input.

The quantity of inputs utilized by the ideal unit and the level of outputs to be produced are expressed as a linear combination of the inputs and outputs associated with the n units being evaluated:

$$x_i^{\text{ideal}} = \sum_{j \in \mathcal{N}} \lambda_j^* x_{ij}, \quad i \in \mathcal{H}, \tag{15.17}$$

$$y_r^{\text{ideal}} = \sum_{j \in \mathcal{N}} \lambda_j^* y_{rj}, \quad r \in \mathcal{K}. \tag{15.18}$$

For each feasible solution (ϑ, λ) to problem (15.13), the slack variables s_i^-, $i \in \mathcal{H}$, and s_r^+, $r \in \mathcal{K}$, can be defined, which represent respectively the quantity of input i used in excess by DMU$_j$ and the quantity of output r produced in shortage by the DMU_j with respect to the ideal unit:

$$s_i^- = \vartheta x_{ij} - \sum_{j \in \mathcal{N}} \lambda_j x_{ij}, \quad i \in \mathcal{H}, \tag{15.19}$$

$$s_r^+ = \sum_{j \in \mathcal{N}} \lambda_j y_{rj} - y_{rj}, \quad r \in \mathcal{K}. \tag{15.20}$$

As with the primal problem, it is possible also for the dual problem to provide a definition of efficiency. DMU_j is efficient if $\vartheta^* = 1$ and if the optimum value of the slack variables is equal to zero: $s_i^{-*} = 0$, $i \in \mathcal{H}$, and $s_r^{+*} = 0$, $r \in \mathcal{K}$. In other words, DMU_j is efficient if it is not possible to improve the level of an input used or the level of an output produced without a deterioration in the level of another input or of another output. If $\vartheta^* < 1$, DMU_j is said to be *technically inefficient*, in the sense that, in order to obtain the same output, the input quantities used could be simultaneously reduced in the same proportion. The maximum reduction allowed by the efficient frontier is defined by the value $1 - \vartheta^*$. If $\vartheta^* = 1$, but some slack variables are different from zero, DMU_j presents a *mix inefficiency* since, keeping the same output level, it could reduce the use of a few inputs without causing an increase in the quantity of other production factors used.

15.3.1 Definition of target objectives

In real-world applications it is often desirable to set improvement objectives for inefficient units, in terms of both outputs produced and inputs utilized. Data envelopment analysis provides important recommendations in this respect, since it identifies the output and input levels at which a given inefficient unit may become efficient. The efficiency score of a unit expresses the maximum proportion of the actually utilized inputs that the unit should use in conditions of efficiency, in order to guarantee its current output levels. Alternatively, the inverse of the efficiency score indicates the factor by which the current output levels of a unit should be multiplied for the unit to be efficient, holding constant the level of the productive inputs used. Based on the efficiency values, data envelopment analysis therefore gives a measure for each unit being compared of the savings in inputs or the increases in outputs required for the unit to become efficient.

To determine the target values, it is possible to follow an input- or output-oriented strategy. In the first case, the improvement objectives primarily concern the resources used, and the target values for inputs and outputs are given by

$$x_{ij}^{\text{target}} = \vartheta^* x_{ij} - s_i^{-*}, \quad i \in \mathcal{H}, \tag{15.21}$$

$$y_{rj}^{\text{target}} = y_{rj} + s_r^{+*}, \quad r \in \mathcal{K}. \tag{15.22}$$

In the second case, target values for inputs and outputs are given by

$$x_{ij}^{\text{target}} = x_{ij} - \frac{s_i^{-*}}{\vartheta^*}, \quad i \in \mathcal{H}, \qquad (15.23)$$

$$y_{rj}^{\text{target}} = \frac{y_{rj} + s_r^{+*}}{\vartheta^*}, \quad r \in \mathcal{K}. \qquad (15.24)$$

Other performance improvement strategies may be preferred over the proportional reduction in the quantities of inputs used or the proportional increase in the output quantities produced:

- priority order for the production factors – the target values for the inputs are set in such a way as to minimize the quantity used of the resources to which the highest priority has been assigned, without allowing variations in the level of other inputs or in the outputs produced;

- priority order for the outputs – the target values for the outputs are set in such a way as to maximize the quantity produced of the outputs to which highest priority has been assigned, without allowing variations in the level of other outputs or inputs used;

- preferences expressed by the decision makers with respect to a decrease in some inputs or an increase in specific outputs.

15.3.2 Peer groups

Data envelopment analysis identifies for each inefficient unit a set of excellent units, called a *peer group*, which includes those units that are efficient if evaluated with the optimal system of weights of an inefficient unit. The peer group, made up of DMUs which are characterized by operating methods similar to the inefficient unit being examined, is a realistic term of comparison which the unit should aim to imitate in order to improve its performance.

The units included in the peer group of a given unit DMU_j may be identified by the solution to model (15.9). Indeed, these correspond to the DMUs for which the first and the second member of constraints (15.11) are equal:

$$E_j = \left\{ j : \sum_{r \in \mathcal{K}} u_r^* y_{rj} = \sum_{i \in \mathcal{H}} v_i^* x_{ij} \right\}. \qquad (15.25)$$

Alternatively, with respect to formulation (15.13), the peer group consists of those units whose variable λ_j in the optimal solution is strictly positive:

$$E_j = \left\{ j : \lambda_j^* > 0 \right\}. \qquad (15.26)$$

Notice that within a peer group a few excellent units more than others may represent a reasonable term of comparison. The relative importance of a unit belonging to a peer group depends on the value of the corresponding variable λ_j in the optimal solution of the dual model.

The analysis of peer groups allows one to differentiate between really efficient units and apparently efficient units for which the choice of an optimal system of weights conceals some abnormal behavior. In order to draw this distinction, it is necessary to consider the efficient units and to evaluate how often each belongs to a peer group. One may reasonably expect that an efficient unit often included in the peer groups uses for the evaluation of its own efficiency a robust weights structure. Conversely, if an efficient unit rarely represents a term of comparison, its own system of optimal weights may appear distorted, in the sense that it may implicitly reflect the specialization of the unit along a particular dimension of analysis.

15.4   Identification of good operating practices

By identifying and sharing *good operating practices*, one may hope to achieve an improvement in the performance of all units being compared. The units that appear efficient according to data envelopment analysis certainly represent terms of comparison and examples to be imitated for the other units. However, among efficient units some more than others may represent a target to be reached in improving the efficiency.

The need to identify the efficient units, for the purpose of defining the best operating practices, stems from the principle itself on which data envelopment analysis is grounded, since it allows each unit to evaluate its own degree of efficiency by choosing the most advantageous structure of weights for inputs and outputs. In this way, a unit might appear efficient by purposely attributing a non-negligible weight only to a limited subset of inputs and outputs. Furthermore, those inputs and outputs that receive greater weights may be less critical than other factors more intimately connected to the primary activity performed by the units being analyzed. In order to identify good operating practices, it is therefore expedient to detect the units that are really efficient, that is, those units whose efficiency score does not primarily depend on the system of weights selected. To differentiate these units, we may resort to a combination of different methods: *cross-efficiency analysis*, evaluation of *virtual inputs* and *virtual outputs*, and *weight restrictions*.

15.4.1   Cross-efficiency analysis

The analysis of *cross-efficiency* is based on the definition of the *efficiency matrix*, which provides information on the nature of the weights system adopted by the units for their own efficiency evaluation. The square efficiency matrix contains as many rows and columns as there are units being compared. The generic element θ_{ij} of the matrix represents the efficiency of DMU_j evaluated through the optimal weights structure for DMU_i, while the element θ_{jj} provides the efficiency of DMU_j calculated using its own optimal weights. If DMU_j is efficient (i.e. if $\theta_{jj} = 1$), although it exhibits a behavior specialized along a given dimension with respect to the other units, the efficiency values in the column corresponding to DMU_j will be less than 1.

Two quantities of interest can be derived from the efficiency matrix. The first represents the average efficiency of a unit with respect to the optimal weights systems for the different units, obtained as the average of the values in the jth column. The second is the average efficiency of a unit measured applying its optimal system of weights to the other units. The latter is obtained by averaging the values in the row associated with the unit being examined. The difference between the efficiency score θ_{jj} of DMU_j and the efficiency obtained as the average of the values in the jth column provides an indication of how much the unit relies on a system of weights conforming with the one used by the other units in the evaluation process. If the difference between the two terms is significant, DMU_j may have chosen a structure of weights that is not shared by the other DMUs in order to privilege the dimensions of analysis on which it appears particularly efficient.

15.4.2 Virtual inputs and virtual outputs

Virtual inputs and virtual outputs provide information on the relative importance that each unit attributes to each individual input and output, for the purpose of maximizing its own efficiency score. Thus, they allow the specific competencies of each unit to be identified, highlighting at the same time its weaknesses.

The *virtual inputs* of a DMU are defined as the product of the inputs used by the unit and the corresponding optimal weights. Similarly, *virtual outputs* are given by the product of the outputs of the unit and the associated optimal weights.

Inputs and outputs for which the unit shows high virtual scores provide an indication of the activities in which the unit being analyzed appears particularly efficient. Notice that model (15.9) admits in general multiple optimal solutions, corresponding to which it is possible to obtain different combinations of virtual inputs and virtual outputs.

Two efficient units may yield high virtual values corresponding to different combinations of inputs and outputs, showing good operating practices in different contexts. In this case, it might be convenient for each unit to follow the principles and operating methods shown by the other, aiming at improving its own efficiency on a specific dimension.

15.4.3 Weight restrictions

To separate the units that are really efficient from those whose efficiency score largely depends on the selected weights system, we may impose some restrictions on the value of the weights to be associated with inputs and outputs. In general, these restrictions translate into the definition of maximum thresholds for the weight of specific outputs or minimum thresholds for the weight of specific inputs. Notice that, despite possible restrictions on the weights, the units still enjoy a certain flexibility in the choice of multiplicative factors for inputs and outputs. For this reason it may be useful to resort to the evaluation of virtual inputs and virtual outputs in order to identify the units with the most efficient operating practices with respect to the usage of a specific input resource or to the production of a given output.

15.5 Other models

Model (15.9) is based on the hypothesis that the units being compared operate with constant returns to scale. Recall that the returns to scale express the variation in the quantity of outputs in terms of variations in the quantity of inputs used. When the returns to scale are constant, if the inputs increase in a given proportion then the outputs also increase in the same proportion. The hypothesis of constant returns to scale leads to an efficient frontier like the one shown in Figure 15.1. In particular, if $\mathbf{X}$ denotes the matrix of inputs used by the n units and $\mathbf{Y}$ denotes the corresponding matrix of outputs, in the hypothesis of constant returns to scale we can express the production possibility set as

$$P = \{(\mathbf{x}, \mathbf{y}) : \mathbf{x} \geq \mathbf{X}\lambda, \mathbf{y} \leq \mathbf{Y}\lambda, \lambda \geq \mathbf{0}\}. \tag{15.27}$$

This means that if the point $(\mathbf{x}, \mathbf{y})$ belongs to P, then any other point of the form $(k\mathbf{x}, k\mathbf{y})$, $k > 0$, will also belong to the production possibility set. If the hypothesis of constant returns to scale is not adequate, one may resort to formulations other than model (15.9). For example, the Banker–Charnes–Cooper model is based on the hypothesis of variable returns to scale, and takes the form

$$\min \quad \vartheta, \tag{15.28}$$

$$\text{s.to} \quad \sum_{j \in \mathcal{N}} \lambda_j x_{ij} - \vartheta x_{ij} \leq 0, \quad i \in \mathcal{H}, \tag{15.29}$$

$$\sum_{j \in \mathcal{N}} \lambda_j y_{rj} - y_{rj} \geq 0, \quad r \in \mathcal{K}, \tag{15.30}$$

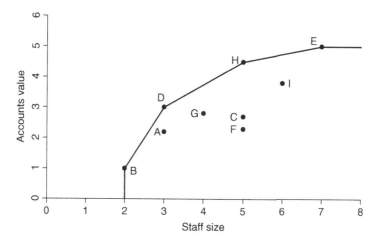

Figure 15.3   Efficient frontier for variable returns to scale

$$\sum_{j \in \mathcal{N}} \lambda_j = 1, \qquad\qquad (15.31)$$

$$\lambda_j \geq 0, \qquad\qquad j \in \mathcal{N}. \qquad (15.32)$$

In model (15.28) the production possibility set is defined as

$$P = \{(\mathbf{x}, \mathbf{y}) : \mathbf{x} \geq \mathbf{X}\lambda, \mathbf{y} \leq \mathbf{Y}\lambda, \mathbf{e}\lambda = 1, \lambda \geq \mathbf{0}\}, \qquad (15.33)$$

where $\mathbf{e}$ is a unit vector. The only difference between model (15.28) and the dual of the CCR model stems from condition (15.31) which, together with constraints (15.32), specifies that the efficient frontier is convex. Figure 15.3 describes the production possibility set for the units of Example 15.1 in the hypothesis of variable returns to scale. The models described above do not exhaust the broad variety of formulations that have been proposed in the framework of data envelopment analysis, although they represent good examples. For more in-depth information, see the suggested references below.

15.6   Notes and readings

The CCR model was first proposed in Charnes *et al.* (1978), based on a previous contribution by Farrell (1957). For an extensive discussion of data envelopment analysis models the reader is referred to Cooper *et al.* (2000) and Thanassoulis (2001). More recent developments are described in Ray (2004). A description of production functions and returns to scale can be found in any microeconomics textbook.

Appendix A

Software tools

This appendix provides readers with references to software tools that can be used to carry out business intelligence analyses based on the methods described in this book. These are basically the same software tools used to process the data and design the majority of graphs presented in the text.

We confine our suggestions and recommendations to *open source*[1] programs, so that readers interested in carrying out some of the analyses suggested in previous chapters can easily download the software from the Internet, freely using the different tools, at least within the restrictions imposed by individual licenses.

The R software for data analysis

R is an environment for data analysis and graph plotting that has a wealth of functions and is constantly evolving. It was jointly developed by researchers in the international community. The R environment offers a programming language, interactive analysis functions and a modular structure extendible by means of libraries (packages) implementing specific algorithms. All the software of which R is composed, both the basic modules and the additional packages, is available for downloading from *The R Project for Statistical Computing* at *http://www.r-project.org/*. Besides the programs, a wide range of documents can also be found on this site.

The functions offered by R are not limited to classical statistics but also cover exploratory data graphical analysis, which we have used in the realization of Chapter 7, time series analysis, supervised methods for classification and regression, unsupervised methods for clustering and association rules. The R environment also includes packages that implement optimization methods.

[1]For more information on open source initiatives, visit the website *http://www.opensource.org/*.

Business Intelligence: Data Mining and Optimization for Decision Making   C. Vercellis
© 2009 John Wiley & Sons, Ltd

The Weka environment for data mining

The Weka environment, also developed by an international community of researchers, comprises a broad collection of inductive learning algorithms and analysis functions that enable the data exploration process. The system is written in Java and has an object-oriented structure that easily allows the algorithms and functions to be extended, thus creating new objects in the hierarchy. Like R, Weka is an open source environment subject to the GNU license. The software can be downloaded from the project site at *http://www.cs.waikato.ac.nz/ ml/weka/*.

Open source database management systems

In many instances, it may be useful to organize the available data within a relational database to create a data mart that facilitates the extraction and the subsequent analyses. Two database environments are worth mentioning, which are also open source and feature excellent functions and performance, comply with the standards of the SQL language and support stored procedures, triggers and concurrency:

- the Firebird system, which can be downloaded from the website *http://www.ibphoenix.com/*;

- the MySQL system, which can be downloaded from the website *http://www.mysql.com/*.

Appendix B

Dataset repositories

In this appendix we wish to briefly describe the datasets used in this book, and provide a few links to repositories that offer datasets in the public domain that can be downloaded to carry out analyses and compare the performance of the learning algorithms.

anscombe. These are four datasets, each consisting of two attributes and 11 observations, originally given in Anscombe (1973). The datasets are characterized by the same values of the major summary statistics (mean, variance, correlation, regression line) although they exhibit a very different graphical behavior. The datasets are contained in the R data package.

diabete. This dataset was created by the National Institute of Diabetes and Digestive and Kidney Disease. It contains 768 observations on women patients aged at least 21 years living in an Indian reservation close to Phoenix, Arizona. Each observation is represented by eight numerical predictive variables (number of times pregnant, plasma glucose concentration, diastolic blood pressure, triceps skin-fold thickness, serum insulin, body mass index, diabetes pedigree function, age) while the target class indicates the state of health of each patient. The dataset can be downloaded from *http://www.ics.uci.edu/mlearn/ MLRepository.html*.

mtcars. This dataset is included in the R data package. It contains information on fuel consumption and 10 features for 32 different models of cars. The variables have the following meaning

- mpg : miles per (US) gallon;
- cyl : number of cylinders;
- disp : displacement (cu. in.);
- hp : gross horsepower;

- drat : rear axle ratio;
- wt : weight (lb/1000);
- qsec : 1/4 mile time;
- vs : engine shape (V/S);
- am : type of transmission (0 = automatic, 1 = manual);
- gear : number of forward gears;
- carb : number of carburetors.

employment. This dataset from the Continuous Workforce Monitoring Survey (RCFL), available from *http://www.istat.it*, provides an analysis by the Italian National Institute of Statistics (Istat) of the labor market over the long, medium and short term for the period between the fourth quarter of 1992 and the fourth quarter of 2003.

eustockmarkets. This dataset, available in the R data package, contains the daily closing prices of the four main European stock exchanges: German DAX (Ibis), Swiss SMI, French CAC and UK FTSE. The multivariate time series consists of 1860 observations and four variables.

wine. This dataset contains the time series of sales of Australian wines between January 1980 and July 1995. The dataset can be downloaded from the *Time Series Data Library* maintained by R.J. Hyndman at *http://www-personal .buseco.monash.edu.au/hyndman/TSDL/*.

Other links of interest providing access to datasets in public domain and of different kinds are:

http://www.ics.uci.edu/ mlearn/MLRepository.html
http://lib.stat.cmu.edu/datasets/
http://www.cs.toronto.edu/ delve/data/datasets.html
http://www-personal.buseco.monash.edu.au/ hyndman/TSDL/.

References

Adamo J. (2001). *Data mining for association rules and sequential patterns: sequential and parallel algorithms*. Springer.

Agrawal R., Shafer J. (1998). Parallel mining of association rules. *IEEE Transactions on Knowledge and Data Engineering*, **8**, 962–969.

Agrawal R., Aggarwal C., Prasad V. (2001). A tree projection algorithm for generation of frequent itemsets. *Journal of Parallel and Distributed Computing*, **61**, 350–371.

Aggarwal C., Yu P. (1998). Mining large itemsets for association rules. *Data Engineering Bulletin*, **21**, 23–31.

Agrawal R., Srikant R. (1994). Fast algorithms for mining association rules in large databases. In: *VLDB '94: Proceedings of the 20th International Conference on Very Large Data Bases*, 487–499. Morgan Kaufmann.

Agrawal R., Srikant R. (1995). Mining sequential patterns. In: P. Yu et A. Chen (eds.), *ICDE '95: Proceedings of the Eleventh International Conference on Data Engineering*, 3–14. IEEE Computer Society.

Agrawal R., Srikant R. (1997). Mining generalized association rules. *Future Generation Computer Systems*, **13**, 161–180.

Agrawal R., Imielinski T., Swami A. (1993a). Database mining: A performance perspective. *IEEE Transactions on Knowledge and Data Engineering*, **5**, 914–925.

Agrawal R., Imielinski T., Swami A. (1993b). Mining association rules between sets of items in large databases. In: *SIGMOD '93: Proceedings of the 1993 ACM SIGMOD international conference on Management of data*, 207–216. ACM Press.

Agrawal R., Mannila H., Srikant R., Toivonen H., Verkamo A. (1996). Fast discovery of association rules. In: *Advances in knowledge discovery and data mining*, 307–328. American Association for Artificial Intelligence.

Aldenderfer M., Blashfield R. (1985). *Cluster Analysis*. Sage Publications.

Allison P. (1998). *Multiple Regression: A Primer*. Pine Forge Press.

Anderberg M. (1973). *Cluster Analysis for Applications*. Academic Press.

Anderson J. (1995). *An Introduction to Neural Networks*. MIT Press.

Anscombe F. (1973). Graphs in statistical analysis. *The American Statistician*, 195–199.

Ayres I. (2007). *Super Crunchers: Why Thinking-by-Numbers Is the New Way to Be Smart*. Bantam.

Bakan J. (2005). *The Corporation: The Pathological Pursuit of Profit and Power*. Free Press.

Baldi P., Brunak S. (2001). *Bioinformatics: the machine learning approach*. MIT Press.

Battistini V., Contini A., Del Prato G., Palopoli G., Valentini D., Vercellis C. (1999). L'ottimizzazione della catena logistica integrata: il caso Barilla alimentare. *Logistica e Management*, **102**, 77–86.

Berry M., Linoff G. (1999). *Mastering Data Mining*. Wiley.

Berry M., Linoff G. (2002). *Mining the Web: Transforming Customer Data into Customer Value*. Wiley.

Berry M., Linoff G. (2004). *Data Mining Techniques: For Marketing, Sales, and Customer Relationship Management*. Wiley.

Berson A., Smith S. (1997). *Data Warehousing, Data Mining, and OLAP*. Mcgraw-Hill.

Berson A., Smith S., Thearling K. (1999). *Building Data Mining Applications for CRM*. McGraw-Hill.

Bertsekas D. (2003). *Convex Analysis and Optimization*. Athena Scientific.

Bishop C. (1995). *Neural Networks for Pattern Recognition*. Oxford University Press.

Bolloju N., Khalifa M., Turban E. (2002). Integrating knowledge management into enterprise environments for the next generation decision support. *Decision Support Systems*, **33**, 163–176.

Box G., Jenkins G., Reinsel G. (1994). *Time Series Analysis: Forecasting & Control*. Prentice Hall.

Bradley P., Fayyad U., Mangasarian O. (1999). Mathematical programming for data mining: formulations and challenges. *INFORMS Journal on Computing*, **11**, 217–238.

Breiman L., Friedman J., Olshen R., Stone C. (1984). *Classification and Regression Trees*. Chapman & Hall.

Breslow L., Aha D. (1997). Simplifying decision trees: A survey. *Knowledge Engineering Review*, **12**, 1–40.

Brockwell P., Davis R. (2002). *Introduction to Time Series and Forecasting*. Springer.

Bruhn M. (2002). *Relationship Marketing: Management of Customer Relationships*. Pearson.

Burges C. (1998). A tutorial on support vector machines for pattern recognition. *Data Mining and Knowledge Discovery*, **2**, 121–167.

Cadez I., Heckerman D., Smyth P., Meek C., White S. (2003). Model-based clustering and visualization of navigation patterns on a web site. *Data Mining and Knowledge Discovery*, **7**, 399–424.

Charnes A., Cooper W., Rhodes E. (1978). Measuring the efficiency of decision making units. *European Journal of Operational Research*, **2**, 429–444.

Chatfield C. (2003). *The Analysis of Time Series: An Introduction*. Chapman & Hall.

Cherkassky V., Mulier F. (1998). *Learning from data, concepts, theory and methods*. Wiley.

Chopra S., Meindl P. (2003). *Supply Chain Management*. Prentice Hall.

Clemen R. (1997). *Making Hard Decisions: An Introduction to Decision Analysis*. Duxbury Press.

Cleveland W. (1993). *Visualizing Data*. Hobart Press.

Cooper W., Seiford L., Tone K. (2000). *Data Envelopment Analysis. A Comprehensive Text with Models, Applications, References and DEA-Solver Software*. Springer.

Crawley M. (2005). *Statistics: An Introduction using R*. Wiley.

Cristianini N., Shawe-Taylor J. (2000). *An introduction to support vector machines and other kernel-based learning methods*. Cambridge University Press.

Cristianini N., Shawe-Taylor J. (2004). *Kernel Methods for Pattern Analysis*. Cambridge University Press.

Davenport T., Harris J. (2007). *Competing on Analytics: The New Science of Winning*. Harvard Business School Press.

Dhar V., Stein R. (1997). *Intelligent Decision Support Methods: The Science of Knowledge*. Prentice-Hall.

Domingos P., Pazzani M. (1997). On the optimality of the simple bayesian classifier under zero-one loss. *Machine Learning*, **29**, 103–130.

Drucker H., Vapnik V., Wu D. (1999). Support vector machines for spam categorization. *IEEE Transactions on Neural Networks*, **10**, 1048–1054.

Duda R., Hart P., Stork D. (2001). *Pattern Classification*. Wiley.

Dumoulin A., Vercellis C. (2000). Tactical models for hierarchical capacitated lot-sizing problems with setups and changeovers. *International Journal of Production Research*, **38**, 51–67.

Durbin J., Koopman S. (2001). *Time Series Analysis by State Space Methods*. Oxford University Press.

Dyche J. (2000). *e-Data: Turning Data Into Information with Data Warehousing*. Addison-Wesley.

Dyche J. (2001). *The CRM Handbook: A Business Guide to Customer Relationship Management*. Addison-Wesley.

Egan J. (2001). *Relationship Marketing: Exploring Relational Strategies in Marketing*. Prentice Hall.

Eliashberg J., Lilien G. (1993). *Marketing*. Elsevier Science.

Elmaghraby S. (1978). *Activity networks*. Wiley.

Everitt B., Landau S., Leese M. (2001). *Cluster Analysis*. Arnold Publishers.

Evgeniou T., Poggio T., Pontil M., Verri A. (2002). Regularization and statistical learning theory for data analysis. *Journal of Computational Statistics and Data Analysis*, **38**, 421–432.

Farrell M. (1957). The measurement of productive efficiency. *Journal of the Royal Statistical Society*, **120**, 253–290.

Fausett L. (1994). *Fundamentals of Neural Networks*. Prentice Hall.

Fayyad U., Grinstein G., Wierse A. (eds.) (2002). *Information visualization in data mining and knowledge discovery*. Morgan Kaufmann.

Fisher D. (1996). Iterative optimization and simplification of hierarchical clusterings. *Journal of Artificial Intelligence Research*, **4**, 147–179.

Fisher R. (1936). The use of multiple measurements in taxonomic problems. *Annals of Eugenics*, **7**, 179–188.

Fraley C., Raftery A. (2002). Model-based clustering, discriminant analysis, and density estimation. *Journal of the American Statistical Association*, **97**, 611–631.

Freitas A. (2000). Understanding the crucial differences between classification and discovery of association rules - a position paper. *SIGKDD Explorations Newsletter*, **2**, 65–69.

Fukunaga K. (1990). *Introduction to Statistical Pattern Recognition*. Academic Press.

Fumero F., Vercellis C. (1997). Integrating distribution, lot-sizing and machine loading via lagrangean relaxation. *International Journal of Production Economics*, **49**, 45–54.

Fumero F., Vercellis C. (1999). Synchronized development of production, inventory and distribution schedules. *Transportation Science*, **33**, 330–340.

Giovinazzo W. (2002). *Internet-Enabled Business Intelligence*. Prentice Hall.

Giudici P. (2003). *Applied Data Mining: Statistical Methods for Business and Industry*. Wiley.

Gorry G., Scott Morton M. (1971). A framework for management information systems. *Sloan Management Review*, **13**, 55–70.

Graves S., De Kok A. (2003). *Supply Chain Management: Design, Coordination and Operation*. Elsevier.

Graves S., Rinnooy Kan A., Zipkin P. (2002). *Logistics of Production and Inventory*. North Holland.

Halkidi M., Batistakis Y., Vazirgiannis M. (2002a). Cluster validity methods: part i. *SIGMOD Record*, **31**, 40–45.

Halkidi M., Batistakis Y., Vazirgiannis M. (2002b). Cluster validity methods: part ii. *SIGMOD Record*, **31**, 19–27.

Hamilton J. (1994). *Time Series Analysis*. Princeton University Press.

Han J., Fu Y. (1995). Discovery of multiple-level association rules from large databases. In: *VLDB '95: Proceedings of the 21th International Conference on Very Large Data Bases*, 420–431. Morgan Kaufmann.

Han J., Kamber M. (2005). *Data Mining: Concepts and Techniques*. Morgan Kaufmann.

Hand D., Mannila H., Smyth P. (2001). *Principles of Data Mining*. MIT Press.

Hartigan J. (1975). *Clustering Algorithms*. Wiley.

Harvey A. (1993). *Time Series Models*. Harvester Wheatsheaf.

Hastie T., Tibshirani R., Friedman J. (2001). *The Elements of Statistical Learning: Data Mining, Inference, and Prediction*. Springer.

Haykin S. (1998). *Neural Networks: A Comprehensive Foundation*. Prentice Hall.

Hipp J., Guntzer U., Nakhaeizadeh G. (2000). Algorithms for association rule mining - a general survey. *SIGKDD Explorations Newsletter*, **2**, 58–64.

Hogg R., Craig A., McKean J. (2004). *Introduction to Mathematical Statistics*. Prentice Hall.

Holt C. (1957). Forecasting seasonals and trends by exponentially weighted moving averages. Technical report, Carnegie Institute.

Inmon W. (2002). *Building the Data Warehouse*. Wiley.

Jain A., Dubes R. (1988). *Algorithms for Clustering Data*. Prentice Hall.

Jain A., Murty M., Flynn P. (1999). Data clustering: A review. *ACM Computing Surveys*, **31**, 264–323.

Jensen D., Cohen P. (2000). Multiple comparisons in induction algorithms. *Machine Learning*, **38**, 309–338.

Joachims T. (2002). *Learning to Classify Text using Support Vector Machines*. Kluwer.

Jones R. (1980). Maximum likelihood fitting of arma models to time series with missing observations. *Technometrics*, **20**, 389–395.

Kaufman L., Rousseeuw P. (1990). *Finding Groups in Data: An Introduction to Cluster Analysis*. Wiley.

Keen P., Scott Morton M. (1978). *Decision support systems: an organizational perspective*. Addison-Wesley.

Keys P. (1995). *Understanding the process of operational research*. Wiley.

Kimball R. (1996). *The Data Warehouse Toolkit*. Wiley.

Kimball R., Ross M. (2002). *The Data Warehouse Toolkit: The Complete Guide to Dimensional Modeling*. Wiley.

Kimball R., Thornthwaite W., Reeves L., Ross M. (1998). *The Data Warehouse Lifecycle Toolkit*. Wiley.

Klein M., Methlie L. B. (1995). *Knowledge-based Decision Support Systems with Applications in Business*. Wiley.

Kleinbaum D., Kupper L., Muller K., Nizam A. (1997). *Applied Regression Analysis and Multivariable Methods*. Duxbury Press.

Kleinrock L. (1975). *Queueing Systems*. Wiley-Interscience.

Kohavi R. (1995). A study on cross-validation and bootstrap for accuracy estimation and model selection. In: *Proceedings of the Fourteenth International Conference on Artificial Intelligence*, 1137–1143.

Kohavi R., John G. (1997). Wrappers for feature subset selection. *Artificial Intelligence*, **97**, 273–324.

Kudyba S., Hoptroff R. (2001). *Data Mining and Business Intelligence: A Guide to Productivity*. Idea Group.

Kulkarni S., Lugosi G., Venkatesh S. (1998). Learning pattern classification - a survey. *IEEE Transactions on Information Theory*, **44**, 2178–2206.

Lawrence R., Almasi G., Kotlyar V., Viveros M., Duri S. (2001). Personalization of supermarket product recommendations. *Data Mining and Knowledge Discovery*, **5**, 11–32.

Levine D., Krehbiel T., Berenson M. (2003). *Business Statistics - A First Course*. Prentice Hall.

Makridakis S., Wheelwright S. (1989). *Forecasting Methods for Management*. Wiley Interscience.

Makridakis S., Wheelwright S., McGee V. (1983). *Forecasting: Methods and Applications*. Wiley-Hamilton.

Makridakis S., Wheelwright S., Hyndman R. (1998). *Forecasting: Methods and Applications*. Wiley.

Mallach E. (2000). *Decision Support and Data Warehouse Systems*. McGraw-Hill.

Malone T., Crowston K., Herman A. (2003). *Organizing Business Knowledge: The MIT Process Handbook*. The MIT Press.

Mangasarian O. (1997). Mathematical programming in data mining. *Data Mining in Knowledge Discovery*, **1**, 183–201.

Mannila H., Toivonen H., Verkamo A. (1994). Efficient algorithms for discovering association rules. In: U. Fayyad et R. Uthurusamy (eds.), *AAAI Workshop on Knowledge Discovery in Databases*, 181–192. AAAI Press.

Marshall B., McDonald D., Chen H., Chung W. (2004). Ebizport: collecting and analyzing business intelligence information. *Journal of the American Society for information Science and Technology*, **55**, 873–891.

Mendenhall W., Sincich T. (2003). *A Second Course in Statistics: Regression Analysis*. Prentice Hall.

Mendenhall W., Beaver R., Beaver B. (2000). *A Brief Course in Business Statistics*. South-Western College Pub.

Miller H., Han J. (2000). *Geographic Data Mining and Knowledge Discovery*. Taylor and Francis.

Mirkin B. (1996). *Mathematical Classification and Clustering*, volume 11. Kluwer.

Montgomery D., Peck E., Vining G. (2001). *Introduction to Linear Regression Analysis*. Wiley-Interscience.

Moss L., Atre S. (2003). *Business Intelligence Roadmap: The Complete Project Lifecycle for Decision-Support Applications*. Addison-Wesley.

Murtagh F. (1985). *Multidimensional Clustering Algorithms*. Physica-Verlag.

Murthy S. (1998). Automatic construction of decision trees from data: a multi-disciplinary survey. *Data Mining and Knowledge Discovery*, **2**, 345–389.

Murthy S., Kasif S., Salzberg S. (1994). A system for induction of oblique decision trees. *Journal of Artificial Intelligence Research*, **2**, 1–33.

Nemati H., Steiger D., Iyer L., Herschel R. (2002). Knowledge warehouse: an architectural integration of knowledge management, decision support, artificial intelligence and data warehousing. *Decision Support Systems*, **33**, 143–161.

Nemhauser G., Rinnooy Kan A., Todd M. (1994). *Optimization*. Elsevier Science.

Nigam K., McCallum A., Thrun S., Mitchell T. (2000). Text classification from labeled and unlabeled documents using em. *Machine Learning*, **39**, 103–134.

Orsenigo C., Vercellis C. (2003). Multivariate classification trees based on minimum features discrete support vector machines. *IMA Journal of Management Mathematics*, **14**, 221–234.

Orsenigo C., Vercellis C. (2004). Discrete support vector decision trees via tabu-search. *Journal of Computational Statistics and Data Analysis*, **47**, 311–322.

Orsenigo C., Vercellis C. (2006). Rule induction through discrete support vector decision trees. In: E. Triantaphyllou et G. Felici (eds.), *Data Mining and Knowledge Discovery Approaches Based on Rule Induction Techniques*, 305–325. Springer.

Orsenigo C., Vercellis C. (2007). Accurately learning from few examples with a polyhedral classifier. *Computational optimization and applications*, **38**, 235–247.

Orsenigo C., Vercellis C. (2009). Multicategory classification via discrete support vector machines. *Computational Management Science*, **1**, 101–114.

Pampel F. (2000). *Logistic Regression: A Primer*. SAGE Publications.

Park J., Chen M., Yu P. (1995). An effective hash based algorithm for mining association rules. In: M. Carey et D. Schneider (eds.), *Proceedings of the 1995 ACM SIGMOD international conference on Management of data*, 175–186. ACM Press.

Parr Rud O. (2000). *Data Mining Cookbook: Modeling Data for Marketing, Risk and Customer Relationship Management*. Wiley.

Pasquier N., Bastide Y., Taouil R., Lakhal L. (1999). Discovering frequent closed itemsets for association rules. In: *ICDT '99: Proceeding of the 7th International Conference on Database Theory*, 398–416. Springer.

Pei J., Han J., Mortazavi-Asl B., Zhu H. (2000). Mining access patterns efficiently from web logs. In: *PADKK '00: Proceedings of the 4th Pacific-Asia Conference on Knowledge Discovery and Data Mining, Current Issues and New Applications*, 396–407. Springer.

Peppers D., Rogers M. (1996). *The One to One Future*. Currency.

Peppers D., Rogers M. (2004). *Managing Customer Relationships: A Strategic Framework*. Wiley.

Poirier C., Reiter S. (1999). *Supply Chain Optimization*. Berrett-Koehler.

Powell R. (2001). *DM Review: A 10 Year Journey*. DM Review.

Pyle D. (1999). *Data Preparation for Data Mining*. Morgan Kaufmann.

Pyle D. (2003). *Business Modeling and Data Mining*. Morgan Kaufmann.

Quinlan J. (1993). *C4.5: Programs for machine learning*. Morgan Kaufmann Publishers.

Ramoni M., Sebastiani P. (2001). Robust bayes classifiers. *Artificial Intelligence*, **125**, 209–226.

Rasmussen N., Goldy P., Solli P. (2002). *Financial Business Intelligence. Trends, Technology, Software Selection, and Implementation*. Wiley.

Raudys S. (2001). *Statistical and Neural Classifiers*. Springer.

Ray S. (2004). *Data Envelopment Analysis: Theory and Techniques for Economics and Operations Research*. Cambridge University Press.

Ripley B. (1996). *Pattern Recognition and Neural Networks*. Cambridge University Press.

Romesburg C. (1984). *Cluster Analysis for Researchers*. Life Time Learning.

Rosenblatt F. (1962). *Principles of Neurodynamics: Perceptrons and the theory of brain mechanisms*. Spartan Books.

Ryan T. (1997). *Modern Regression Methods*. Wiley.

Safavian S., Landgrebe D. (1998). A survey of decision tree classifier methodology. *IEEE Transactions on Systems, Man and Cybernetics*, **22**, 660–674.

Schafer J., Konstan J., Riedl J. (2001). E-commerce recommendation applications. *Journal of Data Mining and Knowledge Discovery*, **5**, 115–153.

Schölkopf B., Smola A. (2001). *Learning with Kernels, Support Vector Machines, Regularization, Optimization, and Beyond*. MIT Press.

Schurmann J. (1996). *Pattern classification, a unified view of statistical and neural approaches*. Wiley.

Scott Long J. (1997). *Regression Models for Categorical and Limited Dependent Variables*. SAGE Publications.

Shapiro J. (2000). *Modeling the Supply Chain*. Duxbury Press.

Shim J., Warkentin M., Courtney J., Power D., Sharda R., Carlsson C. (2002). Past, present, and future of decision support technology. *Decision Support Systems*, **33**, 111–126.

Silverstein C., Brin S., Motwan R. (1998). Beyond market baskets: Generalizing association rules to dependence rules. *Data Mining and Knowledge Discovery*, **2**, 39–68.

Simchi-Levi D., Kaminsky P., Simchi-Levi E. (2002). *Designing and Managing the Supply Chain*. McGraw-Hill/Irwin.

Simchi-Levi D., Wu S., Shen Z. (2004a). *Handbook of Quantitative Supply Chain Analysis: Modeling in the E-Business Era*. Springer.

Simchi-Levi D., Chen X., Bramel J. (2004b). *The Logic of Logistics: Theory, Algorithms, and Applications for Logistics and Supply Chain Management*. Springer.

Simon A., Shaffer S. (2001). *Data warehousing and business intelligence for e-commerce*. Morgan Kaufmann.

Simon H. (1969). *The sciences of the artificial*. The MIT Press.

Simon H. (1977). *The New Science of Management Decision*. Prentice Hall.

Skinner D. (1999). *Introduction to Decision Analysis*. Probabilistic.

Snapper J. (1998). Responsibility for computer-based decisions in health care. In: K. Goodman (ed.), *Ethics, Computing and Medicine: Informatics and the Transformation of Health Care*, 43–56. Cambridge University Press.

Snedecor G., Cochran W. (1989). *Statistical Methods*. Iowa State University Press.

Spaeth H. (1980). *Cluster Analysis Algorithms for Data Reduction and Classification of Objects*, volume 4. Ellis Horwood.

Sprague R. (1980). A framework for research on decision support systems. In: G. Fick et R. Sprague (eds.), *Decision support systems: issues and challenges*, 1–22. Pergamon Press.

Thanassoulis E. (2001). *Introduction to the Theory and Application of Data Envelopment Analysis: A Foundation Text*. Cambridge University Press.

Toivonen H. (1996). Sampling large databases for association rules. In: T. Vijayaraman, A. Buchmann, C. M., et N. Sarda (eds.), *In Proc. 1996 Int. Conf. Very Large Data Bases*, 134–145. Morgan Kaufman.

Tufte E. (1983). *The Visual Display of Quantitative Information*. Graphics Press.

Tukey J. (1977). *Exploratory Data Analysis*. Addison-Wesley.

Turban E., Aronson J., Liang T. (2005). *Decision Support Systems and Intelligent Systems*. Prentice Hall.

Vapnik V. (1996). *The Nature of Statistical Learning Theory*. Springer.

Vapnik V. (1998). *Statistical learning theory*. Wiley.

Velleman P., Hoaglin D. (1981). *The ABC's of EDA: Applications, Basics, and Computing of Exploratory Data Analysis*. Duxbury Press.

Vercellis C. (2008). *Ottimizzazione. Modelli, metodi, applicazioni*. McGraw-Hill.

Weiss N. (2001). *Introductory Statistics*. Addison Wesley.

Winkler R., Makridakis S. (1983). The combination of forecasts. *Journal of the Royal Statistical Society, A*, **145**, 150–157.

Winston W., Venkataramanan M. (2002). *Introduction to Mathematical Programming: Applications and Algorithms*. Duxbury Press.

Winters P. (1960). Forecasting sales by exponentially weighted moving averages. *Management Science*, **6**, 324–342.

Witten I., Frank E. (2005). *Data Mining: Practical Machine Learning Tools and Techniques*. Morgan Kaufmann.

Zaki M. (1999). Parallel and distributed association mining: A survey. *IEEE Concurrency*, **7**, 14–25.

Zaki M. (2000). Generating non-redundant association rules. In: *KDD '00: Proceedings of the sixth ACM SIGKDD international conference on Knowledge discovery and data mining*, 34–43. ACM Press.

Zaki M., Ogihara M. (1998). Theoretical foundations of association rules. In: *Proceedings of 3rd SIGMOD '98 Workshop on Research Issues in Data Mining and Knowledge Discovery*.

Zoltners A., Sinha P., Zoltners G. (2001). *The Complete Guide to Accelerating Sales Force Performance: How to Get More Sales from Your Sales Force*. American Management Association.

Index

Business Intelligence: Data Mining and Optimization for Decision Making C. Vercellis
© 2009 John Wiley & Sons, Ltd